Endothelin in Renal Physiology and Disease

Contributions to Nephrology

Vol. 172

Series Editor

Claudio Ronco Vicenza

Endothelin in Renal Physiology and Disease

Volume Editors

Matthias Barton Zürich
Donald E. Kohan Salt Lake City, Utah

35 figures, 9 in color, and 3 tables, 2011

KARGER Basel · Freiburg · Paris · London · New York · New Delhi · Bangkok · Beijing · Tokyo · Kuala Lumpur · Singapore · Sydney

Contributions to Nephrology
(Founded 1975 by Geoffrey M. Berlyne)

Matthias Barton
Molecular Internal Medicine
University of Zürich
CH-8057 Zürich (Switzerland)

Donald E. Kohan
Division of Nephrology
University of Utah Health Sciences Center
Salt Lake City, UT 84123 (USA)

Library of Congress Cataloging-in-Publication Data

Endothelin in renal physiology and disease / volume editors, Matthias Barton, Donald E. Kohan.
p. ; cm. -- (Contributions to nephrology, ISSN 0302-5144 ; v. 172)
Includes bibliographical references and index.
ISBN 978-3-8055-9794-4 (hardcover : alk. paper) -- ISBN 978-3-8055-9795-1
(e-ISBN)
1. Endothelins. 2. Kidney--Diseases. I. Barton, Matthias. II. Kohan, Donald Elliott. III.
Series: Contributions to nephrology ; v. 172. 0302-5144
[DNLM: 1. Endothelins--physiology. 2. Kidney--physiopathology. 3. Kidney Diseases--drug therapy.
4. Receptor, Endothelin A--metabolism. W1 CO778UN v.172 2011 / QV 150]
QP552.E54E534 2011
612.4'63--dc23

2011023479

www.karger.com
Printed in Switzerland on acid-free and non-aging paper (ISO 9706) by Reinhardt Druck, Basel
ISSN 0302–5144
ISBN 978–3–8055–9794–4
e-ISBN 978–3–8055–9795–1

Contents

Preface

'Those diseases which medicines do not cure, the knife cures; those which the knife cannot cure, fire cures; and those which fire cannot cure, are to be reckoned wholly incurable.'

Hippokrátēs (460 BC – 370 BC)
ΚΥΡΙΑΚΟΣ, Part VII

The prevalence and incidence of kidney disease has been steadily increasing, mostly due to rising numbers of patients with obesity and/or diabetes, as well as an aging world population. Chronic kidney disease aggravates other conditions, including arterial hypertension, dyslipidemia, and atherosclerosis. Current drug therapies modestly slow progression of chronic kidney disease rather than reversing structural injury. Unfortunately, a large number of patients ultimately require renal replacement therapy through dialysis or kidney transplantation.

Following seminal discoveries in endothelial biology in the early 1980s, several endothelium-derived vasoconstrictor activities were reported, and by 1988 Yanagisawa and his team had succeeded in isolating, cloning, and sequencing a peptide which even today remains the strongest vasoconstrictor known to man. Named 'endothelin (ET)' because of its endothelial cell origin, this peptide activates G protein-coupled receptors and exerts a wide variety of effects on almost every organ system. Within the kidney, ET potently regulates physiologic function and contributes to renal injury through multiple mechanisms. Within the past 10 years the ET field has evolved rapidly; we now know that ET plays a central role in renal sodium and water balance, arterial blood pressure regulation, and the development and maintenance of kidney disease. Exciting new studies even suggest that proteinuric renal disease associated with activation of ET is – at least in part – a reversible condition.

Orally active ET receptor antagonists (ERAs) are now available; this new class of drugs exerts renoprotective activity in numerous models of renal disease. In the late 1990s, the first clinical studies suggested therapeutic potential for ERAs in the treatment of kidney injury. Since then, clinical drug development by a number of pharmaceutical companies has led to studies demonstrating remarkable antiproteinuric effects of ERAs, even in patients already on current standard antiproteinuric therapy.

This book offers timely reviews by acclaimed experts in pharmacology, molecular biology, physiology, cardiovascular medicine, and nephrology covering (1) pharmacology, molecular biology, and physiology of ET in the kidney; (2) preclinical evidence for the role of ET and ERAs in renal disease; and (3) the current state of clinical development of ERA therapy in renal medicine. We hope that this book will be a valuable source of information for nephrologists, clinicians, and other healthcare professionals; renal physiologists and molecular biologists; postdoctoral researchers and students in the life sciences; and scientists and decision-makers in drug research and drug development.

We thank all contributors and reviewers for their efforts which helped us to complete the book in time for the *Twelfth International Conference on Endothelin* in Cambridge, UK in September 2011. We sincerely thank Tanja Sebuk, Angela Weber, Yvonne Rebmann, Peter Roth, and Thomas Nold of Karger Publishers for their most professional support.

Matthias Barton, Zürich, Switzerland
Donald E. Kohan, Salt Lake City, Utah, USA

Barton M, Kohan DE (eds): Endothelin in Renal Physiology and Disease.
Contrib Nephrol. Basel, Karger, 2011, vol 172, pp 1–17

Pharmacology of Renal Endothelin Receptors

Anthony P. Davenport · Janet J. Maguire

Clinical Pharmacology Unit, University of Cambridge, Addenbrooke's Hospital, Cambridge, UK

Abstract

Endothelin (ET)-1 is the major isoform in the human kidney where it interacts with two G protein-coupled receptors, ET_A and ET_B. It contains high densities of ET receptors, but in contrast to most other peripheral organs, the majority (70%) are of the ET_B subtype and largely have a differential distribution to ET_A receptors within renal cells, thus mediating contrasting physiological and pathophysiological actions. ET-1 remains the most potent constrictor of human-isolated vessels, including those of the kidney which are particularly sensitive to the actions of this peptide. The pharmacological response is unusual in being sustained for a considerable period of time and slow to wash out. Smooth muscle cells of the renal vasculature mainly express ET_A receptors, and ET_A-selective antagonists fully block these constrictor responses. The vascular endothelium only expresses ET_B receptors. ET-1 also acts in an autocrine or paracrine manner, and binds to ET_B receptors to stimulate the release of vasodilators. ET-1 is unusual amongst the mammalian bioactive peptides in possessing two disulphide bridges, conferring resistance to enzymatic degradation. However, the plasma half-life of ET-1 is surprisingly short as a result of the second major function of endothelial ET_B receptors in removing ET-1 from the circulation, mainly in the kidneys and lungs. Thus, ET_B receptors have a critical role in protecting target organs such as the heart and may limit the detrimental vasoconstrictor effect caused by upregulation of ET-1 associated with disease. Inhibition of the renal medullary ET_B system causes sodium retention because of its role in systemic fluid and electrolyte homeostasis. ET_A/ET_B antagonists would be expected to block the beneficial vasodilatory, clearing, and natriuretic actions of ET_B receptors. Since many of the deleterious actions of ET-1, vasoconstriction, mesangial cell proliferation, and inflammation occur mainly via ET_A receptors, selective ET_A blockade may be more beneficial in renal disease.

In humans, endothelins (ETs) [1] interact with two G protein-coupled receptors, ET_A [2] and ET_B [3]. The main pharmacological characteristics and tools delineating these receptors are compared in table 1. The key feature of this classification is that all three endogenous isoforms bind to the ET_B receptors with similar affinity, but (importantly) at physiological concentrations. ET-3 displays at least 100-fold lower selectivity and does not activate ET_A. In contrast, ET-1 and ET-2 are equipotent at the ET_B subtype [4, 5]. mRNA encoding all three isoforms are present in human kidney [6, 7], although high performance liquid chromatography combined with radioimmunoassay analysis has revealed that ET-1 is the most abundant isoform, with ET-2 and ET-3 difficult to detect [6]. ET-like immunoreactivity is principally localized within endothelial cells lining all intrarenal vessels throughout the cortex and medulla, consistent with ET-1 on release acting locally as an intrarenal mediator. Immunoreactive ET was also detected within endothelial, but not mesangial, cells of the glomeruli. Epithelial cells lining collecting ducts mediating natriuresis are a second major source (fig. 1) [8, 9]. The peptide has also been detected in podocytes (glomerular or visceral epithelial cells) in Bowman's capsule that wrap around the capillaries of the glomerulus and on contraction decrease glomerular filtration rate by reducing the surface area available for filtration [10]. Gauquelin et al. [11] identified ET-1 receptors in membranes prepared from rat glomeruli, and Rebibou et al. [12] characterized the binding of labeled ET-1 to isolated human glomerular mesangial cells and to podocytes (which are glomerular epithelial cells as opposed to parietal epithelial cells). Ortmann et al. [13] showed that human podocytes express mRNA encoding for ET-1, ET_A, and ET_B receptors. ET-1, acting via ET_A receptors, contributes to podocyte injury through cytoskeleton disruption and apoptosis. ET_A, but not ET_B, antagonists are effective in preventing podocyte injury. The role of glomerular podocytes has also been discussed in two recent reviews by Barton [14, 15].

These sources are consistent with the main physiological actions of ET-1 acting in an autocrine or paracrine manner to cause vasoconstriction mainly via ET_A receptors which are opposed by ET_B-mediated vasodilatation as well as natriuresis by acting on ET_B receptors in the renal medulla. In renal disease, many of the deleterious actions of ET-1, vasoconstriction, extracellular matrix production, mesangial cell proliferation, and inflammation occur mainly via ET_A receptors [16, 17]. Our aim is to summarize the pharmacology of renal ET receptors with the main focus in humans. More extensive reviews should be consulted for receptor function, particularly in animal models [18, 19].

Characterizing Endothelin Receptors with Selective Agonists and Antagonists

The human kidney contains high densities of ET receptors compared with other peripheral tissues, with Bmax values ranging in different regions from 50–350 fmol • mg^{-1} of protein. In marked contrast to most other organs where

Table 1. The main characteristics of ET_A and ET_B receptors and tools delineating these subtypes

	ET_A	ET_B
Gene name	*EDNRA*	*EDNRB*
Potency	ET-1 = ET-2 > ET-3	ET-1 = ET-2 = ET-3
Desensitization	slow	fast
Trafficking	endosomes for recycling	lysosomes for degradation
G protein-coupling	Gq/G11 (also Gi/o, Gi2, Gs)	Gq/G11, Gs, Gq/13, Gi2.
Peptide agonists	–	Sarafotoxin S6C
		BQ3020
		IRL1620
Peptide antagonists	BQ123	BQ788
	FR139317	
Radiolabeled ligands	$[^{125}I]$-PD151242	$[^{125}I]$-BQ3020
	$[^{3}H]$-BQ123	
Renal function		
Vasculature	cortical vasoconstriction	medullary vasodilatation
	conduit vessel vasoconstriction	conduit vessel vasodilatation
	afferent arteriolar constriction	afferent arteriolar dilatation
	efferent arteriolar constriction	efferent arteriolar dilatation
Glomerulus	mesangial cell contraction	
	mesangial cell proliferation	
	podocyte injury	
Epithelium		endothelial function not known
Inner medullary collecting duct		natriuresis
	extracellular matrix accumulation	
	interstitial fibrosis	

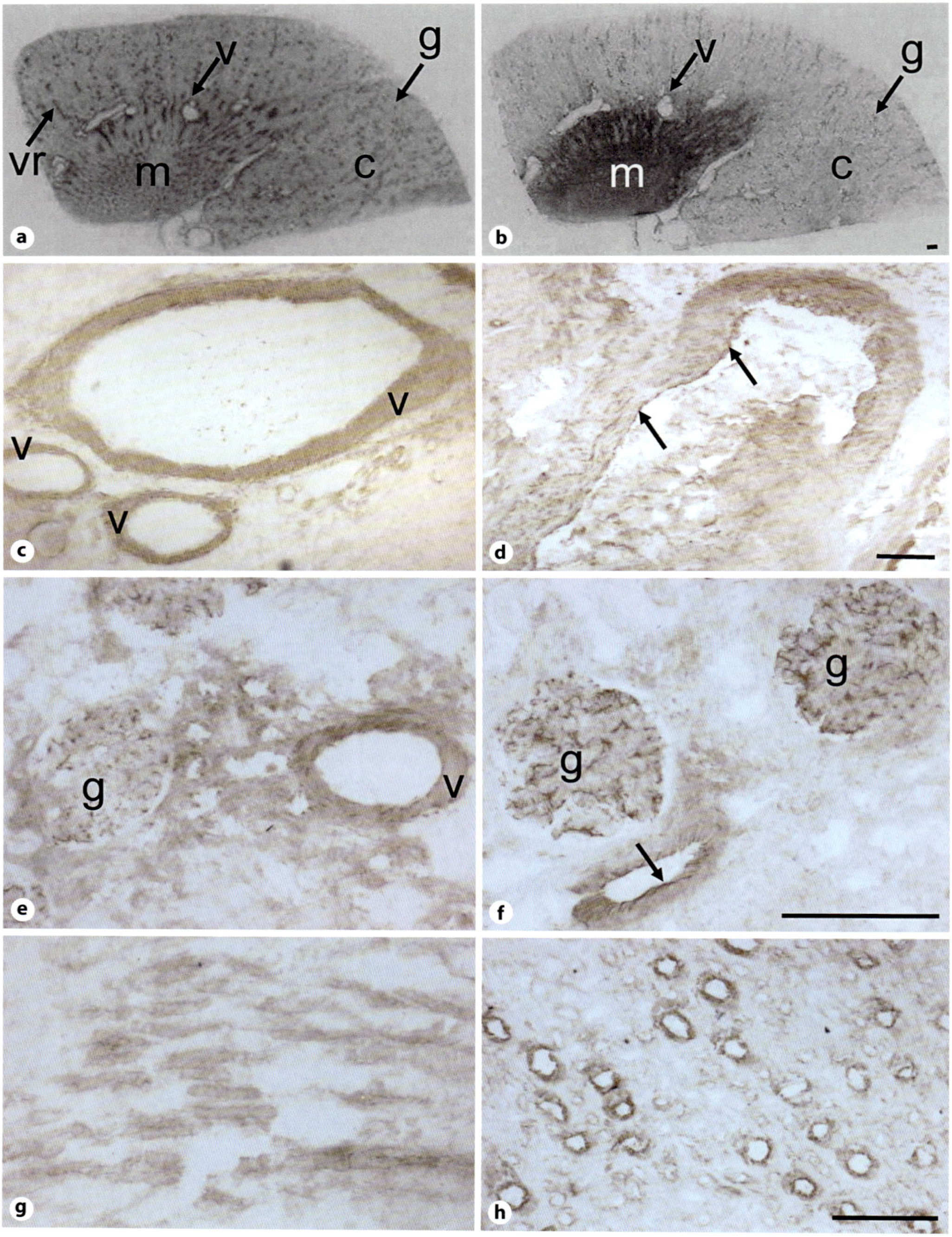

Fig. 1. Microautoradiographical images of radioligand binding using [^{125}I]-PD156707 to label ET_A receptors predominantly in the renal vasculature (**a**) and ET_B receptors labeled with [^{125}I]-BQ3020 mainly to tubules in human kidney (**b**). Immunocytochemical localiza-

ET_A receptors predominate, such as the heart, measurement of the total content of ET receptors reveals that the human kidney also has the highest density of ET_B in the periphery, comprising about 70% [20].

To date, no ET_A-selective agonist has been developed, but truncated linear peptides based on the C-terminal amino acids of ET-1, such as BQ3020, continue to be used as a pharmacological tool to characterize this subtype (table 1) together with a radiolabeled analogue. Sarafotoxin S6c is a moderately selective ET_B agonist which has been used on volunteers and belongs to a family of snake venoms which includes nonselective agonists S6a, S6b, and S6d. They are the only peptides with a high degree of sequence similarity to ETs. The symptoms of envenomation can be lethal with systemic and coronary vasoconstriction, changes in ECG, or direct inotropic actions on the heart, giving insight into the damaging effect of activating ET receptors. The widely used cyclic peptides BQ123 and BQ788 have the advantage that they are highly selective antagonists for the ET_A and ET_B receptors, respectively; therefore, the precise role of the two subtypes can be distinguished pharmacologically with a high degree of confidence [4, 5]. These compounds have been used in vitro and in vivo in a range of species and, importantly, in experimental clinical studies in both patients and volunteers.

Both receptors belong to class 1 G protein-coupled receptors, which are attractive 'druggable' targets, but have proved challenging for the development of nonpeptide small molecule antagonists. From over 500 preclinical compounds, only a limited number have been tested in clinical trials (fig. 2) [21]. The development of receptor antagonists for clinical use has focused on two strategies. ET_A antagonists are classified as having at least 100-fold selectivity for this receptor over the ET_B, which represent the theoretical minimum that can be used in vivo to achieve selective ET_A blockade with little action on ET_B. Mixed antagonists block both receptor subtypes, displaying less than 100-fold selectivity.

There are important caveats in interpreting the results of preclinical or clinical trials using some of these compounds as to whether true selectivity was achieved. The measurement selectivity is usually made by the manufacturer with no agreed upon pharmacological measurement to permit comparison. In practice, over 500- to 1,000-fold are likely to be needed in vivo to be confident of blocking only the ET_A subtype [22]. Two antagonists currently approved for use in the clinic are the mixed antagonist bosentan and ambrisentan. The latter has also gained approval for the treatment of pulmonary arterial hypertension.

tion of ET_A receptors to smooth muscle of arcuate arteries and adjacent veins (**c**), glomeruli and arterioles (**e**), and vasa recta (**g**). ET_B receptors localized to endothelial cells of vessels (**d** and **f**) as well as glomeruli (**f**) and to epithelial cells lining the renal tubules (**h**). c = Cortex; g = glomerulus; m = medulla; v = vessel; vr = vasa recta; arrows indicate endothelial cells. Scale bar, 200 μm. Modified from Kuc and Davenport [8].

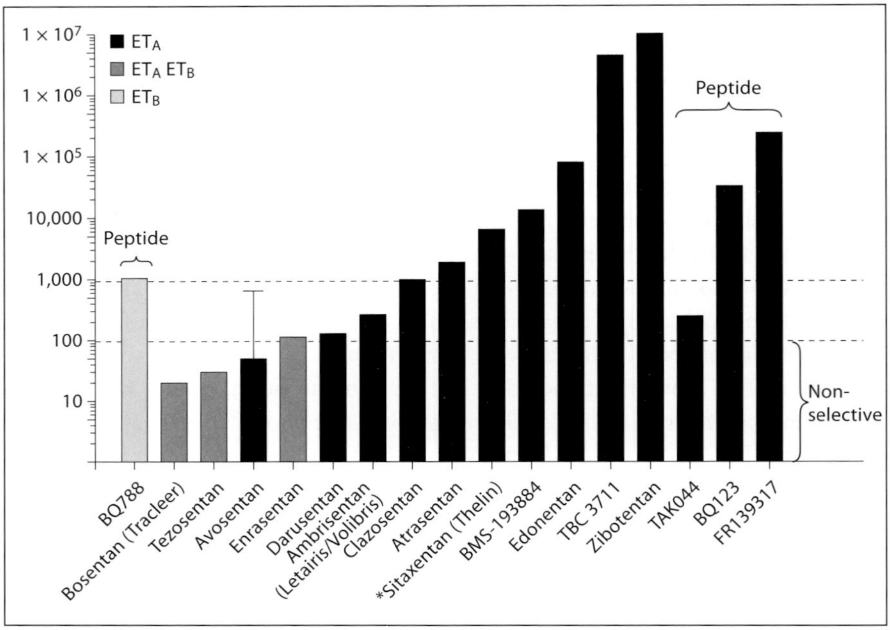

Fig. 2. Comparison of the degree of selectivity of ET receptor antagonists used for human ET_A vs. ET_B receptors. Data for the classification of small molecule antagonists either currently used in the clinic or in clinical trials as reported by the manufacturers. Selectivities of peptide antagonists (BQ788, TAK-044, BQ123) used experimentally in the clinic as well as for laboratory studies are shown for comparison. * Sitaxentan has now been withdrawn from clinical use.

Ambrisentan displays modest selectivity for the human ET_A receptor versus ET_B expressed in CHO cells, although higher values have been reported in the literature depending on the method used [17]; further studies are needed to clarify the precise selectivity. Unfortunately, the highly selective ET_A antagonist sitaxentan [22] was withdrawn in 2010, but when compared with bosentan both compounds had a similar benefit in patients with pulmonary arterial hypertension [23]. The optimum pharmacological profile of ET antagonists continues to be debated and a definitive answer has yet to emerge, but information about the precise localization and function of each subtype in the human kidney can inform the selection of the required profile for chronic kidney disease.

Endothelin A Receptors and Renal Vasculature Vasoconstriction

ET-1 plays a key role in the maintenance of basal vasomotor tone and blood pressure in humans [24]. This is achieved through a balance between ET-1 and other endothelium-derived vasoconstrictors, and the vasodilatory effects of nitric oxide, prostacyclin, and other endothelium-derived relaxing factors. The two ET receptor subtypes exhibit contrasting actions in the cardiovascular system under normal physiological conditions. ET_A receptors are widely expressed on vascular smooth muscle cells throughout and mediate vasoconstriction. Under pathophysiological conditions, ET_A activation is thought to contribute to proliferation, apoptosis, and fibrosis within the vessel wall. The primary transduction system for ET_A receptors is mainly via the G protein Gq/11, but there is evidence for activation of Gi/o, Gi2, and Gs. The major enzyme systems activated include phospholipase C, A2, and D, with calcium signaling being nearly the universal response for ET-1-induced vasoconstriction via inositol triphosphate and DAG generation.

Expression of ET receptors in the human renal vasculature is similar to vessels from other vascular beds [25]. ET_A receptors quantified by receptor autoradiography and confirmed by the use of selective antisera were localized to the smooth muscle layer of intrarenal vessels throughout the cortex and medulla including the large arcuate arteries and adjacent arcuate veins at the corticomedullary junction. ET_B receptors could not be detected. Receptors were also expressed on arterioles and, importantly, only ET_A receptors were visualized in the afferent and efferent vessels of the glomerulus [8, 26].

The pharmacological action of ET-1 in isolated human vessels is unusual if not unique, and has stimulated considerable research interest since the peptide was discovered [1]. In human tissues, radiolabeled ET-1 binds with high, subnanomolar affinity (range of approx. 0.17 nM in human kidney), although in functional assays using human isolated vessels the measured EC_{50} (the concentration producing a half-maximal response) is usually an order of magnitude lower in the nanomolar range (1–10 nM) [27]. The constrictor response to a single concentration of ET-1 including renal vessels develops slowly, compared with other vasoconstrictors such as angiotensin II, reaching a maximum typically after 10 min. However, this is not thought to be the result of a slow association rate constant since for ET-1 this is usually about $1 \times 10^6\ M^{-1} \bullet s^{-1}$, which is comparable to other constrictors. In human vascular smooth muscle preparations which predominantly express ET_A receptors, the contractions can be sustained for a considerable period of time and are slow to wash out, which is consistent with a slow dissociation rate for ET-1. This mechanism is thought to contribute to sustained hypertension and/or ET-induced vasospasm associated with pathophysiological conditions including chronic kidney disease and hypertension following renal transplantation. To be clinically effective, ET antagonists must be able to relax vessels preconstricted with ET-1. Despite the slow dissociation rate, this has been shown. The mechanism of action is not proven, but it is likely by rapid internalization of the ligand receptor complex for ET_A receptors (table 1) and recycling to the membrane.

A low density of ET_B receptors (less than 15%) was measured in human main stem renal artery and vein, where ET-1 potently constricted both vessels with EC_{50} values of 4 and 1 nM, respectively. ET-3 either had no affect on renal artery or they were much less potent than ET-1 where vasoconstrictor responses occurred in vein, consistent with the pharmacological profile of an ET_A receptor. In agreement that the ET-1 response in human kidney is mediated only by ET_A receptors, BQ123 fully reversed the constrictor response with an equilibrium dissociation constant (K_B) of 160 nM for artery and vein [28]. Thus, there is a similar pattern in healthy vessels from the kidney, which is consistent throughout the human vasculature with only ET_A receptors detected in arterioles and resistance vessels. While a low density of ET_B receptors can be detected in conduit arteries and veins, constrictor responses are fully reversed by ET_A-selective antagonists [28, 29].

Endothelial Endothelin B Receptors and Renal Vasculature Vasodilatation

In human kidney, as elsewhere in the peripheral vasculature, only the ET_B subtype is expressed by the vascular endothelium and has been visualized by immunocytochemistry to the main stem renal arteries and veins as well as intrarenal vessels (fig. 1) [8]. While most ET-1 is released from the vascular endothelium towards the underlying smooth muscle to cause vasoconstriction, some ET-1 acts in an autocrine or paracrine manner and binds to ET_B to stimulate cyclic GMP and release vasodilators, including nitric oxide and prostacyclin. This has been shown in detail in the rat regulation of systemic and renal hemodynamics in vivo where endogenous ET acts as a physiological brake for the ET_A-mediated renal vasoconstriction [30].

In volunteers, using venous occlusion plethysmography, local intra-arterial infusions of low doses of exogenous ET-1 into the brachial artery caused an initial vasodilatation by interacting with endothelial ET_B receptors. At higher doses, as the dilator response was overwhelmed, the peptide caused sustained ET_A-mediated vasoconstriction of the forearm vascular bed [31]. In agreement, using the same technique but using BQ788 to block ET_B receptors, a rise in systemic vasoconstriction was measured. Combining the two studies, the results showed that the main consequence of activation of endothelial ET_B receptors by tonically secreted ET-1 was the physiological basal release of nitric oxide [32]. Selective ET_B receptor antagonism by BQ788 also caused arteriolar vasoconstriction in the peripheral circulation of patients with chronic heart failure, demonstrating in disease that the dominant action of endogenous ET-1 is also at endothelial ET_B receptors with no evidence for a contribution by vascular smooth muscle ET_B receptors. Selective ET_B blockade may therefore cause potentially deleterious vasoconstriction in conditions of chronic heart failure [32, 33].

Renal Vascular Responses in Human

In human volunteers, systemic infusion of ET_A receptor antagonists such as BQ123 had no significant effect on renal blood flow, renal vascular resistance, or glomerular filtration rate [34, 35]; however, the ET_B antagonist BQ788 reportedly increased renal vascular resistance, suggesting the presence of basal ET_B-mediated dilator tone [35]. This is consistent with the observation that infusion of L-NAME decreases renal blood flow and glomerular filtration rate, but interestingly these effects were attenuated by BQ123. This implies that both ET_A-mediated vasoconstriction and nitric oxide release via ET_B receptor activation contribute to normal renal vascular function in human kidney in vivo [36]. In agreement with the in vitro data, systemic infusion of ET-1 produced a marked decrease in renal blood flow and glomerular filtration rate and an increase in renal vascular resistance [34, 36] that were all attenuated by ET_A-selective antagonists and augmented by co-infusion of an ET_B antagonist. As would be expected for a predominantly ET_A-mediated renal constrictor response, infusion of ET-3 results in a detectable increase in plasma ET-3, but has no marked effect on renal vascular parameters [36].

Renal Vascular Responses in Rat and Mouse

In mice, ET-1-mediated responses in renal arteries have been compared with other vascular beds. Interestingly, vessels distant from the heart (renal and femoral artery) show a much stronger response to ET-1 than arteries close to the heart (aorta and carotid artery) [37]. In the anesthetized or chronically instrumented rats, infusion of ET-1 produces an initial systemic depressor response followed by sustained constriction and, as expected, a reduction in renal blood flow, glomerular filtration rate, and increased renal vascular resistance. However, the relative contribution of the ET_A and ET_B receptors to these responses is less clear in rats than in humans. Many studies suggest that ET_B receptors contribute substantially to the systemic and renal vasoconstriction to ET-1, whereas others suggest ET_B vasoconstriction is less important than ET_A-mediated responses [19]. It has been reported by some that ET_A antagonism alone is insufficient to completely block ET-1-mediated renal vascular effects, whereas combined ET_A/ET_B blockade is more effective. Comparable to human studies, it is clear that renal ET-mediated vasoconstriction, whether via ET_A, ET_B, or both receptors, is counterbalanced by ET_B-mediated vasodilator tone with, for example, infusion of nitric oxide synthase inhibitors augmenting the renal constrictor response to both infused ET-1 and S6c [30].

The abnormal regulation of renal vascular tone has been identified as one of the main factors of salt-sensitive hypertension in Dahl rats [38]. Hypertension in these rats was associated with renal artery hypertrophy, increased vascular and renal ET-1 content, and glomerulosclerosis. In the renal artery, there was enhanced prostanoid-mediated endothelium-dependent contractions and attenuation of basal nitric oxide release. Rats failed to increase nitric oxide synthase activity in response to salt loading, predisposing the animals to activation of the tissue ET-1 system, abnormal renal vasoconstriction, and renal injury. ET_A blockade normalized salt-induced changes in the renal artery and reduced glomerular injury, suggesting therapeutic potential for ET antagonists in salt-sensitive forms of hypertension [39].

Renal Vasoconstriction and Second Messengers

In isolated smooth muscle cells from preglomerular microvessels, fluorescence microscopy showed that ET-1 and ET-2 produced a biphasic increase in intracellular calcium: an initial peak followed by a sustained (or oscillating) plateau. Responses to ET-3 were smaller and monophasic. Removal of extracellular calcium did not affect the peak of the response, but abolished the later plateau phase, suggesting that ET_A receptor activation leads to a transient release of intracellular calcium with subsequent influx of extracellular calcium in renal vascular smooth muscle [40]. Renal arterial myocytes have been shown to express both delayed rectifying potassium channels and Ca^{2+}-activated potassium channels which are inhibited by ET-1 to promote depolarization that would then result in L-type channel activity and calcium influx [41]. However, the observed effect of calcium channel blockers such as nifedipine and diltiazem was limited, implying a greater importance on release of intracellular stores in contraction of the renal vasculature. In isolated afferent arterioles or the anesthetized rat, ET-1-induced increases in cellular calcium or acute reductions in renal flow, respectively, were diminished by inhibitors of ryanodine receptors or inhibitors of ADP ribosyl cyclase [42]. These data suggest that in addition to the well-established ET-1-induced inositol triphosphate-mediated release of calcium from the sarcoplasmic reticulum, release of calcium from cytosolic stores is mediated at least in part via activation of ryanodine receptors by the signaling molecule cyclic ADP ribose.

In mice, ET-1 constricts afferent arterioles more potently than efferent arterioles, but comparing responses in the presence of BQ123 or in the ET_B receptor-deficient mouse suggests that whereas the afferent response is predominantly ET_A, the efferent response is mediated via both ET_A and ET_B receptors [43]. The more pronounced responsiveness of the afferent vessels is consistent with the ET-1-mediated fall in glomerular filtration rate that has been observed in all species investigated.

Does the Ratio of Endothelin Receptors Change in the Vasculature with Disease?

A key question is whether there is evidence for changes in renal vascular receptors. Although not extensively studied, thus far only upregulation of ET_A receptors in human renal disease has been demonstrated. In kidneys removed from patients with autosomal-dominant polycystic kidney disease, ET_A mRNA was 5- to 10-fold higher in cystic kidneys compared to healthy controls. Increased expression of ET_A receptor protein was found overlying glomeruli and cysts, and was markedly increased in medium-sized renal arteries by microautoradiography. Immunoreactive ET was detected in cyst epithelia and, interestingly, in mesangial and vascular smooth muscles, which do not usually express the peptide, suggesting ET-1 synthesis may be increased in the cystic kidney [44]. Evidence from other human vessels showed no increase in either ET_B expression or functional response in coronary arteries with atherosclerotic lesions compared with healthy vessels, for example. [45]. ET_B receptors linked to nitric oxide have been proposed to be protective in chronic kidney disease [46].

Endothelin B-Clearing Receptors and Natriuresis

ET-1 is unusual amongst the mammalian bioactive peptides in possessing two disulphide bridges, one of which links the N-terminus cysteine with residue 15, conferring resistance to enzymatic degradation in the plasma and tissues, a common route for inactivating peptides. Despite this structural modification, the plasma half-life of ET-1 is surprisingly short as a result of endothelial ET_B receptors removing ET-1 from the circulation [47–49] by the kidneys, lungs, and liver. This rapid clearance has been visualized in vivo using $[^{18}]$F-labeled ET-1 and positron emission tomography in the anesthetized rat (fig. 2) [50], confirming rapid uptake into these organs. Infusion of BQ788 prior to injecting $[^{18}F]$-ET-1 significantly reduced $[^{18}F]$-ET-1 clearing in the kidney and lungs by 85%, although the amount of binding significantly increased in the liver as the label was no longer cleared by ET_B receptors and now bound to the ET_A subtype. Infusion of BQ788 after injecting $[^{18}F]$-ET did not displace the label consistent with internalization of ET-1 by ET_B receptors and degradation in the lysosome (table 1). Although a high density of ET_A receptors are present in the rat heart, binding of $[^{18}F]$-ET-1 could not be visualized in the control animal, but was visualized in this organ when ET_B receptors were blocked by BQ788. ET_B receptors in the kidney and lung have a critical role in protecting target organs such as the heart and may limit the detrimental vasoconstrictor effect caused by upregulation of ET-1 in the vascular system associated with disease. Interestingly, the vasculature of the lung and liver are major sites for conversion of $[^{18}F]$-big ET-1 to the mature peptide $[^{18}F]$-ET-1 which binds to ET_A receptors, whereas in the

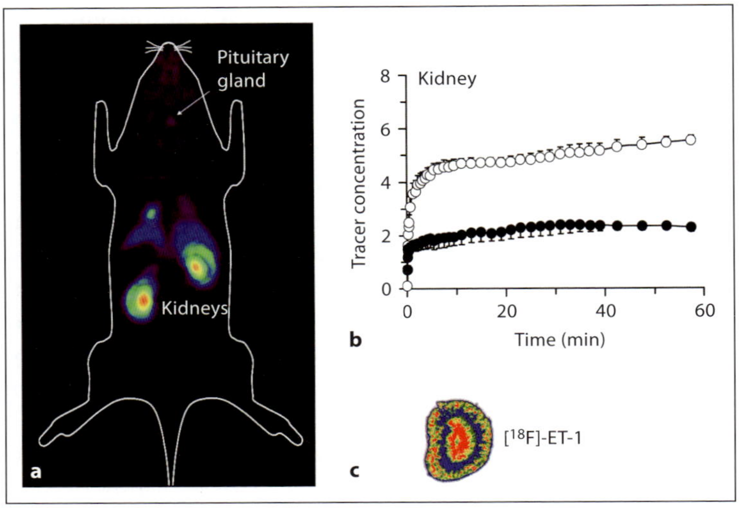

Fig. 3. **a** Positron emission tomography (PET) image showing the whole-body distribution in a coronal plane of [^{18}F]-ET-1 binding to ET receptors in a live rat. Binding was predominantly to liver, lungs, and kidneys where delineation of substructures was detected. Images are color-coded and show the highest levels in red according to the colors of the rainbow with black below the level for detection. **b** Time-activity curves for the amount of [^{18}F]-ET-1 binding measured at time intervals over 60 min within the kidneys of control rats (open circles) compared with rats pretreated with ET_B-selective antagonist rats BQ788 (filled circles). Each value is the mean ± SEM for 3 individuals. **c** Binding of [^{18}F]-ET-1 to ET receptors in sections of kidney removed at the end of the 60 min following PET and visualized using phosphor imaging. Binding was detected in the medulla and cortex with high levels in the vasa recta, papilla and to the glomeruli. Modified from Johnstrom et al. [50].

kidney [^{18}F]-big ET-1 is excreted without conversion [51]. Selectively deleting endothelial ET_B receptors in mice but leaving functioning ET_B receptors in all other cell types demonstrated clearance of an intravenous bolus of labeled ET-1 was significantly reduced, providing clear evidence that endothelial ET_B receptors are mainly, if not exclusively, responsible for ET clearance from the circulation [52–54].

Clinical studies in patients with chronic kidney disease support this concept. In a randomized, placebo-controlled, double-blind, crossover study, the cyclic peptide TAK-044 was used at a low concentration (100 mg intravenous infusions) to selectively block the ET_A, but not the ET_B, receptor [54]. No change in plasma ET-1 level was found. However, at the highest concentration (750 mg), which blocked both receptors, plasma ET-1 doubled as a result of reduced clearance of the peptide by ET_B receptors. Both doses of TAK-044 altered renal dynamics by lowering the effective filtration fraction and caused a beneficial systemic vasodilatation.

Goddard et al. [55] carried out the first clinical study to directly compare the effect of BQ123 or BQ788 alone or in combination in hypertensive patients with chronic renal failure. As expected from in vitro studies, ET_A receptor antagonism alone significantly reduced blood pressure in patients with chronic renal failure with or without ET_B receptor antagonism, but the magnitude of change was significantly higher in the BQ123-only group. Similarly, BQ788 produced both the expected systemic and renal vasoconstriction in patients and volunteers. In this small study (8 individuals), no change was found in sodium excretion despite evidence from animal studies that ET_B receptors present on epithelial cells throughout the tubular epithelium, particularly the inner medullary collecting duct, promote beneficial natriuresis. However, BQ123 alone also increased renal blood flow and renal vascular resistance and reduced proteinuria, consistent with a renoprotective action. These effects were not seen when BQ123 was combined with BQ-788, which supports the idea that ET_B receptors maintain tonic renal vasodilatation. In a larger study of 22 patients with nondiabetic proteinuric chronic kidney disease, BQ123 did produce significant natriuresis, resulting from increased renal blood flow with ET_A receptor antagonism as well as the expected reduced blood pressure, proteinuria, and a new finding of reduced arterial stiffness which is a key determinant of cardiovascular risk [56].

Conclusions and Perspectives

A limitation of the clinical research described above is that they were acute studies using peptide antagonists. Ideally, longer-term head-to-head trials comparing mixed versus ET_A-selective antagonists in the same patient groups to address this question are required. The effects of the mixed ET_A/ET_B antagonist avosentan have been evaluated on progression of diabetic nephropathy in a multicenter, double-blind, placebo-controlled trial. Avosentan reduces albuminuria when added to standard treatment in people with type 2 diabetes and overt nephropathy, but induced significant fluid overload and congestive heart failure. The study was terminated prematurely because of an excess of cardiovascular events [57]. Atrasentan, an ET_A selective antagonist, has been tested in a randomized, double-blind, placebo-controlled trial of subjects with diabetic nephropathy already receiving stable doses of renin-angiotensin system inhibitors. At the doses tested, atrasentan was effective in reducing residual albuminuria [58]. Interestingly, in clinical trials for other conditions, atrasentan improved coronary microvascular endothelial function in patients with early atherosclerosis without changes in renal parameters or edema (measured 2 months after the study was completed) [59]. Since deleterious actions of ET-1 are mainly via ET_A in the kidney and beneficial effects via ET_B, evidence is emerging of the therapeutic benefit of nonpeptide ET_A-selective antagonists in patients with chronic renal failure.

Acknowledgement

We thank the British Heart Foundation for support, grant No. PG/09/050/27734.

References

1 Yanagisawa M, Kurihara H, Kimura S, Tomobe Y, Kobayashi M, Mitsui Y, Yazaki Y, Goto K, Masaki T: A novel potent vasoconstrictor peptide produced by vascular endothelial cells. Nature 1988;332:411–415.
2 Arai H, Hori S, Aramori I, Ohkubo H, Nakanishi S: Cloning and expression of a cDNA encoding an endothelin receptor. Nature 1990;348:730–732.
3 Sakurai T, Yanagisawa M, Takuwa Y, Miyazaki H, Kimura S, Goto K, Masaki T: Cloning of a cDNA encoding a non-isopeptide-selective subtype of the endothelin receptor. Nature 1990;348:732–735.
4 Davenport AP: International Union of Pharmacology. XXIX. Update on endothelin receptor nomenclature. Pharmacol Rev 2002;54:219–226.
5 Davenport AP, Maguire JJ: Endothelin. Handbk Exp Pharmacol 2006;176:295–329.
6 Karet FE, Davenport AP: Localization of endothelin peptides in human kidney. Kidney Int 1996;49:382–387.
7 Ong AC, Jowett TP, Firth JD, Burton S, Karet FE, Fine LG: An endothelin-1 mediated autocrine growth loop involved in human renal tubular regeneration. Kidney Int 1995;48:390–401.
8 Kuc R, Davenport AP: Comparison of endothelin-A and endothelin-B receptor distribution visualized by radioligand binding versus immunocytochemical localization using subtype selective antisera. J Cardiovasc Pharmacol 2004;44(Suppl 1):S224–S226.
9 Herman WH, Emancipator SN, Rhoten RL, Simonson MS: Vascular and glomerular expression of endothelin-1 in normal human kidney. Am J Physiol 1998;275:F8–F17.
10 Morigi M, Buelli S, Angioletti S, Zanchi C, Longaretti L, Zoja C, Galbusera M, Gastoldi S, Mundel P, Remuzzi G, Benigni A: In response to protein load podocytes reorganize cytoskeleton and modulate endothelin-1 gene: implication for permselective dysfunction of chronic nephropathies. Am J Pathol 2005;166:1309–1320.
11 Gauquelin G, Thibault G, Garcia R: Characterization of renal glomerular endothelin receptors in the rat. Biochem Biophys Res Commun 1989;164:54–57.
12 Rebibou JM, He CJ, Delarue F, Peraldi MN, Adida C, Rondeau E, Sraer JD: Functional endothelin 1 receptors on human glomerular podocytes and mesangial cells. Nephrol Dial Transplant 1992;7:288–292.
13 Ortmann J, Amann K, Brandes RP, Kretzler M, Munter K, Parekh N, Traupe T, Lange M, Lattmann T, Barton M: Role of podocytes for reversal of glomerulosclerosis and proteinuria in the aging kidney after endothelin inhibition. Hypertension 2004;44:9749–9781.
14 Barton M: Reversal of proteinuric renal disease and the emerging role of endothelin. Nat Clin Pract Nephrol 2008;4:490–501.
15 Barton M: Therapeutic potential of endothelin receptor antagonists for chronic proteinuric renal disease in humans. Biochim Biophys Acta 2010;1802:1203–1213.
16 Neuhofer W, Pittrow D: Endothelin receptor selectivity in chronic kidney disease: rationale and review of recent evidence. Eur J Clin Invest 2009;39(Suppl 2):50–67.
17 Vachiery JL, Davenport A: The endothelin system in pulmonary and renal vasculopathy: les liaisons dangereuses. Eur Respir Rev 2009;18:260–271.
18 Kohan DE: The renal medullary endothelin system in control of sodium and water excretion and systemic blood pressure. Curr Opin Nephrol Hypertens 2006;15:34–40.

19 Kohan DE, Rossi NF, Inscho EW, Pollock DM: Regulation of blood pressure and salt homeostasis by endothelin. Physiol Rev 2011;91:1–77.
20 Karet FE, Kuc RE, Davenport AP: Novel ligands BQ123 and BQ3020 characterize endothelin receptor subtypes ET_A and ET_B in human kidney. Kidney Int 1993;44:36–42.
21 Palmer MJ: Endothelin receptor antagonists: status and learning 20 years on. Prog Med Chem 2009;47:203–37.
22 Maguire JJ, Davenport AP: Importance of sub-type selectivity for endothelin receptor antagonists in the human vasculature; in Handler C, Abraham D, Dashwood M, Coghlan G, (eds): Translational Vascular Medicine: Pathogenesis, Diagnosis, and Treatment. Berlin, Springer, 2011, pp 1–34.
23 Barst RJ, Langleben D, Badesch D, Frost A, Lawrence EC, Shapiro S, Naeije R, Galie N: Treatment of pulmonary arterial hypertension with the selective endothelin-A receptor antagonist sitaxsentan. J Am Coll Cardiol 2006;47:2049–2056.
24 Haynes WG, Webb DJ: Contribution of endogenous generation of endothelin-1 to basal vascular tone. Lancet 1994;344:852–854.
25 Davenport AP, O'Reilly G, Kuc RE: Endothelin ET_A and ET_B mRNA and receptors expressed by smooth muscle in the human vasculature: majority of the ET_A subtype. Br J Pharmacol 1995;114:1110–1116.
26 Kuc RE, Karet FE, Davenport AP: Characterization of peptide and nonpeptide antagonists in human kidney. J Cardiovasc Pharmacol 1995;26(Suppl 3):S373–S375.
27 Maguire JJ, Kuc RE, O'Reilly G, Davenport AP: Vasoconstrictor endothelin receptors characterized in human renal artery and vein in vitro. Br J Pharmacol 1994;113:49–54.
28 Davenport AP, Kuc RE, Hoskins SL, Karet FE, Fitzgerald F: [125I]-PD151242: a selective ligand for endothelin ET_A receptors in human kidney which localizes to renal vasculature. Br J Pharmacol 1994;113:1303–1310.
29 Maguire JJ, Davenport AP: ET_A receptor-mediated constrictor responses to endothelin peptides in human blood vessels in vitro. Br J Pharmacol 1995;115:191–197.
30 Matsuura T, Miura K, Ebara T, Yukimura T, Yamanaka S, Kim S Iwao H: Renal vascular effects of the selective endothelin receptor antagonists in anaesthetized rats. Br J Pharmacol 1997;122:81–86.
31 Kiowski W, Luscher TF, Linder L, Buhler FR: Endothelin-1-induced vasoconstriction in humans. Reversal by calcium channel blockade but not by nitrovasodilators or endothelium-derived relaxing factor. Circulation 1991;83:469–475.
32 Love MP, Ferro CJ, Haynes WG, et al: Endothelin receptor antagonism in patients with chronic heart failure. Cardiovasc Res 2000;47:166–172.
33 Strachan FE, Spratt JC, Wilkinson IB, Johnston NR, Gray GA, Webb DJ: Systemic blockade of the endothelin-B receptor increases peripheral vascular resistance in healthy men. Hypertension 1999;33:581–585.
34 Montanari A, Biggi A, Carra N, Fasoli E, Calzolari M, Corsini F, Perinotto P, Novarini A: Endothelin-A blockade attenuates systemic and renal hemodynamic effects of L-NAME in humans. Hypertension 2000;35:518–23.
35 Bohm F, Pernow J, Lindstrom J, Ahlborg G: ET_A receptors mediate vasoconstriction, whereas ET_B receptors clear endothelin-1 in the splanchnic and renal circulation of healthy men. Clin Sci (Lond) 2003;104:143–151.
36 Kaasjager KA, Shaw S, Koomans HA, Rabelink TJ: Role of endothelin receptor subtypes in the systemic and renal responses to endothelin-1 in humans. J Am Soc Nephrol 1997;8:32–39.
37 Widmer CC, Mundy AL, Kretz M, Barton M: Marked heterogeneity of endothelin-mediated contractility and contraction dynamics in mouse renal and femoral arteries. Exp Biol Med (Maywood) 2006;231:777–781.
38 Simchon S, Manger WM, Carlin RD, Peeters LL, Rodriguez J, Batista D, Brown T, Merchant NB, Jan KM, Chien S: Salt-induced hypertension in Dahl salt-sensitive rats. Hemodynamics and renal responses. Hypertension 1989;13:612–621.

39 Barton M, Vos I, Shaw S, Boer P, D'Uscio LV, Grone HJ, Rabelink TJ, Lattmann T, Moreau P, Luscher TF: Dysfunctional renal nitric oxide synthase as a determinant of salt-sensitive hypertension: mechanisms of renal artery endothelial dysfunction and role of endothelin for vascular hypertrophy and Glomerulosclerosis. J Am Soc Nephrol 2000;11:835–845.
40 Schroeder AC, Imig JD, LeBlanc EA, Pham BT, Pollock DM, Inscho EW: Endothelin-mediated calcium signaling in preglomerular smooth muscle cells. Hypertension 2000;35:280–286.
41 Betts LC, Kozlowski RZ: Electrophysiological effects of endothelin-1 and their relationship to contraction in rat renal arterial smooth muscle. Br J Pharmacol 2000;130:787–796.
42 Thai TL, Arendshorst WJ: ADP-ribosyl cyclase and ryanodine receptors mediate endothelin ET_A and ET_B receptor-induced renal vasoconstriction in vivo. Curr Opin Nephrol Hypertens 2009;18:40–49.
43 Schildroth J, Rettig-Zimmermann J, Kalk P, et al: Endothelin type A and B receptors in the control of afferent and efferent arterioles in mice. Nephrol Dial Transplant. Am J Physiol Renal Physiol 2008;295:F360–F368.
44 Ong AC, Newby LJ, Dashwood MR: Expression and cellular localisation of renal endothelin-1 and endothelin receptor subtypes in autosomal-dominant polycystic kidney disease (abstract). Nephron Exp Nephrol 2003;93:e80.
45 Maguire JJ, Davenport AP: No alteration in vasoconstrictor endothelin-B-receptor density or function in human coronary artery disease. J Cardiovasc Pharmacol 2000;36:S380–S381.
46 Ohkita M, Takaoka M, Matsumura Y: Drug discovery for overcoming chronic kidney disease (CKD): the endothelin ET B receptor/nitric oxide system functions as a protective factor in CKD. J Pharmacol Sci 2009;109:7–13
47 Fukuroda T, Fujikawa T, Ozaki S, Ishikawa K, Yano M, Nishikibe M: Clearance of circulating endothelin-1 by ET_B receptors in rats. Biochem Biophys Res Commun 1994;199:1461–1465.
48 Gasic S, Wagner OF, Vierhapper H, Nowotny P, Waldhausl W: Regional hemodynamic effects and clearance of endothelin-1 in humans: renal and peripheral tissues may contribute to the overall disposal of the peptide. J Cardiovasc Pharmacol 1992;19:176–180.
49 Pumpton C, Ferro CJ, Haynes WG, Webb DJ, Davenport AP: The increase in human plasma immunoreactive endothelin but not big endothelin-1 or its C-terminal fragment induced by systemic administration of the endothelin antagonist TAK-044. Br J Pharmacol 1996;119:311–314.
50 Johnstrom P, Fryer TD, Richards HK, et al: Positron emission tomography using 18F-labelled endothelin-1 reveals prevention of binding to cardiac receptors owing to tissue-specific clearance by ET B receptors in vivo. Br J Pharmacol 2005;144:115–122.
51 Johnstrom P, Fryer TD, Richards HK, Maguire JJ, Clark JC, Pickard JD, Davenport AP: Positron emission tomography of [18F]-big endothelin-1 reveals renal excretion but tissue-specific conversion to [18F]-endothelin-1 in lung and liver. Br J Pharmacol;159:812–819.
52 Kelland NF, Kuc RE, McLean DL, Azfer A, Bagnall AJ, Gray GA, Gulliver-Sloan FH, Maguire JJ, Davenport AP, Kotelevtsev YV, Webb DJ: Endothelial cell-specific ET_B receptor knockout: autoradiographic and histological characterisation and crucial role in the clearance of endothelin-1. Can J Physiol Pharmacol;88:644–651.
53 Bagnall AJ, Kelland NF, Gulliver-Sloan F, Davenport AP, Gray GA, Yanagisawa M, Webb DJ, Kotelevtsev YV: Deletion of endothelial cell endothelin B receptors does not affect blood pressure or sensitivity to salt. Hypertension 2006;48:286–293.
54 Dhaun N, Ferro CJ, Davenport AP, Haynes WG, Goddard J, Webb DJ: Haemodynamic and renal effects of endothelin receptor antagonism in patients with chronic kidney disease. Nephrol Dial Transplant 2007;22:3228–3234.

55 Goddard J, Johnston NR, Hand MF, Cumming AD, Rabelink TJ, Rankin AJ, Webb DJ: Endothelin-A receptor antagonism reduces blood pressure and increases renal blood flow in hypertensive patients with chronic renal failure: a comparison of selective and combined endothelin receptor blockade. Circulation 2004;109:1186–1193.
56 Dhaun N, Macintyre IM, Melville V, Lilitkarntakul P, Johnston NR, Goddard J, Webb DJ: Blood pressure-independent reduction in proteinuria and arterial stiffness after acute endothelin-a receptor antagonism in chronic kidney disease. Hypertension 2009;54:113–119.
57 Mann JF, Green D, Jamerson K, Ruilope LM, Kuranoff SJ, Littke T, Viberti G: Avosentan for overt diabetic nephropathy. J Am Soc Nephrol 2010;21:527–535.
58 Kohan DE, Pritchett Y, Molitch M, Wen S, Garimella T, Audhya P, Andress DL: Addition of atrasentan to renin-angiotensin system blockade reduces albuminuria in diabetic nephropathy. J Am Soc Nephrol 2011;22:763–772.
59 Reriani M, Raichlin E, Prasad A, Mathew V, Pumper GM, Nelson RE, Lennon R, Rihal C, Lerman LO, Lerman A: Long-term administration of endothelin receptor antagonist improves coronary endothelial function in patients with early atherosclerosis. Circulation 2010;122:958–966.

Dr. A.P. Davenport
Clinical Pharmacology Unit, University of Cambridge, Addenbrooke's Hospital
Cambridge, CB2 0QQ (UK)
Tel. +44 1223 336899
E-Mail apd@medschl.cam.ac.uk

Barton M, Kohan DE (eds): Endothelin in Renal Physiology and Disease.
Contrib Nephrol. Basel, Karger, 2011, vol 172, pp 18–34

Renal Function and Blood Pressure: Molecular Insights into the Biology of Endothelin-1

Nicolas Vignon-Zellweger[a] · Susi Heiden[a] · Noriaki Emoto[a,b]

[a]Department of Clinical Pharmacy, Kobe Pharmaceutical University, and [b]Division of Cardiovascular Medicine, Department of Internal Medicine, Kobe University Graduate School of Medicine, Kobe, Japan

Abstract

The therapeutic implications of the actions of endothelin (ET)-1 upon renal and cardiovascular function are evident. Among other diseases, ET-1 is recognized to be involved in hypertension and renal failure and, in a rush to develop novel treatments, has been extensively studied. However, given the broad localization of the two receptors (ET_A and ET_B) and the diverse effects resulting from their activation, analysis of the role of ET-1 in kidney-regulated blood pressure remains complicated. Moreover, the actions of ET-1 depend upon the cell type and physiological situation. To add to the complexity, both receptors often activate opposing signaling pathways within a single cell. Thus, until recently, reliable insights into the respective involvement of both receptors in the physiology and pathology of the kidney were eagerly awaited. These have been obtained using mice that are genetically modified for different members of the ET system. In this article, the molecular biology of ET-1 and its receptors in the control of renal vasculature tonicity, glomerular function, and management of water and salt reabsorption is discussed. The role of renal ET-1 in the context of blood pressure regulation will be discussed, and the potential of utilizing ET receptor antagonism in the treatment and prevention of glomerular and proteinuric diseases is also outlined.

Endothelin (ET)-1 is an autocrine- and paracrine-signaling hormone involved in various physiological functions, as well as in pathological alterations of the kidney. It has proven implications in many diseases, from hypertension and renal failure to diabetes-induced end-organ damage. At the cellular level, ET-1 influences proliferation, contraction, metabolism, and apoptosis – particularly in the kidney.

The active 21-amino-acid ET-1 is processed from a prohormone by ET-converting enzymes, and produces its biological effects after interaction with two G protein-coupled receptors, ET_A and ET_B.

Genetically manipulated animals represent a powerful tool for understanding gene function in vivo. Therefore, this review pays particular attention to genetic mouse models of the ET system within the scope of kidney function and hypertension. In mice, systemic knockout (KO) of any of the ET system components leads to death at birth or early in life, demonstrating that a functioning ET system is a prerequisite for correct embryogenesis. Nevertheless, diverse genetic technologies have been used to obtain living adult KO mice. For instance, transgenic expression of ET_B in enteric nervous system precursors in mice with otherwise systemic inactivation of the ET_B gene could be used to prevent potentially lethal megacolon formation [1]. In addition, tissue-specific KO of ET-1 and its receptors in renal cells have led to viable offspring [2], and KO mice heterozygous for ET-1 have been generated [3]. These strategies have allowed the analysis of ET-1 function and the opportunity to characterize its receptors in vivo.

How Endothelin-1 Regulates Blood Pressure through Kidney Function

In the physiological state, ET-1 participates in the maintenance of vascular tone, and is often described as one of the most powerful vasoconstrictor agents. Indeed, infusion of exogenous ET-1 leads to a strong elevation of blood pressure [4]. In mice, ET-1 deficiency (restricted to the endothelial cells) reduces blood pressure, confirming the constrictive role of ET-1 upon the vasculature [5]. Nevertheless, mice overexpressing ET-1 in the endothelial cells are normotensive, and develop only moderate salt-sensitive hypertension [6]. Moreover, heterozygous ET-1 KO mice are slightly hypertensive [3]. Even though ET-1 participates in maintaining a positive vascular tone, this phenotype indicates that ET-1 plays a minor role in controlling blood pressure through its vasoconstrictive effect, at least in mice with a chronic ET-1 deficiency. The net effect of ET-1 on blood pressure seems to be rather negative. In line with this are findings that systemic and endothelial-cell-specific overexpressions of ET-1 do not influence blood pressure [7, 8]. The hypertensive phenotype in heterozygous ET-1 KO mice was first attributed to the central action of ET-1 on cardiovascular function and possible compensatory mechanisms, like ET-1-induced activation of endothelial nitric oxide synthase (eNOS) in endothelial cells [9]. Given that the kidney is the organ with the strongest expression of ET-1, we believe that the renal effects of ET-1 play a predominant role in the control of blood pressure.

Endothelin-1 Modulates Renal Function through Its Role as a Vasoconstrictor in the Renal Vasculature

ET-1 is typically produced by the endothelial cells of the renal vasculature. As in the rest of the vasculature, ET_A is present on both the smooth muscle and endothelial cells, while ET_B expression is restricted to endothelial cells [10]. Physiological doses of ET-1 contribute to the renal vascular tone, and thereby regulate glomerular pressure [11].

The glomerular filtration rate (GFR) depends greatly on the constriction of afferent and efferent arterioles. Whether ET-1 has a contractive effect predominantly upon afferent or efferent arterioles remains under dispute. Seminal studies have shown that ET-1 sensitivity is relatively greater in afferent than in efferent arterioles, indicating that ET-1 may reduce the GFR [12]. Using isolated cortical arterioles from ET_B-deficient mice, Schildroth et al. [13] elegantly showed that ET_A participates in the contraction of afferent and efferent arterioles, whereas ET_B activation has no effect on afferent arterioles, but induces NO-mediated vasodilatation in efferent arterioles. This study suggests that ET-1 preferentially constricts afferent arterioles and diminishes the GFR. In line with these observations, infusion of ET-1 also slightly reduces the GFR in humans [14]. Denton et al. [15] showed that infusion of ET-1 constricts both afferent and efferent arterioles in rabbits. However, this did not modify the GFR, but did increase the filtration fraction – thereby suggesting increased glomerular pressure. They explained these results by the fact that the glomerular efferent arterioles have smaller diameters than afferent ones, and therefore ET-1 constriction may have a bigger effect on the resistance of efferent vessels. Moreover, ET_A-specific blockade elevates renal plasma flow and reduces filtration fraction, but does not affect GFR in CKD patients. This may be accompanied by a reduced glomerular pressure, again suggesting that ET-1 preferentially constricts efferent arterioles [16, 17]. The inconsistent observations regarding the constrictor actions of ET-1 in afferent and efferent arterioles may be explained by different experimental set-ups (isolated arterioles vs. in vivo infusion studies) and species-related differences (mouse, rabbit and human).

ET_B on endothelial cells counters the effect of ET_A, and participates in tonic vasodilatation via activation of eNOS and the release of prostaglandins [18]. However, ET_B antagonists have little effect on renal blood flow and GFR in rodents and humans [17, 19, 20].

The impact of ET_B differs according to whether it is present on the microvasculature of the cortical or juxtamedullary glomeruli. In contrast to cortical ET_B, ET_B on both afferent and efferent arterioles of perfused juxtamedullary nephrons mediate constriction under normal conditions [21]. In the same in vitro system, ET_B can nevertheless mediate vasodilatation in afferent arterioles of mice fed a high-salt diet [22]. The authors ascribe this vasodilatory response to the increased ET_B expression induced by the diet.

Renal infusion of ET-1 reduces cortical blood flow, yet has no effect on medullar blood flow, even though the diameter of the juxtamedullary arterioles is reduced. This can be explained by the fact that ET-1 constriction has a lower impact on the resistance of juxtamedullary vessels compared to cortical arterioles, due to their larger resting diameter. Moreover, juxtamedullary arterioles may only marginally affect medullary blood flow [15].

The vasa recta are branches of efferent arterioles that perfuse the renal medulla. Therefore, regulation of vasa recta contraction plays a major role in medullary blood flow and water/sodium retention. In the vasa recta, ET_B is expressed on the endothelial cells and ET_A on the pericytes [10]. ET-1 is a potent constrictor of the isolated vasa recta, mostly through an ET_A pathway [23]. This occurs through the depolarization of pericytes following inhibition of the potassium channel. In this way, ET-1 could reduce renal medullary blood flow [24]. Nevertheless, after intravenous injection, ET-1-induced production of NO and prostaglandins dilates the vasa recta and elevates medullary blood flow – thus promoting water reabsorption [25]. In conclusion, ET_A contributes to the constriction of the glomerular arterioles, and in this way controls renal function. ET_B participates in tonic renal vascular dilatation.

Endothelin-1/Endothelin B Axis Regulates Fluid Volume in the Tubular System: Endothelin-1 as a Antihypertensive Agent

Endothelin-1 Controls Channel Activity and Inhibits Salt Reabsorption

Whereas ET_A is the predominant receptor type in the vasculature, ET_B is overrepresented in the tubular system of the kidney. In the proximal tubule, ET_B (not ET_A) is expressed and it participates in sodium excretion. Proximal tubular Na^+/K^+-ATPase is inhibited by ET-1 [26]. In vitro studies have shown that this effect is due to activation of ET_B and the consequent enhancement of calcium recruitment [27].

In cells from the medullary thick ascending limb (TAL), ET-1 expression increases with osmolality. ET_B is the only receptor for ET-1 represented in the TAL, and inhibits sodium reabsorption. The mechanism involves activation of PI3K, AKT and eNOS, and a consecutive elevation in NO production that inhibits the $Na^+/K^+/Cl^-$ cotransporter [28]. In the cortical TAL of rats, the ET_B activation elevates the Ca^{2+} concentration and NO production, which participates in the inhibition of chloride reabsorption [29] (fig. 1).

ET-1 expression is relatively low in the thin limb segments of Henle's loop [30]. In the ascending and descending thin limbs, the expression of ET_A and ET_B is disputed depending on the species. In rats, recent immunohistochemical studies have shown no expression at all in this part of the nephron [10, 31].

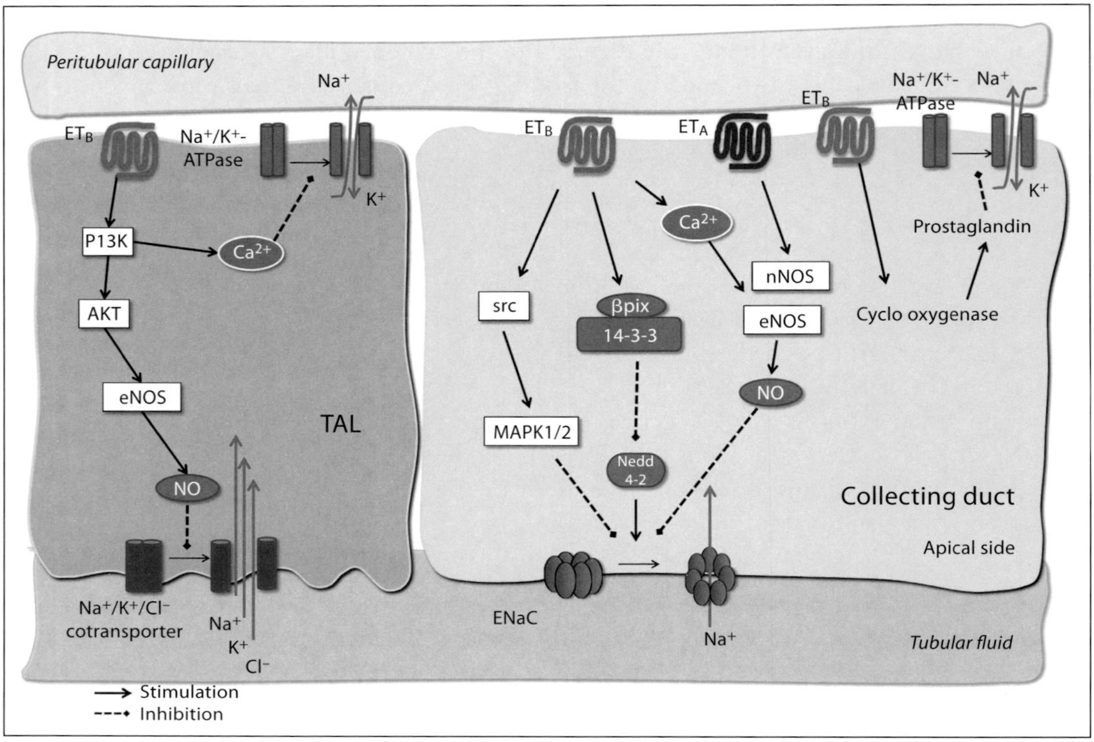

Fig. 1. Actions of ET-1 on sodium reabsorption. In the TAL, ET_B activation inhibits the Na^+/K^+/Cl^- cotransporter function. The opening of sodium channels (ENaC) in the collecting duct is reduced after ET_B activation. In both cell types, Na^+/K^+-ATPase activity is reduced by ET-1. PI3K = phosphatidylinositol 3-kinase; AKT = α-serine/threonine-protein kinase; src = proto-oncogene tyrosine-protein kinase; MAPK = mitogen-activated protein kinase.

The collecting duct (CD) shows the highest density of ET_B and the strongest ET-1 expression compared to other cell types [30]. The CD expresses a low amount of ET_A, if any, depending on the species.

Activation of ET_B receptors on the epithelial cells of the CD induces sodium reabsorption (fig. 1). The capacity of ET-1 to regulate Na^+/K^+-ATPase in the CD was known shortly after the discovery of ET-1 itself [32]. The underlying mechanisms presumably involve ET_B-induced cyclooxygenase activity and prostaglandin production [32]. Moreover, ET-1 regulates the activity of epithelial sodium channels (ENaC). ET-1 inhibits the opening of ENaC, possibly through different signaling pathways (fig. 1). In the long-term inactivation of ENaC by ET-1, the guanine exchange factor βPix plays a role by binding to the 14-3-3 protein. This prevents the binding of the 14-3-3 protein with the Nedd4-2 protein, which is required for ENaC activation [33]. In a rapid fashion,

ET-1 induces an *src*-kinase-dependent MAPK1/2 activation via ET_B, which may directly modify the phosphorylation status and thus reduce the activity of EnAC [34]. This mechanism does not require phospholipase C and protein kinase C (PKC). Finally, eNOS and iNOS are both involved in the inhibition of sodium reabsorption mediated by ET-1. In cells from the inner medullary CD, ET-1 enhances nNOS activity via ET_B [35]. eNOS and nNOS activity is reduced in the CD isolated from ET_B KO mice, but ET_A may be able to promote nNOS expression in these mice [36]. Finally, the activity of both eNOS and nNOS is lower in CD-specific ET-1-deficient mice and this is accompanied by sodium retention [37]. Taken together, the respective roles of ET_A and ET_B in the CD are distinct, but both seem to act in concert towards increasing sodium reabsorption.

The importance of the ET system in the CD for the control of blood pressure has been clearly demonstrated over the past decade by the use of genetically modified animals (table 1). Rescued ET_B KO mice (with ET_B transgenic expression specific to the enteric nervous system) develop salt-loading-dependent hypertension; importantly this phenotype is not due to endothelial dysfunction [1]. In the same way, ET_B deficiency in endothelial cells does not modify blood pressure on either a normal or high-salt diet [46]. Interestingly, the elevation of blood pressure whilst receiving a high-salt diet due to systemic ET_B deficiency is much more pronounced in female rats than in males, whereas female ET_B KO rats do not become hypertensive whilst receiving a normal diet [48]. The protective ET_B/eNOS pathway has a stronger effect in females than in males, and may account for this phenotype [49]. It is important to note that renal ET_A expression is diminished in ET_B KO mice. This may represent an adaptive mechanism to an enhanced tissue level of ET-1 consecutive to reduced ET-1 clearance by ET_B [13]. Using mice expressing Cre recombinase under the control of the aquaporin 2 promotor, Kohan and colleagues [2] developed mice deficient for ET-1, ET_B, ET_A, and combined ET_A/ET_B specifically in the CD. The main finding of these experiments is that CD ET-1 is necessary for maintaining normal blood pressure in vivo. With the exception of CD ET_A KO, all strains were hypertensive, suggesting that ET_B alone is responsible for blood pressure regulation. These results confirmed the predominant role of ET_B in the inhibition of sodium reabsorption. Nevertheless, deletion of ET_B alone leads to an elevation of 10 mm Hg in arterial blood pressure, while CD ET-1 KO and combined ET_A/ET_B KO mice show an elevation of pressure twice as high (almost 20 mm Hg higher than wild-type) [2]. Under the high-salt diet, blood pressure further increases in CD ET-1 KO mice (15 mm Hg higher than under a normal diet) [43] and, if rather modestly, also in CD ET_B KO mice (5 mm Hg higher than under normal diet) [45], while blood pressure remains unchanged in wild-type and CD ET_A KO mice. In combined CD ET_A/ET_B KO mice, blood pressure reaches the level of CD ET-1 KO mice after 4 days on a high-salt diet [2]. Using ET_B-deficient rats, Nakano et al. [50] revealed that ET_A can induce natriuresis independently of its hemodynamic effect through the activation of nNOS in CD cells. Note that ET_B

Table 1. Genetically modified mouse and rat lines for the endothelin system and their phenotype regarding fluid management, vascular function, blood pressure and kidney function

Target gene	Genetic manipulation	Model	Phenotype and main findings	Ref. no.
ET-1	systemic overexpression		normotensive, renal fibrosis, glomerulosclerosis	8
	endothelial overexpression		normotensive	7
	endothelial overexpression	high salt	hypertensive +, vascular inflammation	6
	endothelial overexpression	cerebral artery occlusion	loss of brain-blood barrier integrity	38
	astrocyte overexpression	acute water intoxication	water accumulation, brain edema	39
	systemic KO (heterozygous)		hypertensive	3
	CD KO	water loading/ vasopressin treatment	reduced acute diuresis/enhanced acute vasopressin response	40
			enhanced adenylate cyclase abundance/cAMP accumulation	41
		high salt	impaired water excretion, compensation by prostaglandin E	42
			NOS involved	37
			Na retention, hypertensive on normal ++ and high-salt diets +++	43
	vascular endothelial cells KO		low blood pressure	5
ET_A	CD	high salt	normotensive, normal Na excretion, enhanced acute water excretion	44
ET_B	CD KO	high salt	hypertensive on normal and high-salt diets +	45
	endothelial cells KO	high salt	normotensive	46
	systemic KO (rescued rats)	high salt	hypertensive on high-salt diet +++, prevention by amiloride	47
			ET_B is involved in vasodilative tone and diuresis	20
		high salt	females normotensive on normal diet, more hypertensive on high-salt diet than males	48

Table 1. Continued

Target gene	Genetic manipulation	Model	Phenotype and main findings	Ref. no.
ET_B	systemic KO (rescued mice)	high salt	hypertensive on high-salt diet	1
ET_A/ ET_B	CD KO	high salt	hypertensive on normal ++ and high-salt diets +++, Na retention	2

+ = 10–15 mm Hg; ++ = 15–25 mm Hg; +++ = > 25 mm Hg elevation of blood pressure vs. WT control.

has a much higher density compared to ET_A in the CD, and that absence of the latter alone may not change CD physiology.

Taken together, ET_A is involved in the regulation of sodium reabsorption in the CD to a lesser extent than ET_B or at least in the absence of ET_B. Whether this effect is biologically relevant or results from compensatory effects due to chronic gene deletion is not clear so far.

Because the response to the high-salt diet of CD ET-1 KO mice differs from that of combined CD ET_A/ET_B KO mice, and because single-receptor KO have lower blood pressures than CD ET-1 KO mice, CD ET-1 may act on adjacent structures for the control of blood pressure [2, 45]. Amiloride treatment reduces hypertension in CD ET-1 KO and ET_B rescued rats, confirming the role of ET-1/ ET_B on the activity of ENaC channels [43, 47]. The hypotensive effect of furosemide in CD ET-1 KO mice additionally confirmed the impact of ET-1 on the Na^+-K^+-Cl^- cotransporter in vivo [43].

In conclusion, ET_B is responsible for inhibiting salt reabsorption. The exact role of ET_A in handling salt in the CD remains unclear.

Endothelin-1 Antagonizes Vasopressin Actions and Regulates Blood Volume

ET-1 controls water handling in the CD, but not in the TAL. It was first demonstrated in the perfused CD that the arginine vasopressin (AVP)-induced elevation of water permeability was inhibited by ET-1. ET-1 reduces cAMP concentration, independently of prostaglandins, but in dependence of PKC; this effect is specific to ET_B activation [51]. Moreover, an interplay between ET-1 and the AVP receptor 2/aquaporin 2 (V2R/AQP2) system exists at the expression level (fig. 2). Using the adenylate cyclase (AC)-cAMP-PKA pathway, activation of V2R leads to a quick phosphorylation of ET_B and consecutive internalization. In the long-term, the same pathway can reduce ET_B mRNA expression [52]. The other way around, the interaction between ET-1 and ET_B increases mRNA expression of the V2R [53]. Furthermore, both AQP2 and V2R expressions are reduced by bosentan treatment in rodents [54].

CD ET-1 KO mice have an impaired ability to excrete an acute water load, suggesting that the net effect of ET-1 is diuretic. An elevated AVP-induced

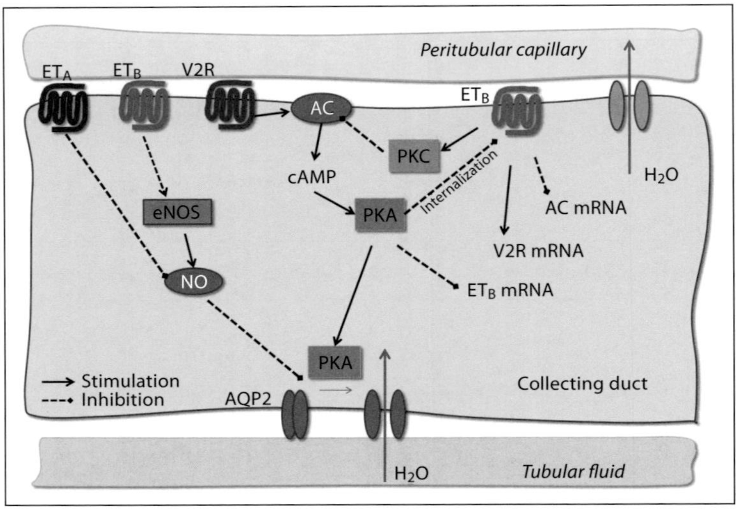

Fig. 2. Actions of ET-1 on water handling in the CD: ET_B activation results in enhanced activity of PKC which inactivates the adenylyl cyclase and thereby counteracts vasopressin actions. ET-1 may act on aquaporin activity through the regulation of NO production by both receptors. V2R = Vasopressin receptor 2; AQP2 = aquaporin 2; AC = adenylate cyclase; cAMP = cyclic adenosine monophosphate; PKA = protein kinase A.

cAMP accumulation in these mice further indicates that ET-1 decreases AVP-induced AC activity [40]. Abundance of AC itself is enhanced in the CD of CD ET-1 KO mice, showing that CD ET-1 exerts a diuretic effect by tonic repression of AC expression [41]. Contrarily, CD ET_A KO mice show a reduced sensitivity to AVP associated with reduced AVP-induced cAMP accumulation, and can handle an acute water load more quickly. Therefore, ET_A seems to increase the response to AVP and reduce diuresis in mice – in which sense, it opposes the effects of ET_B [44]. These data suggest that ET_A could moderate the predominant diuretic action of ET-1, at least during acute water loading. Given that ET-1 is overexpressed in such situations, one could consider the antidiuretic effect of ET_A as a negative feedback to compensate ET-1/ET_B-induced water loss. The authors of this study speculated on possible signaling mechanisms explaining how potential ET_A-induced superoxide could scavenge ET_B-eNOS-produced NO. They proposed a regulatory mechanism intrinsic to the ET system, yet similar to the one observed inside the vascular wall. Nevertheless, whether NO formed after ET_B activation is able to reduce AVP-induced cAMP accumulation is uncertain [35]. ET-1 stimulates NO formation in the CD, and NO has been demonstrated to reduce AVP-stimulated water permeability in isolated CD [55], but ET-1-induced NO has no influence on AVP-induced cAMP accumulation in rat CD cells [35].

In conclusion, ET-1 is an important regulator of water reabsorption. Through inhibition of AVP-induced AQP2 activity, activation of ET_B plays a major role in facilitating diuresis. ET_A may attenuate this effect, at least after hyperhydration.

Pathological Features of Endothelin-1 Related to the Glomerulus

All components of the ET system are present in the mesangial cells and podocytes. Endothelial cells of the glomerulus express only ET_B, but are also a source of ET-1 [10, 30]. ET-1 plays a crucial role in the function of the glomerular filtration barrier in health and disease. Under diabetic conditions, ET-1 expression is enhanced in the glomerulus [56]. The expression of the receptors however remains unchanged [57, 58].

In mesangial cells, ET-1 induces the three main pathophysiological phenomena: excessive contraction, proliferation and extracellular matrix accumulation (collagen I, IV and fibronectin). This leads to glomerulosclerosis and kidney failure. Mice carrying an ET-1 transgene develop glomerulosclerosis and renal fibrosis [8]. Importantly, the effects of AVP and ANG-II on mesangial cell proliferation and extracellular matrix production are in part modulated by ET-1/ET_A [59, 60]. In part because of the convenience of using mesangial cell cultures, the multiple signaling pathways involved have been well documented (fig. 3). Activation of ET_A activates phospholipase C, which produces IP_3 (inositol-3-phosphate) and diacylglycerol. Pathways 1–5 are displayed in figure 3: (1) Diacylglycerol activates PKC which phosphorylates nonreceptor tyrosine kinase *src* and stimulates *ras*, and thus proliferation [61]. (2) IP_3 binds to its sarcomere receptor and promotes the release of calcium, thereby activating the tyrosine kinase Pyk2 which mediates ET-1 signaling via p130Cas/BCAR3/rap1 and disturbs mesangial cell structure and function [62]. (3) ET-1 inhibits iNOS expression through ET_A. ET_B activation in turn enhances NO production [63]. This effect is calcium dependent, which excludes the involvement of iNOS. eNOS is only present in mesangial cells during high-glucose situations [40], indicating that nNOS may be the source of ET_B-induced NO [64]. (4) ET-1 induces EGF receptor transactivation, which generates a proliferative and profibrotic pathway involving $G\alpha_i$, β_1Pix, and ERK1/2 [65]. A newly discovered MAP kinase ERk5 may be involved in the action of ET-1 through transactivation of the EGF receptors [66]. (5) ET-1 also promotes mesangial cell hypertrophy through a pathway implicating PI3K, AKT, $NFAT_4$, and p8 proteins [67].

Podocytes and their foot processes represent a crucial layer of the glomerular filtration barrier. A correct cytoskeletal structure is essential for podocyte function. ET-1 destabilizes f-actin filaments in these cells and promotes nephrin shedding, which causes foot process effacement and podocyte

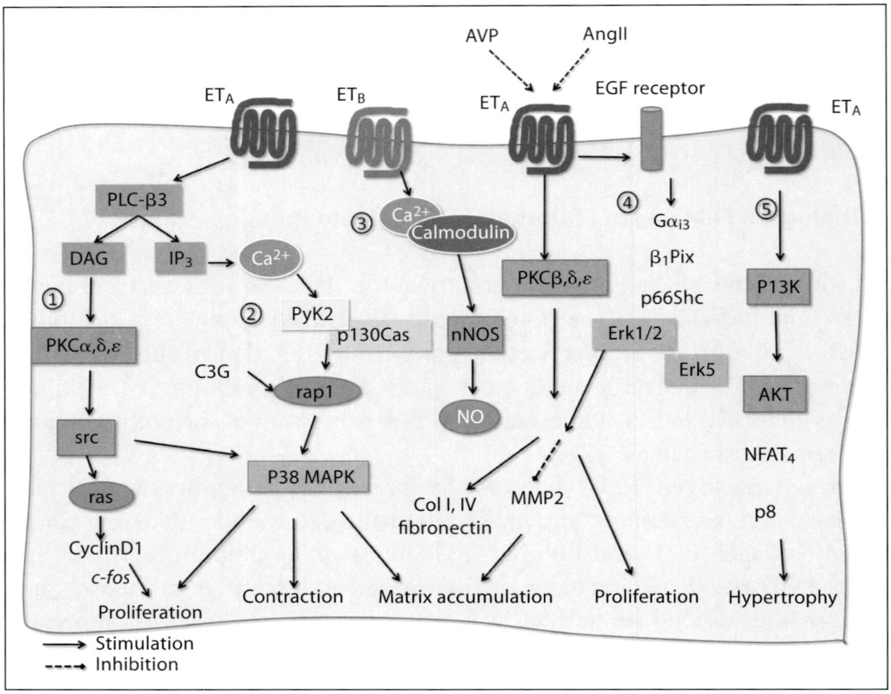

Fig. 3. Actions of ET-1 on mesangial cells. An excess of ET-1 stimulates diverse pathological pathways leading to exaggerated constriction, matrix protein production, proliferation and hypertrophy. These effects are partly under the control of AVP and angiotensin II (AngII). Pathways 1–5 are explained in the main text. PLC = Phospholipase C; DAG = diacylglycerol; IP_3 = inositol-3-phosphate; src = proto-oncogene tyrosine-protein kinase; MAPK and ERK = mitogen-activated protein kinases; ras and rap1 = GTPases; Pyk2 = protein tyrosine kinase 2β; MMP = matrix metalloprotease; AKT = alpha serine/threonine-protein kinase; C3G = guanine nucleotide exchange factor 1; NFAT = nuclear factor of activated T cells.

detachment/loss. This cytopathological situation is strongly linked to proteinuria and glomerular disease. Protein overload promotes cytoskeleton disassembly in podocytes and elevates ET-1 production. Excessive ET-1 can, in turn, further destabilize the cytoskeletal arrangement of podocytes which leads to enhanced permeability of the podocyte monolayer and thus proteinuria [68]. In isolated glomeruli and in rats, an ET_A blockade can prevent the elevation of albumin permeability caused by an ET-1 infusion, independently of vasoconstriction [69]. Particularly in cases of diabetes-induced hyperglycemia, excessive ET-1 is also responsible for podocyte death. In rats, a blockade of ET_A in experimental diabetes prevents the decrease in podocyte numbers [70]. Notably, these mechanisms may be induced by angiotensin, but ET-1 possesses specific features on podocytes since angiotensin-converting enzyme inhibitor and ET_A blockade show additive protective effects [70]. Podocyte damage can also be prevented in a rat model for glomerulosclerosis by darusentan, an ET_A antagonist [71]. ET_A antagonism also prevents adverse changes of the glomerular basement membrane composition in experimental diabetes, and is accompanied by a reduction in proteinuria [72]. The importance of a functional ET-1/ET_A pathway may be underlined by the analogy of the blood-brain barrier, a system similar to the renal filtration barrier and the integrity of which depends upon normal ET-1 expression in astrocytes and endothelial cells [38, 39].

ET_B is particularly present in podocyte foot processes and may therefore regulate filtration [73].

Given the cross-talk between each of the glomerular cells, the paracrine characteristic of ET-1 is certainly relevant in this context. It is possible that podocyte ET-1 adversely affects mesangial cells and modifies the adjacent microcirculation status [68]. Unpublished data from our group support the role of endothelium-derived ET-1 in the structure and function of podocytes in the diabetic mouse.

Unfortunately, to date, no mesangial cell-specific gene is known and it is therefore impossible to generate specific KO mice. Podocyte-specific KO is however possible using overexpression of Cre recombinase under the control of the nephrin promotor [74]. Such genetic mouse models for the ET system would be of particular interest to reveal the therapeutic potential of ET-1 antagonists in proteinuric diseases.

Conclusion

ET-1 is a major contributor to renal physiology. The distinct roles of the two receptors and their wide localization in the different compartments of the kidney render the analysis of ET-1's actions complex. Over the last years, in part thanks to genetically manipulated rodents, it has become clear that the net effect of ET_B activation is a reduction in water and salt reabsorption and that ET_A may participate in fine-tuning these effects. In proteinuric kidney disease, excessive activation of ET_A on podocytes and mesangial cells seems to be part of the etiology. Promising clinical studies using ET antagonists have confirmed these assumptions, but patients often develop fluid accumulation. The natriuretic and diuretic features of ET-1 are responsible for these undesirable side effects. Therefore, specific blockade of ET_A in the glomerulus represents a challenging but rational approach for the treatment of proteinuric diseases.

References

1 Quaschning T, Rebhan B, Wunderlich C, Wanner C, Richter CM, Pfab T, Bauer C, Kraemer-Guth A, Galle J, Yanagisawa M, Hocher B: Endothelin B receptor-deficient mice develop endothelial dysfunction independently of salt loading. J Hypertens 2005;23:979–985.

2 Ge Y, Bagnall A, Stricklett PK, Webb D, Kotelevtsev Y, Kohan DE: Combined knockout of collecting duct endothelin A and B receptors causes hypertension and sodium retention. Am J Physiol Renal Physiol 2008;295:F1635–F1640.

3 Kurihara Y, Kurihara H, Suzuki H, Kodama T, Maemura K, Nagai R, Oda H, Kuwaki T, Cao WH, Kamada N: Elevated blood pressure and craniofacial abnormalities in mice deficient in endothelin-1. Nature 1994;368:703–710.

4 Yanagisawa M, Kurihara H, Kimura S, Tomobe Y, Kobayashi M, Mitsui Y, Yazaki Y, Goto K, Masaki T: A novel potent vasoconstrictor peptide produced by vascular endothelial cells. Nature 1988;332:411–415.

5 Kisanuki YY, Emoto N, Ohuchi T, Widyantoro B, Yagi K, Nakayama K, Kedzierski RM, Hammer RE, Yanagisawa H, Williams SC, Richardson JA, Suzuki T, Yanagisawa M: Low blood pressure in endothelial cell-specific endothelin 1 knockout mice. Hypertension 2010;56:121–128.

6 Amiri F, Ko EA, Javeshghani D, Reudelhuber TL, Schiffrin EL: Deleterious combined effects of salt-loading and endothelial cell restricted endothelin-1 overexpression on blood pressure and vascular function in mice. J Hypertens 2010;28:1243–1251.

7 Amiri F, Virdis A, Neves MF, Iglarz M, Seidah NG, Touyz RM, Reudelhuber TL, Schiffrin EL: Endothelium-restricted overexpression of human endothelin-1 causes vascular remodeling and endothelial dysfunction. Circulation 2004;110:2233–2240.

8 Hocher B, Thone-Reineke C, Rohmeiss P, Schmager F, Slowinski T, Burst V, Siegmund F, Quertermous T, Bauer C, Neumayer HH, Schleuning WD, Theuring F: Endothelin-1 transgenic mice develop glomerulosclerosis, interstitial fibrosis, and renal cysts but not hypertension. J Clin Invest 1997;99:1380–1389.

9 Kuwaki T, Ling GY, Onodera M, Ishii T, Nakamura A, Ju KH, Cao WH, Kumada M, Kurihara H, Kurihara Y, Yazaki Y, Ohuchi T, Yanagisawa M, Fukuda Y: Endothelin in the central control of cardiovascular and respiratory functions. Clin Exp Pharmacol Physiol 1999;26:989–994.

10 Wendel M, Knels L, Kummer W, Koch T: Distribution of endothelin receptor subtypes ET_A and ET_B in the rat kidney. J Histochem Cytochem 2006;54:1193–1203.

11 Qiu C, Samsell L, Baylis C: Actions of endogenous endothelin on glomerular hemodynamics in the rat. Am J Physiol 1995;269(part 2):R469–R473.

12 Edwards RM, Trizna W, Ohlstein EH: Renal microvascular effects of endothelin. Am J Physiol 1990;259:F217–F221.

13 Schildroth J, Rettig-Zimmermann J, Kalk P, Steege A, Fahling M, Sendeski M, Paliege A, Lai EY, Bachmann S, Persson PB, Hocher B, Patzak A: Endothelin type A and B receptors in the control of afferent and efferent arterioles in mice. Nephrol Dial Transplant 2011;26:779–789.

14 Vuurmans JL, Boer P, Koomans HA: Effects of endothelin-1 and endothelin-1-receptor blockade on renal function in humans. Nephrol Dial Transplant 2004;19:2742–2746.

15 Denton KM, Shweta A, Finkelstein L, Flower RL, Evans RG: Effect of endothelin-1 on regional kidney blood flow and renal arteriole calibre in rabbits. Clin Exp Pharmacol Physiol 2004;31:494–501.

16 Dhaun N, Ferro CJ, Davenport AP, Haynes WG, Goddard J, Webb DJ: Haemodynamic and renal effects of endothelin receptor antagonism in patients with chronic kidney disease. Nephrol Dial Transplant 2007;22:3228–3234.

17 Goddard J, Johnston NR, Hand MF, Cumming AD, Rabelink TJ, Rankin AJ, Webb DJ: Endothelin-A receptor antagonism reduces blood pressure and increases renal blood flow in hypertensive patients with chronic renal failure: a comparison of selective and combined endothelin receptor blockade. Circulation 2004;109:1186–1193.
18 Matsuura T, Miura K, Ebara T, Yukimura T, Yamanaka S, Kim S, Iwao H: Renal vascular effects of the selective endothelin receptor antagonists in anaesthetized rats. Br J Pharmacol 1997;122:81–86.
19 Evans RG, Madden AC, Oliver JJ, Lewis TV: Effects of ET(A) – and ET(B)-receptor antagonists on regional kidney blood flow, and responses to intravenous endothelin-1, in anaesthetized rabbits. J Hypertens 2001;19:1789–1799.
20 Pollock DM: Contrasting pharmacological ET_B receptor blockade with genetic ET_B deficiency in renal responses to big ET-1. Physiol Genomics 2001;6:39–43.
21 Inscho EW, Imig JD, Cook AK, Pollock DM: ET_A and ET_B receptors differentially modulate afferent and efferent arteriolar responses to endothelin. Br J Pharmacol 2005;146:1019–1026.
22 Schneider MP, Inscho EW, Pollock DM: Attenuated vasoconstrictor responses to endothelin in afferent arterioles during a high-salt diet. Am J Physiol Renal Physiol 2007;292:F1208–F1214.
23 Pallone TL, Silldorff EP: Pericyte regulation of renal medullary blood flow. Exp Nephrol 2001;9:165–170.
24 Cao C, Lee-Kwon W, Silldorff EP, Pallone TL: KATP channel conductance of descending vasa recta pericytes. Am J Physiol Renal Physiol 2005;289:F1235–F1245.
25 Rubinstein I, Gurbanov K, Hoffman A, Better OS, Winaver J: Differential effect of endothelin-1 on renal regional blood flow: role of nitric oxide. J Cardiovasc Pharmacol 1995;26(suppl 3):S208–S210.
26 Garvin J, Sanders K: Endothelin inhibits fluid and bicarbonate transport in part by reducing Na^+/K^+ ATPase activity in the rat proximal straight tubule. J Am Soc Nephrol 1991;2:976–982.
27 Liu Y, Yang J, Ren H, He D, Pascua A, Armando MI, Yang C, Zhou L, Felder RA, Jose PA, Zeng C: Inhibitory effect of ET_B receptor on Na(+)-K(+) ATPase activity by extracellular Ca(2+) entry and Ca(2+) release from the endoplasmic reticulum in renal proximal tubule cells. Hypertens Res 2009;32:846–852.
28 Herrera M, Hong NJ, Ortiz PA, Garvin JL: Endothelin-1 inhibits thick ascending limb transport via Akt-stimulated nitric oxide production. J Biol Chem 2009;284:1454–1460.
29 Plato CF, Pollock DM, Garvin JL: Endothelin inhibits thick ascending limb chloride flux via ET(B) receptor-mediated NO release. Am J Physiol Renal Physiol 2000;279:F326–F333.
30 Moridaira K, Nodera M, Sato G, Yanagisawa H: Detection of prepro-ET-1 mRNA in normal rat kidney by in situ RT-PCR. Nephron Exp Nephrol 2003;95:e55–e61.
31 Yamamoto T, Suzuki H, Kubo Y, Matsumoto A, Uemura H: Endothelin A receptor-like immunoreactivity on the basal infoldings of rat renal tubules and collecting ducts. Arch Histol Cytol 2008;71:77–87.
32 Zeidel ML, Brady HR, Kone BC, Gullans SR, Brenner BM: Endothelin, a peptide inhibitor of Na(+)-K(+)-ATPase in intact renaltubular epithelial cells. Am J Physiol 1989;257(part 1):C1101–C1107.
33 Pavlov TS, Chahdi A, Ilatovskaya DV, Levchenko V, Vandewalle A, Pochynyuk O, Sorokin A, Staruschenko A: Endothelin-1 inhibits the epithelial Na^+ channel through betaPix/14-3-3/Nedd4-2. J Am Soc Nephrol 2010;21:833–843.
34 Bugaj V, Pochynyuk O, Mironova E, Vandewalle A, Medina JL, Stockand JD: Regulation of the epithelial Na^+ channel by endothelin-1 in rat collecting duct. Am J Physiol Renal Physiol 2008;295:F1063–F1070.
35 Stricklett PK, Hughes AK, Kohan DE: Endothelin-1 stimulates NO production and inhibits cAMP accumulation in rat inner medullary collecting duct through independent pathways. Am J Physiol Renal Physiol 2006;290:F1315–F1319.

36 Sullivan JC, Goodchild TT, Cai Z, Pollock DM, Pollock JS. Endothelin(A) (ET(A)) and ET(B) receptor-mediated regulation of nitric oxide synthase 1 (NOS1) and NOS3 isoforms in the renal inner medulla. Acta Physiol (Oxf) 2007;191:329–336.
37 Schneider MP, Ge Y, Pollock DM, Pollock JS, Kohan DE: Collecting duct-derived endothelin regulates arterial pressure and Na excretion via nitric oxide. Hypertension 2008;51:1605–1610.
38 Leung JW, Ho MC, Lo AC, Chung SS, Chung SK: Endothelial cell-specific over-expression of endothelin-1 leads to more severe cerebral damage following transient middle cerebral artery occlusion. J Cardiovasc Pharmacol 2004;44(suppl 1):S293–S300.
39 Yeung PK, Lo AC, Leung JW, Chung SS, Chung SK: Targeted overexpression of endothelin-1 in astrocytes leads to more severe cytotoxic brain edema and higher mortality. J Cereb Blood Flow Metab 2009;29:1891–1902.
40 Ge Y, Ahn D, Stricklett PK, Hughes AK, Yanagisawa M, Verbalis JG, Kohan DE: Collecting duct-specific knockout of endothelin-1 alters vasopressin regulation of urine osmolality. Am J Physiol Renal Physiol 2005;288:F912–F920.
41 Strait KA, Stricklett PK, Kohan DE: Altered collecting duct adenylyl cyclase content in collecting duct endothelin-1 knockout mice. BMC Nephrol 2007;8:8.
42 Ge Y, Strait KA, Stricklett PK, Yang T, Kohan DE: Role of prostaglandins in collecting duct-derived endothelin-1 regulation of blood pressure and water excretion. Am J Physiol Renal Physiol 2007;293:F1805–F1810.
43 Ahn D, Ge Y, Stricklett PK, Gill P, Taylor D, Hughes AK, Yanagisawa M, Miller L, Nelson RD, Kohan DE: Collecting duct-specific knockout of endothelin-1 causes hypertension and sodium retention. J Clin Invest 2004;114:504–511.
44 Ge Y, Stricklett PK, Hughes AK, Yanagisawa M, Kohan DE: Collecting duct-specific knockout of the endothelin A receptor alters renal vasopressin responsiveness, but not sodium excretion or blood pressure. Am J Physiol Renal Physiol 2005;289:F692–F698.
45 Ge Y, Bagnall A, Stricklett PK, Strait K, Webb DJ, Kotelevtsev Y, Kohan DE: Collecting duct-specific knockout of the endothelin B receptor causes hypertension and sodium retention. Am J Physiol Renal Physiol 2006;291:F1274–F1280.
46 Bagnall AJ, Kelland NF, Gulliver-Sloan F, Davenport AP, Gray GA, Yanagisawa M, Webb DJ, Kotelevtsev YV: Deletion of endothelial cell endothelin B receptors does not affect blood pressure or sensitivity to salt. Hypertension 2006;48:286–293.
47 Gariepy CE, Ohuchi T, Williams SC, Richardson JA, Yanagisawa M: Salt-sensitive hypertension in endothelin-B receptor-deficient rats. J Clin Invest 2000;105:925–933.
48 Taylor TA, Gariepy CE, Pollock DM, Pollock JS: Gender differences in ET and NOS systems in ET_B receptor-deficient rats: effect of a high salt diet. Hypertension 2003;41(part 2):657–662.
49 Chappell MC, Gallagher PE, Averill DB, Ferrario CM, Brosnihan KB: Estrogen or the AT1 antagonist olmesartan reverses the development of profound hypertension in the congenic mRen2. Lewis rat. Hypertension 2003;42:781–786.
50 Nakano D, Pollock DM: Contribution of endothelin A receptors in endothelin 1-dependent natriuresis in female rats. Hypertension 2009;53:324–330.
51 Kohan DE, Padilla E, Hughes AK: Endothelin B receptor mediates ET-1 effects on cAMP and PGE2 accumulation in rat IMCD. Am J Physiol 1993;265(part 2):F670–F676.
52 Wong NL, Wong BP, Tsui JK: Vasopressin regulates endothelin-B receptor in rat inner medullary collecting duct. Am J Physiol Renal Physiol 2000;278:F369–F374.
53 Wang MH, Fok A, Huang MH, Wong NL: Interaction between endothelin and angiotensin II in the up-regulation of vasopressin messenger RNA in the inner medullary collecting duct of the rat. Metabolism 2007;56:1372–1376.

54 Wong NL, Sonntag M, Tsui JK: Attenuation of renal vasopressin V2 receptor upregulation by bosentan, an ET_A/ ET_B receptor antagonist. Metabolism 2003;52:1141–1146.
55 Garcia NH, Stoos BA, Carretero OA, Garvin JL: Mechanism of the nitric oxide-induced blockade of collecting duct water permeability. Hypertension 1996;27(part 2):679–683.
56 Benigni A, Colosio V, Brena C, Bruzzi I, Bertani T, Remuzzi G: Unselective inhibition of endothelin receptors reduces renal dysfunction in experimental diabetes. Diabetes 1998;47:450–456.
57 Fukui M, Nakamura T, Ebihara I, Osada S, Tomino Y, Masaki T, Goto K, Furuichi Y, Koide H: Gene expression for endothelins and their receptors in glomeruli of diabetic rats. J Lab Clin Med 1993;122:149–156.
58 Watson AM, Li J, Schumacher C, de Gasparo M, Feng B, Thomas MC, Allen TJ, Cooper ME, Jandeleit-Dahm KA: The endothelin receptor antagonist avosentan ameliorates nephropathy and atherosclerosis in diabetic apolipoprotein E knockout mice. Diabetologia 2010;53:192–203.
59 Tahara A, Tsukada J, Tomura Y, Suzuki T, Yatsu T, Shibasaki M: Vasopressin stimulates the production of extracellular matrix by cultured rat mesangial cells. Clin Exp Pharmacol Physiol 2008;35:586–593.
60 Kuo HT, Shin SJ, Kuo MC, Chen HC: Effects of specific endothelin-1 receptor antagonists on proliferation and fibronectin production of glomerular mesangial cells stimulated with angiotensin II. Kaohsiung J Med Sci 2006;22:371–376.
61 Mishra R, Wang Y, Simonson MS: Cell cycle signaling by endothelin-1 requires Src nonreceptor protein tyrosine kinase. Mol Pharmacol 2005;67:2049–2056.
62 Rufanova VA, Alexanian A, Wakatsuki T, Lerner A, Sorokin A: Pyk2 mediates endothelin-1 signaling via p130Cas/BCAR3 cascade and regulates human glomerular mesangial cell adhesion and spreading. J Cell Physiol 2009;219:45–56.
63 Owada A, Tomita K, Terada Y, Sakamoto H, Nonoguchi H, Marumo F: Endothelin (ET)-3 stimulates cyclic guanosine 3′,5′-monophosphate production via ET_B receptor by producing nitric oxide in isolated rat glomerulus, and in cultured rat mesangial cells. J Clin Invest 1994;93:556–563.
64 Zhang W, Meng H, Li ZH, Shu Z, Ma X, Zhang BX: Regulation of STIM1, store-operated Ca^{2+} influx, and nitric oxide generation by retinoic acid in rat mesangial cells. Am J Physiol Renal Physiol 2007;292:F1054–F1064.
65 Chahdi A, Sorokin A: Endothelin-1 induces p66Shc activation through EGF receptor transactivation: role of beta(1)Pix/Galpha(i3) interaction. Cell Signal 2010;22:325–329.
66 Dorado F, Velasco S, Esparis-Ogando A, Pericacho M, Pandiella A, Silva J, Lopez-Novoa JM, Rodriguez-Barbero A: The mitogen-activated protein kinase Erk5 mediates human mesangial cell activation. Nephrol Dial Transplant 2008;23:3403–3411.
67 Goruppi S, Kyriakis JM: The pro-hypertrophic basic helix-loop-helix protein p8 is degraded by the ubiquitin/proteasome system in a protein kinase B/Akt- and glycogen synthase kinase-3-dependent manner, whereas endothelin induction of p8 mRNA and renal mesangial cell hypertrophy require $NFAT_4$. J Biol Chem 2004;279:20950–20958.
68 Morigi M, Buelli S, Zanchi C, Longaretti L, Macconi D, Benigni A, Moioli D, Remuzzi G, Zoja C: Shigatoxin-induced endothelin-1 expression in cultured podocytes autocrinally mediates actin remodeling. Am J Pathol 2006;169:1965–1975.
69 Saleh MA, Boesen EI, Pollock JS, Savin VJ, Pollock DM: Endothelin-1 increases glomerular permeability and inflammation independent of blood pressure in the rat. Hypertension 2010;56:942–949.
70 Gagliardini E, Corna D, Zoja C, Sangalli F, Carrara F, Rossi M, Conti S, Rottoli D, Longaretti L, Remuzzi A, Remuzzi G, Benigni A: Unlike each drug alone, lisinopril if combined with avosentan promotes regression of renal lesions in experimental diabetes. Am J Physiol Renal Physiol 2009;297:F1448–F1456.

71 Ortmann J, Amann K, Brandes RP, Kretzler M, Munter K, Parekh N, Traupe T, Lange M, Lattmann T, Barton M: Role of podocytes for reversal of glomerulosclerosis and proteinuria in the aging kidney after endothelin inhibition. Hypertension 2004;44:974–981.
72 Hocher B, Schwarz A, Reinbacher D, Jacobi J, Lun A, Priem F, Bauer C, Neumayer HH, Raschack M: Effects of endothelin receptor antagonists on the progression of diabetic nephropathy. Nephron 2001;87:161–169.
73 Yamamoto T, Hirohama T, Uemura H: Endothelin B receptor-like immunoreactivity in podocytes of the rat kidney. Arch Histol Cytol 2002;65:245–250.
74 Zhang S, Ji Y, Liu X, Lu X, Su W, Zhang D, Hao F, Yi F, Guo L, Li X, Zheng Y: Podocyte-specific VEGF down-regulation and pathophysiological development. IUBMB Life 2010;62:677–683.

Noriaki Emoto, MD, PhD
Division of Cardiovascular Medicine, Department of Internal Medicine
Kobe University Graduate School of Medicine
7–5–1 Kusunoki-cho, Chuo-ku
Kobe 650-0017 (Japan)
Tel. +81 78 382 5846, E-Mail emoto@med.kobe-u.ac.jp

Barton M, Kohan DE (eds): Endothelin in Renal Physiology and Disease.
Contrib Nephrol. Basel, Karger, 2011, vol 172, pp 35–49

Endothelin and the Renal Vasculature

Zhengrong Guan · Edward W. Inscho

Department of Physiology, Georgia Health Sciences University, Augusta, Ga., USA

Abstract

The endothelin (ET) system comprises a family of three isopeptides (ET-1, ET-2, and ET-3) involved in diverse physiological and pathophysiological events. ET-1 is the major renal peptide that exerts its biological activity by binding to ET_A and ET_B receptors. Both ET_A and ET_B receptors are expressed by renal microvascular smooth muscle cells, where activation causes vasoconstriction. ET_B receptors are also expressed by microvascular endothelial cells, where activation leads to vasodilator responses. ET-1 influences preglomerular and postglomerular microvascular tone and thus can significantly influence renal hemodynamics. Alteration of renal ET-1 synthesis and receptor expression has been reported in cardiovascular diseases, and could contribute to renal injury by altering renal microvascular reactivity. In this brief review, we will try to summarize what is known about ET control of renal microvascular function.

Since endothelin (ET) was discovered in 1988 [1], three distinct ET isopeptides have been classified: ET-1, ET-2, and ET-3 [2]. Initially, ET was considered to be of endothelial origin. It is now clear that ET is synthesized and released from many vascular and nonvascular cell types, including renal epithelial cells [3–5]. ET-1 is the predominant isoform produced by the vasculature and is constitutively secreted from the endothelium, where it acts in a paracrine/autocrine manner on adjacent endothelial or vascular smooth muscle cells (VSMC). ET-1 is produced from big ET-1 catalyzed by ET-converting enzymes present in endothelial cells and is reportedly the most potent and long-lasting vasoconstrictor [1]. Approximately 80% of secreted ET-1 is released into the basolateral compartment, whereas only a limited amount of ET-1 enters the circulation [6]. Generally, plasma ET-1 concentrations range between 0.5 and 5 pM in healthy human subjects [4, 7]. Interestingly, levels of immunoreactive ET are higher in the kidney inner medulla than any other organ or tissue [8], implying an important physiological role for ET-1 in regulating renal function.

ET influences diverse physiological and pathophysiological mechanisms by activation of distinct ET_A and ET_B receptor subtypes [3–5]. In the vasculature, ET_A and ET_B receptors are expressed by VSMC, while endothelial cells mainly express ET_B receptors. Activation of ET_A and ET_B receptors on VSMC increases intracellular Ca^{2+} concentration ($[Ca^{2+}]_i$), leading to increased vascular tone [4, 5]. Activation of endothelial ET_B receptors stimulates release of nitric oxide (NO) and prostacyclin leading to vasodilatation [9]. Additionally, ET_B receptors also serve as 'clearance receptors' to sequester ET-1 from the plasma [10, 11]. Accordingly, blockade or deletion of ET_B receptors tends to elevate ET-1 levels. Competition-binding studies reveal that ET_A and ET_B receptors have different binding affinities for the three ET isoforms. ET_A receptors have a higher affinity for ET-1 and ET-2, and a weaker affinity for ET-3. ET_B receptors, on the other hand, display similar affinities for all three ET isopeptides [3]. These receptor-binding characteristics have been widely used to identify tissue distribution of receptors.

In the kidney, evidence indicates that ET-1 is produced locally and regulates renal hemodynamics by acting on preglomerular and postglomerular microvascular reactivity. ET-1 is also involved in sodium and water transport, and cell proliferation [4, 5]. Alteration of renal ET-1 levels and receptor expression are implicated in cardiovascular disease development. This review will focus on the role of ET peptides in the renal microcirculation.

Endothelin Receptor Expression in Kidney

Early studies using receptor-binding, immunocytochemical, and pharmacological tools revealed that almost every renal vascular and tubular cell type expresses ET_A and/or ET_B receptors, but the expression levels vary greatly between different regions of the kidney and across species [4, 5]. In the canine kidney, ET_A and ET_B receptor ratios average 22:78, 40:60, and 50:50 in cortical, medullary, and papillary membranes, respectively [12], while in rat kidneys, the regional ET_A/ET_B receptor ratios were 50:50, 30:70, and 90:10, respectively [13]. ET receptors are expressed in high levels in the renal medulla, particularly in the inner medulla [14, 15]. Positron emission tomography makes it possible to quantify receptor concentration by measuring receptor-bound radioligands in vivo. This approach revealed that medullary ET_B receptor expression is double that of the cortex in rabbit kidney [15]. Similarly, ET_B receptor expression is nearly four times greater in the human renal medulla than in the cortex (33:9) [15], but the ET_A/ET_B receptor distribution is similar between the cortex and medulla (30:70) [16].

Little is known regarding how ET_A and ET_B receptors differentially influence renal vascular reactivity to ET-1. Most studies have employed pharmacological tools to determine the contribution of specific ET receptors in modulating renal microvascular function. Several studies conducted to localize ET receptor expression in renal vessels indicate that both ET_A and ET_B receptors are present in preglomerular and postglomerular microvessels [17–21]. ET_A receptors are present in VSMC, while ET_B receptors are mainly expressed in vascular and glomerular endothelial cells and are barely detectable in VSMC [17]. Radioligand-binding studies show that the proportion of ET_A and ET_B receptors in membranes of preglomerular vessels is 40:60 in rabbits [18] and 50:50 in rats [19]. In human kidneys, ET_A receptors are present in the vasculature throughout the cortex and medulla, with low binding densities in vasa recta and glomeruli [20]. Renal and arcuate arteries express a high density of ET_A receptors at 90–95% of total ET receptors [21]. Additionally, both ET_A and ET_B receptors are detected in rat vasa recta but they are located in pericytes and endothelial cells, respectively [17], implying that they might play different roles in regulating medullary perfusion.

Effects of Endothelin on Renal Hemodynamics

The renal microcirculatory system is a unique portal circulation characterized by two independently regulated resistance vessels (afferent and efferent arterioles) in series straddling the glomerular capillaries. The endothelial cells release a variety of vasodilators and vasoconstrictors in response to physical and chemical stimuli, shear stress, or stretch. These vasoactive substances modulate the glomerular microcirculation through autocrine and/or paracrine mechanisms to influence afferent and efferent arteriolar resistance and hence renal hemodynamics. Thus, the balance of vascular resistance conferred by afferent and efferent arterioles is a crucial factor in determining overall glomerular hemodynamics.

A role for ET in regulating the renal microcirculation has been substantiated by numerous in vivo and in vitro studies in different animal models. The kidney is highly sensitive to exogenous ET-1 compared to other organs [22]. Almost all in vivo studies show that infusion of exogenous ET-1 produces a marked and prolonged renal vasoconstriction marked by profound reductions in renal blood flow (RBF) and glomerular filtration rate (GFR) reflecting increased renal vascular resistance [12, 23–31]. However, the relative contribution of ET receptors to the ET-1-mediated vasoconstriction is variable among species and results are controversial. In canine kidneys, ET-1-induced reduction of RBF and GFR was abolished by ET_A receptor blockade [12], implicating ET_A receptors in regulating renal vascular reactivity. Furthermore, infusion of a highly selective ET_B receptor agonist, sarafotoxin 6c (S6c), markedly increased urine flow and sodium excretion, but had little effect on RBF and GFR [12], indicating that ET_B receptors primarily regulate tubular transport compared to vascular function in dog kidneys. ET-1 is also a powerful vasoconstrictor in rabbit kidneys [25, 26].

Interestingly, however, ET-1-mediated decreases in rabbit RBF reflect cortical vasoconstriction because medullary perfusion reportedly either increased [26] or did not change [32] during ET-1 infusion. ET-1-induced reduction of cortical perfusion was attenuated by ET_A receptor blockade, but enhanced by ET_B receptor blockade [26]. These studies indicate that ET-1-induced renal vasoconstriction in dogs and rabbits is primarily mediated by ET_A receptors.

In contrast to dogs and rabbits, many in vivo studies in rats indicate that both ET_A and ET_B receptors mediate renal vasoconstriction induced by ET-1, but ET_B receptors may play a more prominent role [27–31]. Intravenous infusion of ET-1 in anesthetized rats usually decreases RBF and GFR. This is accompanied by a transient reduction in blood pressure followed by a sustained pressor response. ET_A receptor antagonists abrogated the pressor effects of ET-1, but had little effect on the renal hemodynamic response [27, 31]. Others have shown that ET-1 and the ET_B receptor ligand or agonist (ET-3 or S6c) elicited a similar reduction in RBF [28, 30, 31]. The ET-1-induced reduction of RBF was partially blocked by ET_A receptor antagonists and completely abolished by combined $ET_{A/B}$ receptor blockade [30, 31]. Similar to observations in rabbits, ET-1-mediated reduction of RBF results from cortical vasoconstriction because medullary blood flow remained unchanged [33] or increased [34, 35]. Collectively, these findings indicate that both ET_A and ET_B receptors are involved in the renal microvascular response to ET-1, but ET_B receptors play a predominant role in rats. Moreover, the ability of ET-1 to decrease RBF was exaggerated by selective ET_B receptor blockade with BQ-788 [30], suggesting that ET-1 elicits both vasoconstrictor and vasodilator effects. The vasodilator component could reflect ET_B receptor-dependent release of vasodilators or reduced ET-1 clearance due to ET_B receptor blockade.

Effects of Endothelin on Renal Microvascular Reactivity in vitro

While the whole kidney studies provide in vivo evidence supporting ET-1 as a regulator of renal vascular reactivity, the data are often confounded by concomitant changes in mean arterial pressure and released circulating factors. Furthermore, the distribution of ET_A and ET_B receptors in different microvascular segments compromises interpretation of which ET receptors are responsible for ET-1's overall effect and to what degree they contribute to ET-1-induced alteration of renal hemodynamics. Accordingly, many in vitro models have been used to provide site-specific information. Studies have frequently utilized partially resected kidney models, with or without hydronephrosis, or isolated renal microvessels to directly visualize ET's effects on afferent and efferent arteriolar diameters.

The in vitro blood-perfused juxtamedullary nephron preparation is frequently used for direct assessment of renal microvascular reactivity while

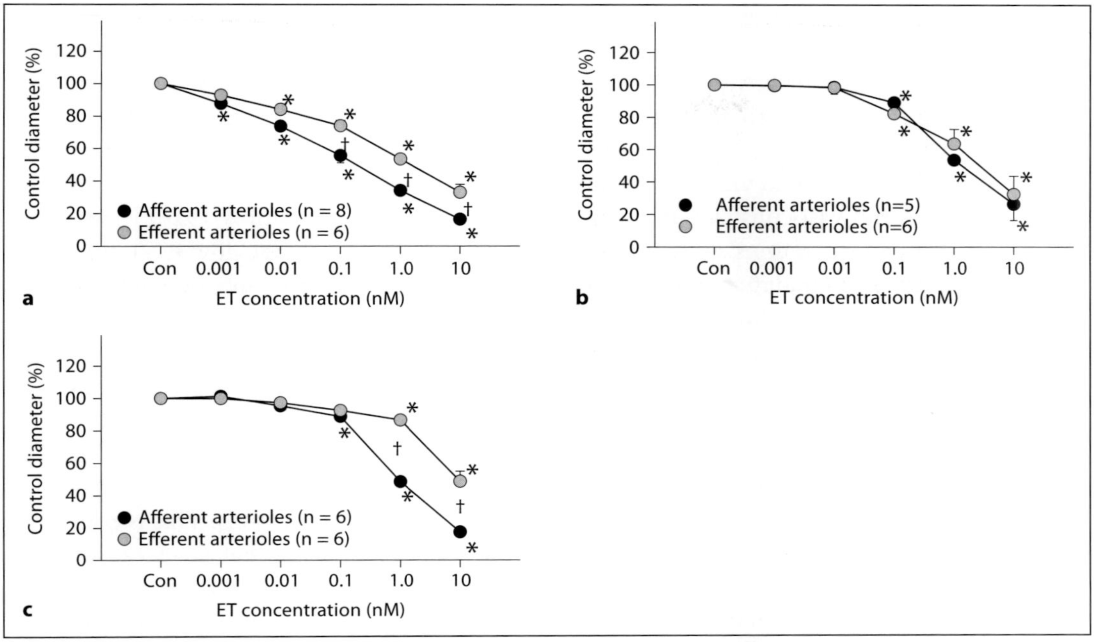

Fig. 1. Renal microvascular responses to ET peptides. Kidneys were superfused with increasing concentrations of ET-1 (**a**), ET-2 (**b**), or ET-3 (**c**) from 1.0 pM to 10 nM (5 min for each concentration), while perfusion pressure was maintained at 100 mm Hg. Afferent arteriolar diameters (black symbols) and efferent arteriolar diameters (gray symbols) were measured at 12-second intervals and calculated from the average of all measurements obtained during the final 2 min of each 5-min treatment period. Data are expressed as percent of the control diameter. Values are means ± SEM. *$p < 0.05$ vs. control diameter in same group; † $p < 0.05$ vs. control rats at same perfusion pressure. Adapted from Inscho et al. [38].

maintaining an intact association between the vascular and tubular elements [36]. ET-1, ET-2, and ET-3 all evoke concentration-dependent vasoconstriction of afferent and efferent arterioles (fig. 1) [37–40]. ET-1 vasoconstricts afferent and efferent arterioles at concentrations of 1 and 10 pM, respectively [38]. ET-1 vasoconstricted afferent arterioles more effectively than efferent arterioles such that the afferent diameter declined by 83% of the control compared to 67% for efferent arterioles at a concentration of 10 nM. Similar to ET-1, ET-3 also yielded a greater vasoconstrictor response in afferent than efferent arterioles. ET-2, however, elicited similar degrees of vasoconstriction for afferent and efferent arterioles. The ET-1 concentration curve in afferent arterioles was significantly shifted to the right by pretreatment with ET_A or ET_B receptor blockers and was completely abolished by combined $ET_{A/B}$ blockers, indicating the vasoconstrictor property of both ET receptors at preglomerular microvessels. In contrast, ET_A receptor blockade in efferent arterioles converted ET-1-induced vasoconstriction to vasodilation at low concentrations (0.01–0.1 nM) while vasoconstriction was retained at higher concentrations (1–10 nM), suggesting that ET_B receptors can elicit vasodilation and vasoconstriction of efferent arterioles. The dual effect of ET_B receptors in efferent arterioles was confirmed using S6c, which yielded efferent arteriolar vasodilation under control conditions but vasoconstriction during ET_B receptor blockade with A-192621 [38]. These observations suggest that during ET_A receptor blockade, low ET-1 concentrations mainly stimulate endothelial ET_B receptors, probably by releasing NO or prostacyclin, while higher concentrations of ET-1 shift to activating ET_B receptors on VSMC in efferent arterioles.

Preferential ET-1-mediated vasoconstriction of rat afferent arterioles has been demonstrated using different experimental approaches [24, 31, 41–43]. For example, micropuncture studies in Munich-Wistar rats showed that ET-1 reduced single nephron GFR with significant increases in afferent and efferent arteriolar resistance, but particularly in afferent arterioles [41]. This is consistent with studies in hydronephrotic rat kidneys showing that ET-1 caused substantial afferent arteriolar vasoconstriction but only modest efferent arteriolar vasoconstriction [24, 42]. Alternatively, other studies have indicated that efferent arterioles are more sensitive to ET-1 than afferent arterioles [23, 44]. Glomerular capillary pressure, as assessed by measuring stop-flow pressure, was significantly increased in response to ET, which was associated with a greater increase in efferent arteriolar resistance than in afferent arteriolar resistance [23]. Studies in isolated-perfused microvessels showed that the EC_{50} of ET-1 was lower in efferent arterioles (5.7 pM) compared to afferent arterioles (52 pM) [44], suggesting that ET-1 is a more potent vasoconstrictor of efferent arterioles. The explanation for the discrepancies among these studies is unclear, but may reflect differences between experimental conditions.

The aforementioned studies of ET-1 on renal microvascular reactivity mostly employed pharmacological tools to establish the contribution of ET receptors. Genetic approaches provide useful strategies to determine specific gene or protein functions. Although several ET-1 or ET_B receptor knockout mouse models have been developed, only one study has been conducted to assess the distribution of ET receptors in ET-1-induced renal microvascular responsiveness in vascular ET_B receptor-deficient mice and wild-type control mice [45]. The results obtained from isolated-perfused renal microvessels indicate that ET-1 more potently vasoconstricts afferent arterioles compared to efferent arterioles. Furthermore, ET-1-induced vasoconstriction involves activation of ET_A receptors in afferent arterioles, but involves both ET_A and ET_B receptors in efferent arterioles [45]. The relative contribution of ET receptors in ET-1-induced responses in mouse renal microvessels is different from rats. Interestingly, endothelium-specific ET-1 knockout mice exhibit 10- to 12-mm Hg lower mean arterial pressure [46], suggesting a pivotal role of endogenous endothelium-derived ET receptor system in control of blood pressure and maintenance of vascular tone under physiological conditions. How these changes affect renal hemodynamics remains to be determined.

Endothelin in Medullary Microcirculation

Medullary perfusion is supplied by descending vasa recta (DVR) arising from juxtamedullary efferent arterioles [47]. Given the potent vasoconstrictor effect of ET-1 in juxtamedullary nephron microvasculature and expression of ET_A and ET_B receptors in vasa recta [17], it is not surprising that ET is also involved in regulating medullary microcirculation. Elegant studies using isolated-perfused DVR establish ET as a potent vasoconstrictor of DVR [47]. ET-1, ET-2, and ET-3 elicit potent vasoconstriction with a rank order of potency of ET-1>ET-2>ET-3. DVR diameter started to decrease with 0.01 pM ET-1 and almost completely collapsed at 100 pM. ET-1-induced vasoconstriction was attenuated by an ET_A antagonist or a mixed $ET_{A/B}$ receptor antagonist, whereas ET-3-induced vasoconstriction was only blocked by an ET_B receptor antagonist, suggesting the dual involvement of ET_A and ET_B receptors. These observations, however, seem to contradict rabbit studies where infusion of ET-1 increases medullary blood flow with concurrent reduction in total RBF and cortical blood flow [26]. Conflicting outcomes probably reflect species variation (rats vs. rabbits) and different experimental conditions (in vitro vs. in vivo). The intense vasoconstriction of DVR by ET-1 suggests a potentially important role of ET-1 for the development of renal injury in cases where the renal ET system is upregulated [47].

Endogenous Endothelin-1 in Regulating Renal Microvascular Tone

While both in vivo and in vitro studies provide compelling evidence that ET-1 regulates renal hemodynamics by influencing preglomerular and postglomerular microvascular tone, there is conflicting evidence establishing that ET-1 regulates renal vascular tone under basal conditions. Infusion of ET_A receptor antagonists did not change RBF or GFR in conscious rats [9, 27, 48], implying that ET_A receptors may not regulate renal hemodynamics under basal conditions. In contrast, combined $ET_{A/B}$ receptor blockade with bosentan markedly reduced glomerular capillary pressure with a significant increase in preglomerular resistance while postglomerular resistance remained unchanged [48]. ET_B receptor blockade with RES-701-1 or BQ-788 significantly decreased RBF without affecting arterial blood pressure [9, 29]. Reduction of RBF was prevented by pretreatment with L-NAME (NO synthase inhibitor) and ibuprofen (cyclooxygenase inhibitor). These findings suggest that endogenous ET-1 contributes to basal renal vascular resistance by ET_B receptor-dependent release of NO and/or prostacyclin. Overall, these studies imply that ET_B receptor-mediated vasodilation may influence renal microcirculatory function under physiological conditions. For example, exogenous ET-1 or S6c normally vasoconstricts normal rat afferent arterioles [38]; however, when rats are fed a high-salt diet, ET-1 and S6c-mediated vasoconstriction is attenuated and even converted to vasodilation at low ET-1 concentrations. ET_B receptor expression in preglomerular microvessels is increased during salt loading [40]. Accordingly, salt-induced upregulation of ET_B receptor expression may counteract ET_A and ET_B receptor activation in VSMC and facilitate sodium excretion by increasing RBF. The physiological mechanisms behind salt-induced increases in ET_B receptor expression need further investigation.

Endothelin in Renal Autoregulation

Autoregulation is an important mechanism protecting selected organs against injury from variations in arterial pressure. This protective mechanism is accomplished in the kidney by precise pressure-mediated adjustments of preglomerular resistance via myogenic and tubuloglomerular feedback responses [49]. Although earlier studies indicated that ET-1 was not involved in renal autoregulation [50, 51], recent studies suggest that ET-1 might interact with NO in modulating or resetting renal autoregulation [52, 53]. L-NAME augmentation of autoregulation was eliminated by the nonselective $ET_{A/B}$ receptor antagonist bosentan or the ET_B receptor blocker BQ-788 [52]. A subsequent study by Shi et al. [53] showed that myogenic reactivity was significantly impaired by pretreatment with BQ-788, suggesting that ET-1 may be involved in resetting renal autoregulation by release of NO via ET_B receptor stimulation. Interestingly, exposure of macula densa cells to ET-1 caused a rapid increase of signal-nephron GFR, a measure of tubuloglomerular feedback responsiveness, without changes in systemic and renal hemodynamics [54]. This study suggests that ET-1 may influence tubuloglomerular feedback responses, but the mechanism(s) underlying this is (are) unclear.

Endothelin-Induced Intracellular Signaling in Renal Microvessels

Both ET_A and ET_B receptors belong to the G protein-coupled receptor family. ET receptor activation is coupled to many second messenger pathways including phospholipase C, phospholipase D, protein kinase C, mitogen-activated protein kinases, cytosolic Ca^{2+}, and tyrosine kinase [4, 5]. Application of ET-1 to VSMC increases $[Ca^{2+}]_i$ via influx of extracellular Ca^{2+} and mobilization of Ca^{2+} from intracellular stores [55]. ET_A receptors are coupled with activation of

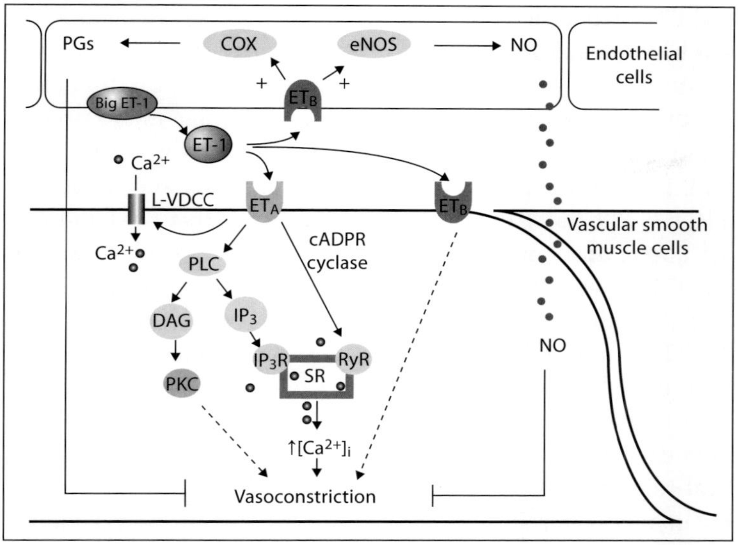

Fig. 2. Intracellular ET pathway in renal preglomerular microvessels. ET-1 is produced from big ET-1 catalyzed by the ET-converting enzyme from endothelial cells. Activation of ET_A receptors in VSMC increases intracellular calcium concentration ($[Ca^{2+}]_i$) via influx of extracellular Ca^{2+} through activating L-type voltage-dependent Ca^{2+} channels (L-VDCC) and mobilization of Ca^{2+} from sarcoplasmic reticulum (SR) through activating inositol triphosphate (IP_3) pathway and cyclic adenine diphosphate ribose (cADPR) cyclase/ryanodine receptor (RyR) pathways. Activation of endothelial ET_B receptors releases NO and prostaglandins (PGs), causing vasodilation. Activation of ET_B receptors in vascular smooth muscle cells mediates vasoconstriction via as yet undefined intracellular mechanisms (dashed line).

phospholipase C to generate diacylglycerol and inositol triphosphate, which in turn stimulates Ca^{2+} release from intracellular stores, ultimately causing vasoconstriction (fig. 2). In contrast, activation of ET_B receptors in endothelial cells stimulates release of NO and/or prostacyclin, which increase cGMP and cAMP formation, leading to vasodilation. Downstream signaling of ET_B receptor activation in VSMC still remains uncertain.

In the kidney, L-type voltage-dependent Ca^{2+} channels (L-VDCC) are important for regulating basal vascular tone and preglomerular microvascular responses to many vasoconstrictors [49]. However, there are conflicting results whether ET-1-induced vasoconstriction involves L-VDCC signaling. Several in vivo and in vitro studies indicate that L-VDCC blockade has little effect on ET-1 or ET_B receptor-mediated renal vasoconstriction [39, 55, 56], while others found that ET-1-induced renal vasoconstriction is via L-VDCC-dependent influx of extracellular Ca^{2+} [57–59], specifically in afferent arterioles but not in efferent arterioles [24, 60]. Studies in freshly isolated preglomerular microvascular

VSMC found that ET-1 evoked a biphasic increase in $[Ca^{2+}]_i$, with a rapid initial increase followed by a sustained plateau [55]. The sustained plateau was eliminated by superfusion with Ca^{2+}-free medium while the initial peak was unaltered, indicating that ET-1 increases $[Ca^{2+}]_i$ in preglomerular microvascular VSMC by stimulating Ca^{2+} release from intracellular stores and Ca^{2+} influx from extracellular fluid. In contrast, ET-3 or S6c only evoked small and monophasic increases in $[Ca^{2+}]_i$ [39, 55]. Subsequent studies revealed that ET-1-induced increases in $[Ca^{2+}]_i$ were unchanged by blockade of L-VDCC with diltilazem or Ni^{2+}. This was confirmed with both in vivo and in vitro studies. Nifedipine only slightly attenuated the reduction of RBF when ET-1 or S6c were applied at very high doses, while ET-1-induced juxtamedullary afferent arteriolar vasoconstriction was slightly inhibited by diltilazem only at very low ET-1 concentrations (1 and 10 pM) [39]. Although L-VDCC inhibition led to slightly different observations in these two experimental settings, the results suggest that L-VDCC do not play a critical role in ET-1 and ET_B receptor-mediated afferent arteriolar vasoconstriction in rats. Alternatively, studies in hydronephrotic rat kidney showed that nifedipine completely eliminated ET-1-induced afferent arteriolar vasoconstriction, indicating a prominent role for L-VDCC in ET-1-mediated afferent arteriolar responses [24].

Recent studies indicate that cyclic adenine diphosphate ribose (cADPR) cyclase/ryanodine receptor-mediated Ca^{2+} signaling pathways contribute to ET-1-mediated vasoconstriction in afferent arterioles [61, 62]. Blockade of ribosyl cyclase with nicotinamide reduced ET-1-mediated $[Ca^{2+}]_i$ signaling by 60% in rat preglomerular microvascular VSMC. This effect is largely due to stimulation of NADPH oxidase by ET-1 via ET_A receptor activation because application of tempol, apocynin, or a specific cADPR cyclase inhibitor attenuated $[Ca^{2+}]_i$ signaling induced by ET-1 but not by S6c [61]. The finding of cADPR and RyR signaling in ET-1-induced renal microvasoconstriction was further demonstrated in vivo as well as in the mice lacking the major mammalian ADPR cyclase (CD38–/–) [62]. These studies suggest that the superoxide-NADPH oxidase pathway contributes to the preglomerular microvascular response to ET-1 through cADPR.

Additionally, other studies suggest that activation of ET_A and ET_B receptors in the renal microvasculature are also coupled with a variety of second messenger pathways such as cytochrome *P*-450 metabolites [37], RhoA/Rho-kinase pathway [63], reactive oxygen species [64], and NO [65]. The role of prostacyclin in ET-1 responses is controversial. Some have reported that ET receptor activation is coupled to prostacyclin [66, 67], whereas others found that ET-1-mediated vasoconstriction was unaffected by COX inhibition [65, 68, 69] or that ET-1 stimulated a vasoconstrictor cyclooxygenase metabolite [70]. Interested readers are referred to some recent excellent reviews which provide more comprehensive discussions of the ET-1-mediated intracellular pathways in renal and nonrenal vascular beds [4, 5, 71, 72].

Impact of Endothelin-1 on Endothelial Function in Hypertension

Since discovery of this powerful vasoconstrictor released from the endothelium, activation of the ET system has been reported in a number of models of endothelial dysfunction, including hypertension [5]. Impaired renal vasodilation has been suggested as one of the mechanisms initiating salt-sensitive hypertension [73]. Studies in Dahl salt-sensitive hypertensive rats showed increased vascular and renal ET-1 protein content and impaired endothelium-dependent renal vasodilation. Treatment with an ET_A receptor antagonist not only prevented activation of tissue ET-1, but also prevented renal artery injury and endothelial dysfunction, suggesting that ET_A receptor-dependent increases in renal ET-1 level contributes to renal vascular injury in Dahl salt-sensitive hypertensive rats [74]. Together, these studies indicate an important role of ET-1 in mediating endothelial injury and development of hypertensive renal injury. The precise roles of ET-1 in hypertension, however, are still under investigation.

Conclusion

Collectively, ET-1-mediated vasoconstriction of afferent arterioles is coupled with activation of ET_A and ET_B receptors with a predominant role of ET_A receptors in dogs, rabbits, and mice, and a predominant role of ET_B receptors in rats. ET-1-mediated vasoresponsiveness in efferent arterioles seems to be exclusively coupled to ET_B receptors, except in mice. Low ET-1 concentrations activate endothelial ET_B receptors leading to vasodilation, while high ET-1 concentrations activate ET_B receptors expressed in VSMC evoking vasoconstriction. Additionally, endogenous ET-1 may contribute to control of basal renal vascular resistance via vasodilator influence of ET_B receptors in rats. The magnitude of ET-1-mediated vasoresponsiveness depends upon the segmental distribution of ET receptor expression as well as species variation. Different responses of afferent and efferent arterioles to ET-1 may certainly have a distinct impact on glomerular capillary pressure and hence alter glomerular hemodynamics. Any changes in ET receptor expression or renal ET-1 levels may play a crucial role in renal injury under pathological conditions by impairing renal hemodynamics.

Acknowledgements

The authors would like to acknowledge grant support from National Institutes of Health (PO1 HL095499) for E.W. Inscho and the American Heart Association (10SDG3770010) for Z. Guan.

References

1 Yanagisawa M, Kurihara H, Kimura S, Tomobe Y, Kobayashi M, Mitsui Y, Yazaki Y, Goto K, Masaki T: A novel potent vasoconstrictor peptide produced by vascular endothelial cells. Nature 1988;332:411–415.

2 Inoue A, Yanagisawa M, Kimura S, Kasuya Y, Miyauchi T, Goto K, Masaki T: The human endothelin family: three structurally and pharmacologically distinct isopeptides predicted by three separate genes. Proc Natl Acad Sci USA 1989;86:2863–2867.

3 Sakurai T, Yanagisawa M, Takuwa Y, Miyazaki H, Kimura S, Goto K, Masaki T: Cloning of a cDNA encoding a non-isopeptide-selective subtype of the endothelin receptor. Nature 1990;348:732–735.

4 Simonson MS: Endothelins: multifunctional renal peptides. Physiol Rev 1993;73:375–411.

5 Kohan DE, Rossi NF, Inscho EW, Pollock DM: Regulation of blood pressure and salt homeostasis by endothelin. Physiol Rev 2011;91:1–77.

6 Wagner OF, Christ G, Wojta J, Vierhapper H, Parzer S, Nowotny PJ, Schneider B, Waldhausl W, Binder BR: Polar secretion of endothelin-1 by cultured endothelial cells. J Biol Chem 1992;267:16066–16068.

7 Treiber FA, Jackson RW, Davis H, Pollock JS, Kapuku G, Mensah GA, Pollock DM: Racial differences in endothelin-1 at rest and in response to acute stress in adolescent males. Hypertension 2000;35:722–725.

8 Kitamura K, Tanaka T, Kato J, Ogawa T, Eto T, Tanaka K: Immunoreactive endothelin in rat kidney inner medulla: marked decrease in spontaneously hypertensive rats. Biochem Biophys Res Commun 1989;162:38–44.

9 Matsuura T, Miura K, Ebara T, Yukimura T, Yamanaka S, Kim S, Iwao H: Renal vascular effects of the selective endothelin receptor antagonists in anaesthetized rats. Br J Pharmacol 1997;122:81–86.

10 Fukuroda T, Fujikawa T, Ozaki S, Ishikawa K, Yano M, Nishikibe M: Clearance of circulating endothelin-1 by ET_B receptors in rats. Biochem Biophys Res Commun 1994;199:1461–1465.

11 Bagnall AJ, Kelland NF, Gulliver-Sloan F, Davenport AP, Gray GA, Yanagisawa M, Webb DJ, Kotelevtsev YV: Deletion of endothelial cell endothelin B receptors does not affect blood pressure or sensitivity to salt. Hypertension 2006;48:286–293.

12 Brooks DP, DePalma PD, Pullen M, Nambi P: Characterization of canine renal endothelin receptor subtypes and their function. J Pharmacol Exp Ther 1994;268:1091–1097.

13 Gellai M, DeWolf R, Pullen M, Nambi P: Distribution and functional role of renal ET receptor subtypes in normotensive and hypertensive rats. Kidney Int 1994;46:1287–1294.

14 Waeber C, Hoyer D, Palacios JM: Similar distribution of [125I]sarafotoxin-6b and [125I]endothelin-1, -2, -3 binding sites in the human kidney. Eur J Pharmacol 1990;176:233–236.

15 Johnstrom P, Rudd JH, Richards HK, Fryer TD, Clark JC, Weissberg PL, Pickard JD, Davenport AP: Imaging endothelin ET_B receptors using [18F]-BQ3020:in vitro characterization and positron emission tomography (microPET). Exp Biol Med 2006;231:736–740.

16 Nambi P, Pullen M, Wu HL, Aiyar N, Ohlstein EH, Edwards RM: Identification of endothelin receptor subtypes in human renal cortex and medulla using subtype-selective ligands. Endocrinology 1992;131:1081–1086.

17 Wendel M, Knels L, Kummer W, Koch T: Distribution of endothelin receptor subtypes ET_A and ET_B in the rat kidney. J Histochem Cytochem 2006;54:1193–1203.

18 Edwards RM, Trizna W: Characterization of 125I-endothelin-1 binding to rat and rabbit renal microvasculature. J Pharmacol Exp Ther 1995;274:1084–1089.

19 De Leon H, Garcia R: Characterization of endothelin receptor subtypes in isolated rat renal preglomerular microvessels. Regul Pept 1995;60:1–8.

20 Kuc R, Davenport AP: Comparison of endothelin-A and endothelin-B receptor distribution visualized by radioligand binding versus immunocytochemical localization using subtype selective antisera. J Cardiovasc Pharmacol 2004;44(Suppl 1):S224–S226.

21 Davenport AP, Kuc RE, Maguire JJ, Harland SP: ET_A receptors predominate in the human vasculature and mediate constriction. J Cardiovasc Pharmacol 1995;26(Suppl 3):S265–S267.
22 Clozel M, Clozel JP: Effects of endothelin on regional blood flows in squirrel monkeys. J Pharmacol Exp Ther 1989;250:1125–1131.
23 King AJ, Brenner BM, Anderson S: Endothelin: a potent renal and systemic vasoconstrictor peptide. Am J Physiol Renal Physiol 1989;256:F1051–1058.
24 Loutzenhiser R, Epstein M, Hayashi K, Horton C: Direct visualization of effects of endothelin on the renal microvasculature. Am J Physiol Renal Physiol 1990;258:F61–F68.
25 Evans RG, Bergstrom G, Cotterill E, Anderson WP: Renal haemodynamic effects of endothelin-1 and the ET_A/ET_B antagonist TAK-044 in anaesthetized rabbits. J Hypertens 1998;16:1897–1905.
26 Evans RG, Madden AC, Oliver JJ, Lewis TV: Effects of ET_A- and ET_B-receptor antagonists on regional kidney blood flow, and responses to intravenous endothelin-1, in anaesthetized rabbits. J Hypertens 2001;19:1789–1799.
27 Pollock DM, Opgenorth TJ: Evidence for endothelin-induced renal vasoconstriction independent of ET_A receptor activation. Am J Physiol Regul Integr Comp Physiol 1993;264:R222–R226.
28 Cristol JP, Warner TD, Thiemermann C, Vane JR: Mediation via different receptors of the vasoconstrictor effects of endothelins and sarafotoxins in the systemic circulation and renal vasculature of the anaesthetized rat. Br J Pharmacol 1993;108:776–779.
29 Gellai M, Fletcher T, Pullen M, Nambi P: Evidence for the existence of endothelin-B receptor subtypes and their physiological roles in the rat. Am J Physiol Regul Integr Comp Physiol 1996;271, R254–R261.
30 Just A, Olson AJ, Arendshorst WJ: Dual constrictor and dilator actions of ET_B receptors in the rat renal microcirculation: interactions with ET_A receptors. Am J Physiol Renal Physiol 2004;286, F660–F668.
31 Wellings RP, Corder R, Warner TD, Cristol JP, Thiemermann C, Vane JR: Evidence from receptor antagonists of an important role for ET_B receptor-mediated vasoconstrictor effects of endothelin-1 in the rat kidney. Br J Pharmacol 1994;111:515–520.
32 Denton KM, Shweta A, Finkelstein L, Flower RL, Evans RG: Effect of endothelin-1 on regional kidney blood flow and renal arteriole calibre in rabbits. Clin Exp Pharmacol Physiol 2004;31:494–501.
33 Vassileva I, Mountain C, Pollock DM: Functional role of ET_B receptors in the renal medulla. Hypertension 2003;41:1359–1363.
34 Brodsky S, Abassi Z, Wessale J, Ramadan R, Winaver J, Hoffman A: Effects of A-192621.1, a specific endothelin-B antagonist, on intrarenal hemodynamic responses to endothelin-1. J Cardiovasc Pharmacol 2000;36:S311–S313.
35 Hercule HC, Oyekan AO: Role of NO and cytochrome P-450-derived eicosanoids in ET-1-induced changes in intrarenal hemodynamics in rats. Am J Physiol Regul Integr Comp Physiol 2000;279:R2132–R2141.
36 Casellas D, Navar LG: In vitro perfusion of juxtamedullary nephrons in rats. Am J Physiol Renal Physiol 1984;246:F349–F358.
37 Imig JD, Pham BT, LeBlanc EA, Reddy KM, Falck JR, Inscho EW: Cytochrome P450 and cyclooxygenase metabolites contribute to the endothelin-1 afferent arteriolar vasoconstrictor and calcium responses. Hypertension 2000;35:307–312.
38 Inscho EW, Imig JD, Cook AK, Pollock DM: ET_A and ET_B receptors differentially modulate afferent and efferent arteriolar responses to endothelin. Br J Pharmacol 2005;146:1019–1026.
39 Pollock DM, Jenkins JM, Cook AK, Imig JD, Inscho EW: L-type calcium channels in the renal microcirculatory response to endothelin. Am J Physiol Renal Physiol 2005;288:F771–F777.
40 Schneider MP, Inscho EW, Pollock DM: Attenuated vasoconstrictor responses to endothelin in afferent arterioles during a high-salt diet. Am J Physiol Renal Physiol 2007;292:F1208–F1214.
41 Kon V, Yoshioka T, Fogo A, Ichikawa I: Glomerular actions of endothelin in vivo. J Clin Invest 1989;83:1762–1767.
42 Takenaka T, Forster H, Epstein M: Protein kinase C and calcium channel activation as determinants of renal vasoconstriction by angiotensin II and endothelin. Circ Res 1993;73:743–750.

43 Saito M, Homma S, Yamatsu I, Sato M, Ohshima N: Visualization of renal microcirculation in isolated Munich-Wistar rat kidneys: effects of endothelin-1 on renal hemodynamic activity. Jpn J Pharmacol 1994;66:221–229.
44 Lanese DM, Yuan BH, McMurtry IF, Conger JD: Comparative sensitivities of isolated rat renal arterioles to endothelin. Am J Physiol Renal Physiol 1992;263:F894–F899.
45 Schildroth J, Rettig-Zimmermann J, Kalk P, Steege A, Fahling M, Sendeski M, Paliege A, Lai EY, Bachmann S, Persson PB, Hocher B, Patzak A: Endothelin type A and B receptors in the control of afferent and efferent arterioles in mice. Nephrol Dial Transplant 2011;26:779–789.
46 Kisanuki YY, Emoto N, Ohuchi T, Widyantoro B, Yagi K, Nakayama K, Kedzierski RM, Hammer RE, Yanagisawa H, Williams SC, Richardson JA, Suzuki T, Yanagisawa M: Low blood pressure in endothelial cell-specific endothelin 1 knockout mice. Hypertension 2010;56:121–128.
47 Pallone TL, Zhang Z, Rhinehart K: Physiology of the renal medullary microcirculation. Am J Physiol Renal Physiol 2003;284:F253–F266.
48 Qiu C, Samsell L, Baylis C: Actions of endogenous endothelin on glomerular hemodynamics in the rat. Am J Physiol Regul Integr Comp Physiol 1995;269:R469–R473.
49 Navar LG, Arendshorst WJ, Pallone TL, Inscho EW, Imig JD, Bell PD: Microcirculation; in Tuma RFD, Wa LK (eds): Handbook of Physiology 2008. San Diego, Elsevier, p.550–683.
50 Takabatake T, Ise T, Ohta K, Kobayashi K: Effects of endothelin on renal hemodynamics and tubuloglomerular feedback. Am J Physiol Renal Physiol 1992;263:F103–F108.
51 Berthold H, Munter K, Just A, Kirchheim HR, Ehmke H: Contribution of endothelin to renal vascular tone and autoregulation in the conscious dog. Am J Physiol Renal Physiol 1999;276:F417–F424.
52 Kramp R, Fourmanoir P, Caron N: Endothelin resets renal blood flow autoregulatory efficiency during acute blockade of NO in the rat. Am J Physiol Renal Physiol 2001;281:F1132–F1140.
53 Shi Y, Lau C, Cupples WA: Interactive modulation of renal myogenic autoregulation by nitric oxide and endothelin acting through ET_B receptors. Am J Physiol Regul Integr Comp Physiol 2007;292:R354–R361.
54 Romano G, Giagu P, Favret G, Bartoli E: Effect of endothelin 1 on proximal reabsorption and tubuloglomerular feedback. Kidney Blood Press Res 2000;23:360–365.
55 Schroeder AC, Imig JD, LeBlanc EA, Pham BT, Pollock DM, Inscho EW: Endothelin-mediated calcium signaling in preglomerular smooth muscle cells. Hypertension 2000;35:280–286.
56 Fretschner M, Endlich K, Gulbins E, Lang RE, Schlottmann K, Steinhausen M: Effects of endothelin on the renal microcirculation of the split hydronephrotic rat kidney. Ren Physiol Biochem 1991;14:112–127.
57 Madeddu P, Yang XP, Anania V, et al: Efficacy of nifedipine to prevent systemic and renal vasoconstrictor effects of endothelin. Am J Physiol 1990;259:F304–F311.
58 Yukimura T, Miura K, Yamashita Y, Shimmen T, Okumura M, Yamanaka S, Saito M, Yamamoto K: Effects of the calcium channel antagonist nicardipine on renal action of endothelin in dogs. Contrib Nephrol 1991;90:105–110.
59 Takahashi K, Katoh T, Fukunaga M, Badr KF: Studies on the glomerular microcirculatory actions of manidipine and its modulation of the systemic and renal effects of endothelin. Am Heart J 1993;125:609–619.
60 Edwards RM, Trizna W, Ohlstein EH: Renal microvascular effects of endothelin. Am J Physiol Renal Physiol 1990;259:F217–F221.
61 Fellner SK, Arendshorst W: Endothelin-A and -B receptors, superoxide, and Ca^{2+} signaling in afferent arterioles. Am J Physiol Renal Physiol 2007;292:F175–F184.
62 Thai TL, Churchill GC, Arendshorst WJ: NAADP receptors mediate calcium signaling stimulated by endothelin-1 and norepinephrine in renal afferent arterioles. Am J Physiol Renal Physiol 2009;297:F510–F516.
63 Cavarape A, Endlich N, Assaloni R, Bartoli E, Steinhausen M, Parekh N, Endlich K: Rho-kinase inhibition blunts renal vasoconstriction induced by distinct signaling pathways in vivo. J Am Soc Nephrol 2003;14:37–45.

64 Just A, Whitten CL, Arendshorst WJ: Reactive oxygen species participate in acute renal vasoconstrictor responses induced by ET_A and ET_B receptors. Am J Physiol Renal Physiol 2008;294:F719–F728.
65 Just A, Olson AJ, Falck JR, Arendshorst WJ: NO and NO-independent mechanisms mediate ET_B receptor buffering of ET-1-induced renal vasoconstriction in the rat. Am J Physiol Regul Integr Comp Physiol 2005;288:R1168–R1177.
66 Hirata Y, Matsuoka H, Kimura K, Sugimoto T, Hayakawa H, Suzuki E, Sugimoto T: Role of endothelium-derived relaxing factor in endothelin-induced renal vasoconstriction. J Cardiovasc Pharmacol 1991;17(Suppl 7): S169–S171.
67 Chou SY, Porush JG: Renal actions of endothelin-1 and endothelin-3: interactions with the prostaglandin system and nitric oxide. Am J Kidney Dis 1995;26:116–123.
68 Cao LQ, Banks RO: Cardiorenal actions of endothelin, part II: effects of cyclooxygenase inhibitors. Life Sci 1990;46:585–590.
69 Duval M, D'Orleans-Juste P: Endothelin-1 induces an endothelin-A-receptor-dependent and L-NAME-sensitive vasoconstriction in the isolated perfused guinea-pig kidney. J Cardiovasc Pharmacol 2000;36:S241–S243.
70 Munger KA, Takahashi K, Awazu M, Frazer M, Falk SA, Conger JD, Badr KF: Maintenance of endothelin-induced renal arteriolar constriction in rats is cyclooxygenase dependent. Am J Physiol 1993;264:F637–F644.
71 Bouallegue A, Daou GB, Srivastava AK: Endothelin-1-induced signaling pathways in vascular smooth muscle cells. Curr Vasc Pharmacol 2007;5:45–52.
72 Kedzierski RM, Yanagisawa M: Endothelin system: the double-edged sword in health and disease. Annu Rev Pharmacol Toxicol 2001;41:851–876.
73 Simchon S, Manger WM, Carlin RD, Peeters LL, Rodriguez J, Batista D, Brown T, Merchant NB, Jan KM, Chien S: Salt-induced hypertension in Dahl salt-sensitive rats. Hemodynamics and renal responses. Hypertension 1989;13:612–621.
74 Barton M, Vos I, Shaw S, Boer P, D'Uscio LV, Grone HJ, Rabelink TJ, Lattmann T, Moreau P, Luscher TF: Dysfunctional renal nitric oxide synthase as a determinant of salt-sensitive hypertension: mechanisms of renal artery endothelial dysfunction and role of endothelin for vascular hypertrophy and Glomerulosclerosis. J Am Soc Nephrol 2000;11:835–845.

Prof. Edward W. Inscho
Department of Physiology, Georgia Health Sciences University
CA Building, CA31371120, 15th Street
Augusta, GA 30912 (USA)
Tel. +1 706 721 5615, E-Mail einscho@georgiahealth.edu

Barton M, Kohan DE (eds): Endothelin in Renal Physiology and Disease.
Contrib Nephrol. Basel, Karger, 2011, vol 172, pp 50–62

Endothelin Signaling and Actions in the Renal Mesangium

Andrey Sorokin

Kidney Disease Center, Division of Nephrology, Department of Medicine, Medical College of Wisconsin, Milwaukee, Wisc., USA

Abstract

Endothelins (ETs), and particularly ET-1, activate a complex network of interconnected signaling cascades in mesangial glomerular cells, which play an important role in the physiology and pathophysiology of the glomerulus. Excessive stimulation of ET-1 production by mesangial cells results in activation of a wide variety of signaling pathways in the renal mesangium, which is at least partially responsible for glomerular damage in the setting of diabetes, hypertension, and glomerulonephritis. Mesangial cells express both types of ET receptors (ET_A-R and ET_B-R), which are G protein-coupled receptors. ET-1 induces mobilization of Ca^{2+}; activation of phospholipases A, C, and D; activation of protein kinase C; GTP-loading of several families of small GTPases; and activation of intracellular tyrosine kinases resulting in protein tyrosine phosphorylation of adaptor, scaffolding, and signaling proteins. ET-1-triggered posttranslational modification of signaling molecules sets the base for the formation of multiunit signaling complexes which define the specificity of ET signaling. Long-term effects of ET-1 are also mediated via increased expression of particular signaling proteins. It is likely that ET-1 acts via ET_A-R to trigger the contraction of mesangial cells, which decreases glomerular filtration area and reduces the glomerular filtration rate, promoting impaired renal function. Proliferation of mesangial cells is observed in the progress of several types of glomerulonephritis. ET-1 is a potent mitogen of mesangial cells and the ability of ET-1 to support mesangial cell proliferation is likely to be associated with both recruitment of cytoplasmic tyrosine kinases which activate the Shc-Sos-Ras-Raf-MEK-ERK signaling pathway and transactivation of the EGF receptor. The guanine nucleotide exchange factor βPix and the adaptor protein $p66^{Shc}$ are important players in Akt-independent inactivation of FOXO3a transcription factor. This results in the depletion of the inhibitor of cell cycle progression $p27^{kip1}$, and promotion of mesangial cell proliferation. Plentiful evidence suggests an essential role of ET-1-signaling and action in the renal mesangium for renal biology and pathobiology.

In the adult glomerulus, interplay of four types of cells, namely endothelial cells of the capillary vessels, mesangial cells, podocytes, and parietal epithelial cells provide efficient filtration of blood and formation of urine. In the mature glomerulus, the tasks of mesangial cells include providing structural support for the glomerular capillary network and regulating filtration area by contraction and removal of apoptotic cells by phagocytosis. In addition, glomerular mesangial cell proliferation is initiated in the progress of experimental diabetic nephropathy and in some forms of glomerulonephritis [1]. Glomerular mesangial cells are also among the major matrix-producing cells responsible for the synthesis of extracellular matrix proteins contributing to renal fibrosis.

To accomplish these tasks, mesangial cells are in constant cross-talk with endothelial cells and podocytes, involving binding and secretion of a number of regulatory molecules, including the potent bioactive peptides of the endothelin (ET) family [2, 3]. The ET family consists of three members, ET-1, ET-2, and ET-3, which play an important role in renal physiology and pathophysiology due to their involvement in the physiological control of blood pressure and salt homeostasis, as well as action as multifunctional bioactive peptides in the renal mesangium [4–6]. ET-1 is the main isoform which is secreted by mesangial cells and which also operates on the renal mesangium. All of ET's effects are elicited by binding to specific G protein-coupled receptors: ET_A-R [7] and ET_B-R [8]. ET_A-R and ET_B-R demonstrate considerable variability in distribution between different tissues and among species [9]. Human and rat renal mesangial cells express substantial levels of both ET_A-R and ET_B-R [4, 10]. A progressive renal injury like inflammation and fibrosis are mediated via both ET_A-R and ET_B-R receptors, while constrictor effects are primarily transduced by ET_A-R receptors [11].

Regulation of ET-1 gene expression in mesangial cells and conversion of proET-1 to Big ET-1 and consequently to ET-1 is well studied and has been reviewed previously [4, 6]. This review will focus on those aspects of ET action and signaling in the renal mesangium which seem to be the most relevant to renal physiology and disease.

ET-1 production by mesangial cells results in activation of a wide variety of signaling pathways regulating contraction, hypertrophy, proliferation, and extracellular matrix accumulation in the renal mesangium. These actions are partially responsible for glomerular damage in the setting of diabetes, hypertension, and glomerulonephritis. ET-1 receptors couple to members of the $G_{\alpha i}$, $G_{\alpha q}$, $G_{\alpha s}$, and $G_{\alpha 12/13}$ G protein families [6]. Correspondingly, ET-1 binding to its receptor evokes multiple intracellular pathways that can be cell-specific, but usually include mobilization of Ca^{2+}; activation of phospholipases A, C, and D; activation of serine-threonine kinases; and increased protein tyrosine phosphorylation. ET_A-R acts through $G_{\alpha q}$ and $G_{\alpha s}$ to stimulate phospholipase C and adenylate cyclase, respectively. ET_B-R stimulates PLC via $G_{\alpha q}$ and inhibits adenylate cyclase through $G_{\alpha i}$. The recent discovery that ET receptors colocalize and interact with caveolin-1 in caveolae adds further complexity since caveolae integrates different receptor and signaling proteins. Activation of ET receptors leads to the compartmentalization and binding of $G_{\alpha q}$ to β_1Pix in caveolae [12].

Endothelin Signaling and Contraction of Mesangial Cells

The essential biological event in mesangial cell contraction is the increase of cytosolic free Ca^{2+} concentration ($[Ca^{2+}]_i$), which may result either from the mobilization of $[Ca^{2+}]_i$ from intracellular stores or due to activation of receptor-operated channels. It is generally accepted that increased $[Ca^{2+}]_i$ binds to calmodulin, stimulating myosin light-chain kinase which phosphorylates myosin light-chains and promotes assembly of myosin monomers into filaments, leading to cell contraction. ET lowers the glomerular ultrafiltration coefficient (Kf) and contracts mesangial cells, possibly through stimulation of inositol triphosphate generation and elevation of $[Ca^{2+}]_i$ (fig. 1) [13]. It is known that ET-1 causes a biphasic rise in $[Ca^{2+}]_i$, consisting of an initial transient increase, followed by a sustained lesser increase. The early transient peak reflects inositol triphosphate-released Ca^{2+} from intracellular sources, which is associated with cell alkalinization through augmented Na/H exchange [5, 14, 15]. The continued plateau of $[Ca^{2+}]_i$ which over time declines towards the initial level is due to the increased influx of Ca^{2+} across the plasma membrane. Cultured glomerular mesangial cells demonstrate L-type Ca^{2+} channel activity [16] and were shown to express canonical transient receptor potential 1 (TRPC1), a Ca^{2+}-permeable cation channel [17]. Different effects of low (0.1–10 pM) and high (100 pM and higher) ET-1 concentrations were demonstrated with low concentrations causing a slow sustained increase in $[Ca^{2+}]_i$ mediated by Ca^{2+} influx through a voltage-channel-independent mechanism, while higher ET-1 concentrations caused a rapid and transient increase of $[Ca^{2+}]_i$ which was dependent on Ca^{2+} release from intracellular stores and required activation of PLC and protein kinase C (PKC) [14]. Activation of one or more PKC isoforms – either PKC-α, -δ, or -ε, but not -ζ is required for ET-1-stimulated contraction of mesangial cells [18].

The TRPC1 channel could be important for ET-1-mediated contraction of mesangial cells since the ET-1-induced decrease in glomerular filtration rate in vivo was abolished by infusion of a TRPC1 antibody that targets an extracellular domain in the pore region of TRPC1 channel [17].

Apparently, $[Ca^{2+}]_i$-dependent molecular signaling mechanisms important for mesangial cell contraction are not limited to stimulation of myosin light-chain kinase by the Ca^{2+}-calmodulin complex. Evidence has been obtained that tyrosine phosphorylation plays an important role in contraction and that the contractile responsiveness of mesangial cells depends on activation of the p38 mitogen-activated protein kinases (MAPKs) [19]. The ability of p38 MAPK

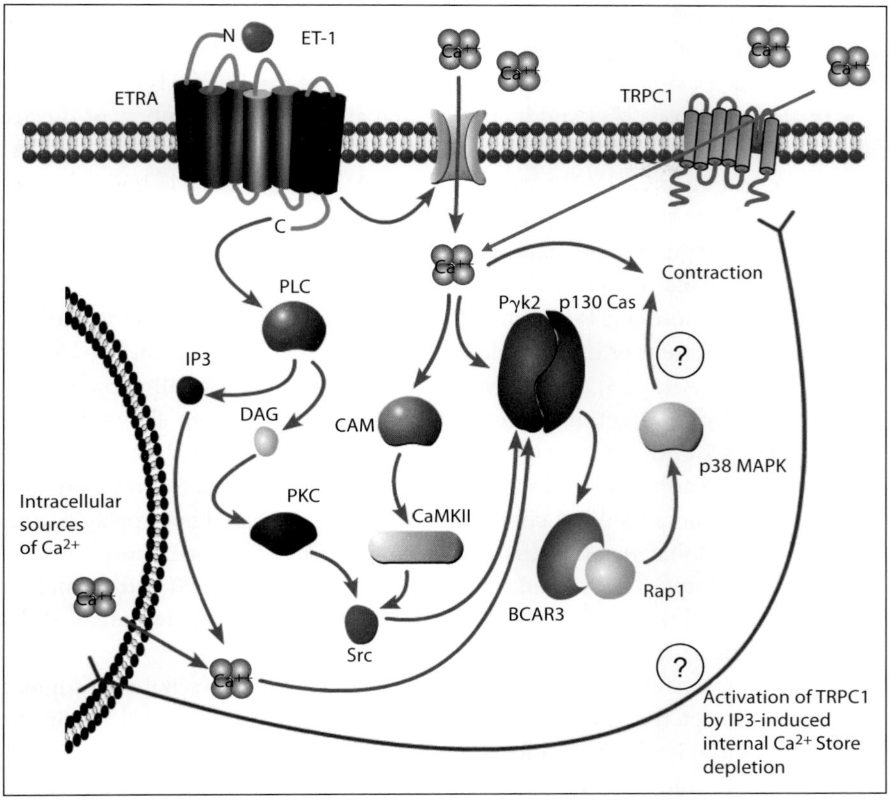

Fig. 1. ET signaling and contraction of renal mesangial cells. Schematic representation of signaling pathways mediated by ET_A-R in renal mesangial cells relevant for contraction. Signaling from ET_B-R causing relaxation is not shown.

to regulate cell contraction is possibly due to its unique capability in vivo to activate MAPK-activated protein kinase 2/3 which phosphorylates small heat shock protein 25. Small heat shock proteins are known to modulate polymerization/depolymerization of F-actin and be involved in cell contraction. Thus, p38 MAPK-mediated phosphorylation of small heat shock proteins provides a potential mechanism of regulation of mesangial cell contractility [19].

ET-1-induced mobilization of $[Ca^{2+}]_i$ activates calcium-dependent intracellular tyrosine kinase Pyk2 in glomerular mesangial cells [20], which appears to be important for activation of p38 MAPK in ET-1-treated mesangial cells. Calcium-regulated cytoplasmic proline-rich tyrosine kinase Pyk2 (also termed related adhesion focal tyrosine kinase, focal adhesion kinase-2, cell adhesion kinase-β, and calcium-dependent tyrosine kinase) provides the link between G protein-coupled ET receptors and the induction of tyrosine phosphorylation via mobilization of intracellular calcium. Activation/autophosphorylation of Pyk2 in ET-1-treated cells depends not only on Ca^{2+} mobilization, but also upon activity of Src kinases since it is sensitive to inhibitors of Src [20]. ET-1 was shown to activate Src kinases [21] and it appears that at least some members of the Src family are recruited to the complex of β-arrestin 1 with ET_A-R in an agonist-dependent manner [22]. Activation of major MAPK cascades is typically preceded by GTP-loading of particular small GTPases achieved by recruitment of a corresponding guanine nucleotide exchange factor (GEF). In human mesangial cells, ET-1 (acting via ET_A-R) induces calcium-dependent activation of Pyk2 which promotes GTP-loading of the small GTPase Rap1 by a mechanism involving recruitment of the complex between the scaffolding protein p130Cas and the GEF BCAR3 [23]. It has not been proven that this signaling pathway is indispensible for contraction of mesangial cells, but it has been shown that the Pyk2/p130Cas/BCAR3/Rap1 signaling cascade activated via ET_A-R is involved in mesangial cell adhesion and spreading regulation [23]. Negative regulation of ET-1-triggered mesangial cell contraction could be mediated by prostaglandins (products of cyclooxygenase activity) and nitric oxide (NO) formation [4]. ET-1 can strongly inhibit cytokine induction of iNO synthase (iNOS) and formation of NO in cultured mesangial cells, and this action is mediated via the ET_A-R by suppressing the expression of iNOS mRNA [24, 25].

Overexpression of the GEF C3G was observed in the renal mesangium after induction of the anti-Thy1 model of glomerulonephritis [26]. Enforced expression of C3G, which mimicked increased C3G expression in vivo, caused enhanced and prolonged Rap1 activation in response to ET-1 in rat mesangial cells [26]. Surprisingly, C3G was not important for Rap1 activation by ET-1 in immortalized human mesangial cells [23]. Whether this difference in C3G requirement for Rap1 activation is species-specific or comes from the difference in nature between cultured primary rat mesangial cells and immortalized human mesangial cells needs further analysis. In primary rat mesangial cells, C3G overexpression led to significant changes in cell spreading and migration patterns in response to ET-1 stimulation and also increased stress fiber formation, phenomena which were mimicked by Rap1A overexpression [26]. An increase in the C3G protein level could contribute to the resolution stage of mesangioproliferative glomerulonephritis by reducing mesangial cells sensitivity to ET-1, modulating cellular motility, and actin dynamics [26].

Endothelin Signaling and Synthesis of Extracellular Matrix Proteins in Mesangial Cells

Accumulation of extracellular matrix proteins is one of the key features of progressive glomerulonephritis [1]. ET-1, acting through ET_A-R, promotes synthesis and accumulation of extracellular matrix in the renal mesangium contributing to the progression of renal fibrosis [6, 27, 28]. ET-1 increases fibronectin, type IV collagen, and type I collagen synthesis by mesangial cells [27–29]. In addition, ET-1 partially mediates angiotensin II-mediated mesangial cell extracellular matrix accumulation [27]. These effects are primarily mediated via ET_A-R, with the resulting activation of PKC-β, -δ, -ε, and -ζ [27, 29]. Analysis of the changes in mRNA transcripts in human mesangial cells identified 122 genes regulated by ET-1 [30]. It must be taken into consideration, however, that extreme caution must be applied when accepting mRNA microarray data as evidence of an increase of protein abundance. A recent study, based on state-of-the-art methods for large-scale protein and mRNA quantification, reported that an unexpectedly large percentage of proteins appear to undergo changes in abundance without corresponding changes in mRNA levels [31]. This could be the reason why some of the proteins shown to be upregulated by ET-1, such as cyclooxygenase-2 (COX-2) [32, 33], C3G [26], or βPix [34], are missing from the list. Among the genes induced by ET-1 in mesangial cells, as detected using Affymetrix U95 microarray platform, are genes encoding macrophage chemoattractant protein -1 (MCP-1) and IL-6 [30]. Both cytokines are capable of increasing collagen accumulation and it was suggested that ET-1, acting via ET_A-R, induces secretion of MCP-1 and IL-6, resulting in their autocrine signaling loop controlling ET-1-induced collagen accumulation in mesangial cells [28]. ET-1 can stimulate mesangial cells to produce TNFα, ICAM-1, and VCAM-1, and thereby induce pro-inflammatory effects [35].

Endothelin Signaling and Proliferation of Mesangial Cells

Glomerular mesangial cell proliferation is an important factor in the development of focal or diffuse glomerulosclerosis as highlighted by data obtained using animal models of chronic renal failure and acute mesangioproliferative glomerulonephritis in rats [36]. The growth factor PDGF is one of the major regulators of mesangial cell proliferation and, notably, ET-1 stimulates PDGF secretion by mesangial cells [37]. Furthermore, multiple studies indicate that ET-1 itself is a potent mitogen for mesangial cells both in vitro and in vivo [19, 27, 38]. Part of this mitogenic activity is probably due to transactivation of the EGF receptor (EGFR) a molecule with strong mitogenic activity (fig. 2). Remarkably, ET-1-triggered transactivation of EGFR in Rat-1 fibroblasts was the first example of transactivation of receptor tyrosine kinase by the ligand of a G protein-coupled receptor [39]. In glomerular mesangial cells, transactivation of EGFR by ET-1 has also been reported [40, 41]. The mechanism of transactivation of EGFR by G protein-coupled receptors is relatively well understood and is based on activation of a disintegrin and metalloprotease (ADAM), a membrane-anchored metalloprotease. This produces mature EGFR ligands [e.g. heparin-binding EGF-like growth factor (HB-EGF)] through the shedding of its transmembrane

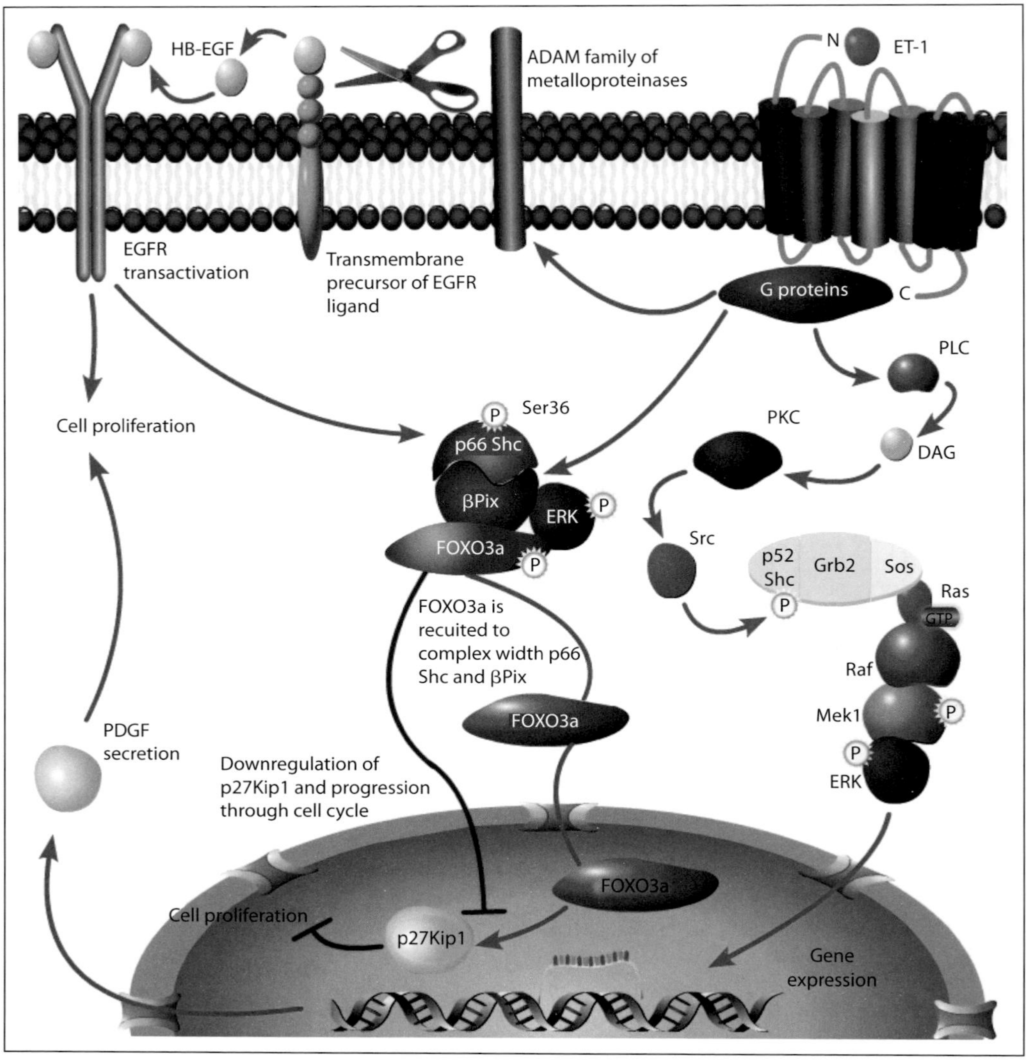

Fig. 2. ET signaling and proliferation of renal mesangial cells. Schematic representation of signaling pathways mediated by ET-1 in renal mesangial cells relevant for proliferation.

precursor [42]. It is of note that upregulation of the transcript for HB-EGF was detected in mesangial cells treated with ET-1 [30].

It is generally accepted that increased GTP-loading of the small GTPase Ras followed by activation of the Raf-MEK-ERK signaling pathway is indispensable for cell proliferation. ERK1 and 2 are members of the MAPKs which are among the most thoroughly studied signaling molecules controlling multiple and diverse cellular functions. In mesangial cells, ET-1 stimulates three major

MAPK: ERK [43], JNK/SAPK [44], and p38 MAPK [20], as well as the less studied ERK5 [45]. The role of these MAPKs in ET-1-mediated cellular effects was evaluated in different cellular systems, including glomerular mesangial cells. Detailed analysis of mechanisms of MAPK activation emphasized the significance of protein-protein interactions as determinants of specificity of signaling initiated by various agonists. Protein-protein interactions are important for signal transduction because (1) they can recruit the signaling molecule to the intracellular location where it can be activated or activate the specific target [46], (2) the direct contact of two or more signaling molecules can facilitate the regulatory modification (such as phosphorylation or conformational change) of one protein by another and contribute to the specificity of signal transmission [47], and (3) the formation and disassembly of multiunit signaling complexes can control the intensity and duration of particular signaling pathways and as a result convert signal transmission mediated by the same signaling molecules into qualitatively different biological responses [19, 47].

ET signaling in glomerular cells is regulated by a number of protein-protein interactions and mediated by several classes of protein-binding domains. The best-characterized protein interaction domains involved in ET mitogenic signaling are the SH2 (Src homology 2) and SH3 (Src homology 3) domains, which are present in many signal transduction proteins [48, 49]. SH2 domains mediate phosphotyrosine-dependent association of proteins, whereas SH3-mediated interactions are based on recognition of proline-rich peptide sequences. Among other well-studied peptide-binding domains involved in ET signaling is a protein module which mediates phosphoserine-mediated interactions and is found in conserved regulatory 14–3–3 proteins which work as dimers representing a particular potent class of adaptor proteins [50, 51]. Studies of protein-protein interactions have emphasized the role of adaptor proteins which contain multiple protein-protein interaction domains. For example, in mesangial cells the $p52^{Shc}$ isoform of adaptor protein Shc (three Shc isoforms are expressed: $p46^{Shc}$, $p52^{Shc}$, and $p66^{Shc}$) plays an important role in ET-1 regulation of Ras and ERK activation. In ET-1-treated glomerular mesangial cells, persistent tyrosine phosphorylation of $p52^{Shc}$ is the cause of the lengthy association of $p52^{Shc}$ with the adaptor protein Grb2. Sustained ET-1-induced interaction between tyrosine-phosphorylated $p52^{Shc}$ and the Grb2 SH2 domain is the basis for biphasic Ras activation by the GEF Sos in mesangial cells [52]. The initial activation of Ras is dependent upon the formation of a Shc/Grb2/Sos1 signaling complex and is followed by rapid deactivation of Ras, as a direct consequence of Sos1 phosphorylation and its release from this trimolecular module. Subsequent to Ras and ERK deactivation, Sos1 reverts to the nonphosphorylated condition, enabling it to bind again to the Grb2/Shc complex, which is stabilized by persistent Shc phosphorylation, resulting in biphasic activation of Ras [52]. The second peak of ERK activation could be attenuated by activation of dual-specificity phosphatases. It appears that biphasic activation of Ras by ET-1 sequentially activates the ERK cascade and phosphatidylinositol 3-kinase in glomerular mesangial cells [52].

The proliferative effect of ET-1 also depends on regulation of proteins in control of the cell cycle and is inhibited by antisense oligonucleotides directed against cyclin D1 and by overexpression of a nonphosphorylatable form of pRb [53]. It was shown that ET-1 induced expression of the cell cycle signaling molecule cyclin D1 by a Src-based mechanism, suggesting a role for Src family cytoplasmic tyrosine kinases in mediating ET-1 mitogenic signaling in mesangial cells [54]. Downregulation of $p27^{kip1}$ results in cell cycle progression into S phase. Expression of $p27^{kip1}$ is under control of forkhead transcription factor FOXO3A which is regulated by Akt. The phosphorylation of FOXO3a by Akt in serum-stimulated cells results in cytoplasmic localization of FOXO3A and downregulation of $p27^{kip1}$. ET-1 employs a different mechanism of FOXO3A inactivation resulting in downregulation of $p27^{kip1}$ and progression of the cell through the cell cycle. This pathway utilized the scaffolding activity of GEF βPix, a molecule which in addition to mediation of ET-1-induced GTP-loading of Cdc42 [55] also participates in regulation ET-1 proliferative activity by assembly with other signaling molecules [34]. Short exposure of mesangial cells to ET-1 induced serine phosphorylation of the $p66^{Shc}$ isoform through $G\alpha_{i3}$ [40, 51] and an increase of β_1Pix GEF activity [55]. Prolonged exposure to ET-1 resulted in the increase of $p66^{Shc}$ and βPix expression and the formation of a multiunit signaling complex between $p66^{Shc}$, β_1Pix, and FOXO3A [34]. Depletion of either βPix or $p66^{Shc}$ inhibited ET-1-induced cell proliferation. The expression of β_1Pix induced FOXO3a phosphorylation through activation of Rac1, ERK1/2, and p66Shc. Using either $p66^{Shc}$ or Akt-depleted cells, we showed that β_1Pix-mediated FOXO3a phosphorylation and $p27^{kip1}$ downregulation required $p66^{Shc}$, but was independent of Akt. Depletion of βPix prevented $p27^{kip1}$ downregulation induced by ET-1 [34].

Endothelin Signaling and Apoptosis of Mesangial Cells

Proliferation of mesangial cells results in the undesirable overpopulation of glomeruli with mesangial cells, but it may be reversed by apoptotic cell death. Mesangial cell apoptosis is considered one of the major mechanisms for resolution of glomerular hypercellularity in experimental mesangial proliferative nephritis [56]. Activation of ERK and phosphoinositide 3-kinase/AKT pathways are usually associated with antiapoptotic signaling, whereas activation of JNK often accompanies apoptotic cell death. Since ET-1 enhanced the activities of JNK in a dose-dependent and time-dependent manner [44] as well as activates ERK and phosphoinositide 3-kinase [52], both pro- and antiapoptotic signaling are triggered by ET-1 in renal mesangial cells. Progression to proliferative glomerulonephritis is accompanied by overexpression of COX-2, increased formation of prostaglandins, and inefficient apoptosis of mesangial cells. Mesangial cells express receptors for TNFα, a cytokine with proapoptotic properties implicated in the resolution of glomerulonephritis. Since COX-2 expression protects cells from apoptosis and ET-1 induces COX-2 expression in mesangial cells, ET-1 is expected to act as an antiapoptotic factor. Indeed, in cultured mesangial cells, COX-2 expression induced by ET-1 inhibited TNFα-mediated apoptosis [57].

In addition to being a target of proapoptotic cytokines, glomerular mesangial cells are themselves capable of taking up apoptotic cells, albeit much less efficiently than peritoneal macrophages [58, 59]. This process is under the control of serum proteins, and even though ET-1 could be one of the regulators due to its ability to modulate cytoskeleton dynamics, the role of ETs in regulation of the phagocytosis activity of mesangial cells has not been studied.

Conclusion

ETs bind to ET_A-R and ET_B-R, which are typical G protein-coupled receptors and activate a wide variety of signaling cascades important for the major tasks of mesangial cells in the glomerulus. ET action in the renal mesangium lies behind multiple renal pathologies. The pathological effects of ETs are mediated in large part through activation of ET_A-R. Correspondingly, selective ET_A-R receptor antagonists should be primarily considered for use in diseases characterized by disordered mesangial cell function. In addition, recent progress in defining molecular mechanisms of ET-mediated mesangial cell contraction, mesangial cell proliferation, and accumulation of extracellular matrix proteins sets the basis for developing novel therapeutic approaches which will target particular signaling molecules in ET-induced signaling cascades to combat kidney pathologies associated with dysfunction of the renal mesangium.

Acknowledgements

Grant support was received from NIH research grant RO1 DK 41684 and a ROTRF grant award. I would also like to thank Dr. Simon Prosser (MCW) for editorial assistance.

References

1 Couser WG, Johnson RJ: Mechanisms of progressive renal disease in glomerulonephritis. Am J Kidney Dis 1994;23:193–198.

2 Barton M: Reversal of proteinuric renal disease and the emerging role of endothelin. Nat Clin Pract Nephrol 2008;4:490–501.

3 Ortmann J, Amann K, Brandes RP, Kretzler M, Munter K, Parekh N, Traupe T, Lange M, Lattmann T, Barton M: Role of podocytes for reversal of glomerulosclerosis and proteinuria in the aging kidney after endothelin inhibition. Hypertension 2004;44:974–981.

4 Kohan DE, Rossi NF, Inscho EW, Pollock DM: Regulation of blood pressure and salt homeostasis by endothelin. Physiol Rev 2011;91:1–77.
5 Simonson MS, Dunn MJ: Endothelin peptides and the kidney. Annu Rev Physiol 1993;55:249–265.
6 Sorokin A, Kohan DE: Physiology and pathology of endothelin-1 in renal mesangium. Am J Physiol Renal Physiol 2003;285:F579–F589.
7 Arai H, Hori S, Aramori I, Ohkubo H, Nakanishi S: Cloning and expression of a cDNA encoding an endothelin receptor. Nature 1990;348:730–732.
8 Sakurai T, Yanagisawa M, Takuwa Y, Miyazaki H, Kimura S, Goto K, Masaki T: Cloning of a cDNA encoding a non-isopeptide-selective subtype of the endothelin receptor. Nature 1990;348:732–735.
9 Pollock DM: Endothelin receptor subtypes and tissue distribution; in Highsmith RF (ed): Endothelin: Molecular Biology, Physiology, and Pathology. Totowa, Humana Press, 1998, pp 1–29.
10 Orth SR, Amann K, Gehlen F, Unger L, Wagner J, Raschack M, Ritz E: Adult human mesangial cells (HMCs) express endothelin-B-receptors which mediate endothelin-1-induced cell growth. J Cardiovasc Pharmacol 2000;36:S232–S237.
11 Neuhofer W, Pittrow D: Role of endothelin and endothelin receptor antagonists in renal disease. Eur J Clin Invest 2006;36(Suppl 3):78–88.
12 Chahdi A, Sorokin A: The role of beta(1) Pix/caveolin-1 interaction in endothelin signaling through Galpha subunits. Biochem Biophys Res Commun 2010;391:1330–1335.
13 Badr KF, Murray JJ, Breyer MD, Takahashi K, Inagami T, Harris RC: Mesangial cell, glomerular and renal vascular responses to endothelin in the rat kidney. Elucidation of signal transduction pathways. J Clin Invest 1989;83:336–342.
14 Simonson MS, Wann S, Mene P, Dubyak GR, Kester M, Nakazato Y, Sedor JR, Dunn MJ: Endothelin stimulates phospholipase C, Na^+/H^+ exchange, c-fos expression, and mitogenesis in rat mesangial cells. J Clin Invest 1989;83:708–712.
15 Simonson MS, Dunn MJ: Ca^{2+} signaling by distinct endothelin peptides in glomerular mesangial cells. Exp Cell Res 1991;192:148–156.
16 McDermott GF, Hurst RD, Whiteside CI: Isolated rat glomerular cells demonstrate L-type Ca(2+)-channel activity. Cell Calcium 1993;14:387–396.
17 Du J, Sours-Brothers S, Coleman R, Ding M, Graham S, Kong DH, Ma R: Canonical transient receptor potential 1 channel is involved in contractile function of glomerular mesangial cells. J Am Soc Nephrol 2007;18:1437–1445.
18 Dlugosz JA, Munk S, Zhou X, Whiteside CI: Endothelin-1-induced mesangial cell contraction involves activation of protein kinase C-alpha, -delta, and -epsilon. Am J Physiol 1998;275:F423–F432.
19 Sorokin A, Foschi M, Dunn MJ: Endothelin signalling and regulation of protein kinases in glomerular mesangial cells. Clin Sci (Lond) 2002;103(Suppl 1):132S–136S.
20 Sorokin A, Kozlowski P, Graves L, Philip A: Protein-tyrosine kinase Pyk2 mediates endothelin-induced p38 MAPK activation in glomerular mesangial cells. J Biol Chem 2001;276:21521–21528.
21 Simonson MS, Herman WH: Protein kinase C and protein tyrosine kinase activity contribute to mitogenic signaling by endothelin-1. Cross-talk between G protein-coupled receptors and pp60c-src. J Biol Chem 1993;268:9347–9357.
22 Imamura T, Huang J, Dalle S, Ugi S, Usui I, Luttrell LM, Miller WE, Lefkowitz RJ, Olefsky JM: beta-Arrestin-mediated recruitment of the Src family kinase Yes mediates endothelin-1-stimulated glucose transport. J Biol Chem 2001;276:43663–43667.
23 Rufanova VA, Alexanian A, Wakatsuki T, Lerner A, Sorokin A: Pyk2 mediates endothelin-1 signaling via p130Cas/BCAR3 cascade and regulates human glomerular mesangial cell adhesion and spreading. J Cell Physiol 2009;219:45–56.
24 Beck KF, Mohaupt MG, Sterzel RB: Endothelin-1 inhibits cytokine-stimulated transcription of inducible nitric oxide synthase in glomerular mesangial cells. Kidney Int 1995;48:1893–1899.

25 Hirahashi J, Nakaki T, Hishikawa K, Marumo T, Yasumori T, Hayashi M, Suzuki H, Saruta T: Endothelin-1 inhibits induction of nitric oxide synthase and GTP cyclohydrolase I in rat mesangial cells. Pharmacology 1996;53:241–249.
26 Rufanova VA, Alexanian A, Ostendorf T, Bokemeyer D, Prosser S, Miller B, Sorokin A: Endothelin signaling via guanine exchange factor C3G in renal glomerular mesangial cells. Can J Physiol Pharmacol 2010;88:808–816.
27 Gomez-Garre D, Ruiz-Ortega M, Ortego M, Largo R, Lopez-Armada MJ, Plaza JJ, Gonzalez E, Egido J: Effects and interactions of endothelin-1 and angiotensin II on matrix protein expression and synthesis and mesangial cell growth. Hypertension 1996;27:885–892.
28 Simonson MS, Ismail-Beigi F: Endothelin-1 increases collagen accumulation in renal mesangial cells by stimulating a chemokine and cytokine autocrine signaling loop. J Biol Chem 2011;286:11003–11008.
29 Whiteside CI, Dlugosz JA: Mesangial cell protein kinase C isozyme activation in the diabetic milieu. Am J Physiol Renal Physiol 2002;282:F975–F980.
30 Mishra R, Leahy P, Simonson MS: Gene expression profile of endothelin-1-induced growth in glomerular mesangial cells. Am J Physiol Cell Physiol 2003;285:C1109–C1115.
31 Khositseth S, Pisitkun T, Slentz DH, Wang G, Hoffert JD, Knepper MA, Yu MJ: Quantitative protein and mRNA profiling shows selective post-transcriptional control of protein expression by vasopressin in kidney cells. Mol Cell Proteomics 2011;10:M110.004036.
32 Hughes AK, Padilla E, Kutchera WA, Michael JR, Kohan DE: Endothelin-1 induction of cyclooxygenase-2 expression in rat mesangial cells. Kidney Int 1995;47:53–61.
33 Pratt PF, Bokemeyer D, Foschi M, Sorokin A, Dunn MJ: Alterations in subcellular localization of p38 MAPK potentiates endothelin-stimulated COX-2 expression in glomerular mesangial cells. J Biol Chem 2003;278:51928–51936.
34 Chahdi A, Sorokin A: Endothelin-1 couples betaPix to p66Shc: role of betaPix in cell proliferation through FOXO3a phosphorylation and p27kip1 down-regulation independently of Akt. Mol Biol Cell 2008;19:2609–2619.
35 Chen Y, Sun Y, Lin J, Zhou A, Wang H: Research on the mechanism of endothelin inflammatory effects on human mesangial cells. Chin Med J (Engl) 1997;110:530–534.
36 Floege J, Johnson RJ, Couser WG: Mesangial cells in the pathogenesis of progressive glomerular disease in animal models. Clin Investig 1992;70:857–864.
37 Jaffer FE, Knauss TC, Poptic E, Abboud HE: Endothelin stimulates PDGF secretion in cultured human mesangial cells. Kidney Int 1990;38:1193–1198.
38 Fukuda K, Yanagida T, Okuda S, Tamaki K, Ando T, Fujishima M: Role of endothelin as a mitogen in experimental glomerulonephritis in rats. Kidney Int 1996;49:1320–1329.
39 Daub H, Weiss FU, Wallasch C, Ullrich A: Role of transactivation of the EGF receptor in signalling by G-protein-coupled receptors. Nature 1996;379:557–560.
40 Chahdi A, Sorokin A: Endothelin-1 induces p66Shc activation through EGF receptor transactivation: role of beta(1)Pix/Galpha(i3) interaction. Cell Signal 2010;22:325–329.
41 Hua H, Munk S, Whiteside CI: Endothelin-1 activates mesangial cell ERK1/2 via EGF-receptor transactivation and caveolin-1 interaction. Am J Physiol Renal Physiol 2003;284:F303–F312.
42 Ohtsu H, Dempsey PJ, Eguchi S: ADAMs as mediators of EGF receptor transactivation by G protein-coupled receptors. Am J Physiol Cell Physiol 2006;291:C1–10.
43 Wang Y, Simonson MS, Pouyssegur J, Dunn MJ: Endothelin rapidly stimulates mitogen-activated protein kinase activity in rat mesangial cells. Biochem J 1992;287:589–594.
44 Araki S, Haneda M, Togawa M, Kikkawa R: Endothelin-1 activates c-Jun NH2-terminal kinase in mesangial cells. Kidney Int 1997;51:631–639.
45 Dorado F, Velasco S, Esparis-Ogando A, Pericacho M, Pandiella A, Silva J, Lopez-Novoa JM, Rodriguez-Barbero A: The mitogen-activated protein kinase Erk5 mediates human mesangial cell activation. Nephrol Dial Transplant 2008;23:3403–3411.

46 Chernoff J: Signal transduction and mammalian cell growth: problems and paradigms; in Golemis E (ed): Protein-Protein Interactions: A Molecular Cloning Manual. Cold Spring Harbor, Cold Spring Harbor Laboratory Press, 2002, pp 7–13.
47 Pawson T, Nash P: Protein-protein interactions define specificity in signal transduction. Genes Dev 2000;14:1027–1047.
48 Feller SM, Ren R, Hanafusa H, Baltimore D: SH2 and SH3 domains as molecular adhesives: the interactions of Crk and Abl. Trends Biochem Sci 1994;19:453–458.
49 Schlessinger J: SH2/SH3 signaling proteins. Curr Opin Genet Dev 1994;4:25–30.
50 Fu H, Subramanian RR, Masters SC: 14–3–3 proteins: structure, function, and regulation. Annu Rev Pharmacol Toxicol 2000;40:617–647.
51 Foschi M, Franchi F, Han J, La Villa G, Sorokin A: Endothelin-1 induces serine phosphorylation of the adaptor protein p66Shc and its association with 14–3–3 protein in glomerular mesangial cells. J Biol Chem 2001;276:26640–26647.
52 Foschi M, Chari S, Dunn MJ, Sorokin A: Biphasic activation of p21ras by endothelin-1 sequentially activates the ERK cascade and phosphatidylinositol 3-kinase. EMBO J 1997;16:6439–6451.
53 Terada Y, Inoshita S, Nakashima O, Yamada T, Tamamori M, Ito H, Sasaki S, Marumo F: Cyclin D1, p16, and retinoblastoma gene regulate mitogenic signaling of endothelin in rat mesangial cells. Kidney Int 1998;53:76–83.
54 Mishra R, Wang Y, Simonson MS: Cell cycle signaling by endothelin-1 requires Src nonreceptor protein tyrosine kinase. Mol Pharmacol 2005;67:2049–2056.
55 Chahdi A, Miller B, Sorokin A: Endothelin 1 induces beta 1Pix translocation and Cdc42 activation via protein kinase A-dependent pathway. J Biol Chem 2005;280:578–584.
56 Baker AJ, Mooney A, Hughes J, Lombardi D, Johnson RJ, Savill J: Mesangial cell apoptosis: the major mechanism for resolution of glomerular hypercellularity in experimental mesangial proliferative nephritis. J Clin Invest 1994;94:2105–2116.
57 Ishaque A, Dunn MJ, Sorokin A: Cyclooxygenase-2 inhibits tumor necrosis factor alpha-mediated apoptosis in renal glomerular mesangial cells. J Biol Chem 2003;278:10629–10640.
58 Cortes-Hernandez J, Fossati-Jimack L, Carugati A, Potter PK, Walport MJ, Cook HT, Botto M: Murine glomerular mesangial cell uptake of apoptotic cells is inefficient and involves serum-mediated but complement-independent mechanisms. Clin Exp Immunol 2002;130:459–466.
59 Savill J, Smith J, Sarraf C, Ren Y, Abbott F, Rees A: Glomerular mesangial cells and inflammatory macrophages ingest neutrophils undergoing apoptosis. Kidney Int 1992;42:924–936.

Andrey Sorokin, PhD
CVC, Kidney Disease Center, Department of Medicine, Medical College of Wisconsin
8701 Watertown Plank Road
Milwaukee, WI 53226, USA
Tel. +1 414 4564438, E-Mail Sorokin@mcw.edu

Barton M, Kohan DE (eds): Endothelin in Renal Physiology and Disease.
Contrib Nephrol. Basel, Karger, 2011, vol 172, pp 63–75

Regulation of Sodium Transport in the Proximal Tubule by Endothelin

Ye Zhang[a,b] · Pedro A. Jose[c] · Chunyu Zeng[a,b]

[a]Department of Cardiology, Daping Hospital, The Third Military Medical University, and [b]Chongqing Institute of Cardiology, Chongqing, P.R. China; [c]Center for Molecular Physiology Research, Children's National Medical Center and Department of Pediatrics, George Washington University School of Medicine and Public Health, Washington, D.C., USA

Abstract

Human essential hypertension and rodent genetic hypertension are associated with increased sodium transport in the renal proximal tubule and medullary thick ascending limb of Henle. The proximal tubule, which secretes endothelin (ET), expresses the ET_B receptor. Low (nM) concentrations of ET, via the ET_B receptor, inhibit sodium and water transport and ATP-driven drug secretion in the proximal tubule. In contrast, very low (pM) and high nM concentrations of ET increase renal proximal sodium transport, but the receptor involved remains to be determined. The natriuretic effect of ET_B receptor stimulation is impaired in spontaneously hypertensive rats, due in part to a defective interaction with D_3 dopamine and angiotensin II type 1 receptors. Impaired ET_B receptor function in the renal proximal tubule may be important in the pathogenesis of genetic hypertension.

The kidney is important in the long-term regulation of blood pressure and is the major organ involved in the regulation of body sodium homeostasis [1–3]. The proximal tubule and medullary thick ascending limb of Henle are pre-eminent in the overall regulation of sodium balance in essential hypertension [4]. Indeed, several studies have shown that human essential hypertension and rodent genetic hypertension are associated with increased sodium transport in the renal proximal tubule and medullary thick ascending limb of Henle [4].

Endothelin (ET) was initially identified as an endothelial cell-derived peptide, with the greatest vasoconstrictor potency of any known endogenous compound [5]. ETs are a family of isopeptides (ET-1, ET-2, and ET-3), with at least two receptors subtypes (ET_A and ET_B). Renal tissue expresses both ET receptor subtypes [6], and ET is secreted by renal tubules, including the renal proximal tubule [7, 8], where it can regulate sodium transport in an autocrine/paracrine manner [9]. In the renal proximal tubules, ET-1 (the major ET expressed in the renal proximal tubule) inhibits ion transport principally through the ET_B receptor [10], an effect opposite that of the ET_A receptor. This review focuses on the regulation of ET on sodium transport in the renal proximal tubule and its role in the pathogenesis of essential hypertension.

Endothelin and Its Metabolism

In 1985, Hickey et al. [11] described the existence of a trypsin-sensitive endothelium-derived constricting factor in cultured bovine endothelial cells, which they named as ET. Subsequently, more ET family members were found. These members include ET-1, ET-2, ET-3, and ET-4 (analogue of human ET-2 in rat and mouse, also named as a vasoactive intestinal contractor), as well as three isoforms of 31-amino-acid ETs ($ET\text{-}1^{1-31}$, $ET\text{-}2^{1-31}$, and $ET\text{-}3^{1-31}$) [12]. Owing to the similarity of actions between $ET\text{-}1^{1-31}$ and ET-1 in vivo, it is possible that some of the effects of ET^{1-31} result from their partial bioconversion into ETs [13]. Among these isoforms, ET-1 is the principal isoform and is the most potent and long-lasting constrictor of human vessels known to date [5].

ET-1 is produced in many cell types in the renal and cardiovascular systems, such as endothelial cells, smooth muscle cells, cardiomyocytes, leucocytes, macrophages, and renal tubular and mesangial cells [14, 15]. Bioactive ETs are the product of post-translational processing of the parent preproET peptide. The transcription and translation of *preproET* result in the formation of a 203-amino-acid peptide which is subsequently cleaved by a furin convertase to the 38-amino-acid peptide big ET^{1-38}. Big ET is processed further into ET^{1-31} by different isoforms of ET-converting enzymes (ECEs), a group of proteases that belong to the metalloprotease family [16], including ECE-1a, ECE-1b, ECE-1c, and ECE-1d, derived from a single gene by the action of alternative promoters.

ET synthesis is regulated by many factors. It is enhanced in response to low-shear stress, turbulent blood flow, hypoxia, cytokines, angiotensin II, epinephrine, and low-density lipoproteins [17]. In contrast, high-shear stress, nitric oxide, vasodilating prostaglandins, and natriuretic peptides suppress ET production [18]. ET synthesis is regulated by sodium diet; a high-sodium diet, independent of blood pressure status, increases renal synthesis of ET [19]. Distal nephron segments synthesize ET-1 to a greater extent than the proximal tubule [8].

Synthesized ET-1 is released in two ways. One way is via a constitutive pathway, producing intense constriction of the underlying smooth muscle, which contributes to the maintenance of endogenous vascular tone. The other way is via release from endothelial cell-specific storage granules (Weibel-Palade bodies) in response to external physiological stimuli, producing further vasoconstriction [20]. Although plasma ET-1 is present in the highest concentration in blood/plasma, compared with ET-2 and ET-3, ET-1 concentrations are still lower in plasma than in endothelial and other cells. It is accepted that ET-1 functions as a locally released, rather than a circulating, hormone.

Endothelin Receptors

Endothelin Receptor Classification

ET receptors are classified as ET_A and ET_B by the International Union of Pharmacology Committee on Receptor Nomenclature and Drug Classification [21]. Both ET_A and ET_B receptors belong to the 7-transmembrane domain or G protein-coupled rhodopsin-type receptor superfamily. Pharmacologically heterogeneous responses seem to be related to the existence of alternatively spliced variants of ET_A and ET_B receptors. A third receptor, named ET_C, which is specific for ET-3 binding, was cloned from dermal melanophores from the amphibian *Xenopus laevis*, but a mammalian homologue has yet to be identified [21] (see below).

The three ET receptors have different affinities to the three isoforms of ET (ET-1, ET-2, and ET-3). ET receptors can bind all ET isoforms, but the ET_A receptor has a much higher affinity for ET-1, the most abundant ET in human plasma, than for ET-3, while the ET_C receptor has a higher affinity for ET-3 than ET-1. In contrast, the ET_B receptor binds to all three ET isoforms with equal affinity.

Localization of Endothelin Receptors

In all species studied, including humans, ET-binding is greater in the renal inner medulla than in the inner cortex [8, 22, 23].

In the rat, there is faint immunostaining of the ET_A receptor in the proximal convoluted and straight tubules, restricted to the basal side, and intense staining in the distal tubule collecting duct [24, 25]. The ET_A receptor is also expressed in mesangial cells, pericytes of descending vasa recta, and vascular smooth muscle cells of veins and arteries, specifically the interlobar, arcuate, and interlobular arteries, as well as efferent and afferent arterioles [25]. The ET_B receptor is the major ET receptor in the kidney; 70–80% of ET receptors in the kidney are ET_B receptors [6, 8], which in the rat are expressed in the proximal tubule, inner medullary collecting duct, glomerular capillaries, vasa recta endothelial cells, and vascular smooth muscle cells of interlobular, efferent, and afferent arteries [25–28].

In the mouse, the ET_A receptor has been shown to be expressed in some proximal tubules and vessels, but not in glomeruli. In agreement with the rat studies, the ET_B receptor is expressed in the proximal tubule and collecting duct [29].

Faint staining of ET_A and ET_B receptors has also been reported in human glomeruli, and proximal and distal tubules. In agreement with the rat studies, the distal nephron expresses more ET receptors than the proximal nephron [30, 31].

Signal Transduction of Endothelin Receptors in the Renal Proximal Tubules

Both ET_A and ET_B receptors are linked to Gq/11, $G\alpha_S$, and $G\alpha_{i/O}$ [32, 33]. In some cells (e.g. Chinese hamster ovary cells), the ET_A receptor is linked to $G\alpha_S$ while the ET_B receptor is linked to $G\alpha_i$ [34]. However, in hepatocytes, the ET_B receptor is also linked to $G\alpha_S$ [35]. The ET_B, but the ET_A, receptor also activates $G\alpha_{13}$ [36].

In rat renal brush border membranes, the ET-mediated increase in phospholipase C activity and protein kinase C translocation to the brush border membrane are mediated by the ET_B but not the ET_A, receptor. In contrast, the ET_C receptor is involved in ET-mediated signaling in basolateral membranes [37]. Both the ET_A and ET_B receptors belong to the class A receptors of the G protein-coupled receptor family because they bind to β-arrestin 1 with a higher affinity than β-arrestin 2, and do not bind to visual arrestin [38, 39].

Effect of Endothelin on Renal Proximal Tubule Transport

In vivo Studies

Low-dose infusion of ET in anesthetized rats has been reported to decrease sodium transport in proximal and distal nephron segments, assessed by lithium clearance, which is associated with an increase in renal blood flow, but not glomerular filtration rate [40]. At a dose that does not alter renal plasma flow, ET also decreases proximal but not distal tubule sodium reabsorption, also assessed by lithium clearance [8, 41, 42]. This shows that ET infusion causes natriuresis without altering glomerular filtration rate or renal blood flow, and suggests a direct inhibitory effect of ET-1 on Na^+ transport along the nephron. However, ET-mediated natriuresis has not been consistently observed [43, 44]. One group reported that the natriuretic effect of exogenous ET-1 is due solely to an increase in blood pressure since renal decapsulation or maintaining renal perfusion pressure at baseline values (ET agonists often increase arterial pressure) with an aortic clamp prevents ET-1-induced natriuresis [43]. The possible explanation for the inconsistent findings on ET agonist-induced natriuresis may be due, at least partly, to differential activation of ET_A and ET_B receptors. For example, when the ET_A receptor, but not the ET_B receptor, is blocked, a natriuretic effect of ET-1 is detected [45]. One study showed that low doses of ET causes a natriuresis by decreasing renal proximal tubular reabsorption that is not due to ET_B receptors [46].

In contrast to the studies on sodium excretion, there is unanimous agreement that systemically administered ET increases urinary water excretion. Even when given at doses that markedly decreased renal blood flow, the renal arterial infusion of ET-1 increased urine volume and free water clearance [47]. This effect may be mediated by the ET_B receptor because the infusion of ET_B-specific agonists (IRL1620 or sarafotoxin 6c) increases urine flow [10, 46].

The ET_B receptor also works as a clearance receptor. The ET_B receptor in endothelial cells removes ET-1 from the circulation [48, 49]. Blockade of the ET_B receptor increases circulating immunoreactive ETs (ET-1 and ET-3), and mice with genetic ablation of the ET_B receptor in endothelial cells have elevated plasma concentrations of ET-1 [48, 49].

In vitro Studies

ET has diverse effects throughout the nephron. In the rat proximal tubule, ET has a biphasic effect on ion and fluid transport [50]. Low concentrations (pM) increase, whereas a high concentration (low nM) of ET-1 decreases fluid transport through protein kinase C-, cyclooxygenase- and lipoxygenase-dependent mechanisms [42, 50]. However, higher concentrations (high nM) of ET also increase renal proximal tubule transport [7]. In agreement with the data on sodium transport, low nM concentrations of ET, via the ET_B receptor, have also been shown to inhibit the secretion of fluorescent substrates in renal proximal tubules of the killifish [51].

ET-1 inhibits fluid and bicarbonate absorption in the isolated-perfused rat proximal straight tubule [52] and fluid transport in the midproximal convoluted tubule measured by the split-drop micropuncture method [53]. A preliminary report also indicates that ET decreases sodium-phosphate cotransport in rat renal brush border membranes [53, 54]. Garvin and Sanders [52] showed that the ability of ET to inhibit fluid and bicarbonate transport in the rat proximal straight tubule is mediated by a reduction in Na^+-K^+ ATPase activity. These effects are probably mediated by the ET_B receptor because stimulation of the ET_B receptor also inhibits Na^+-K^+ ATPase activity in immortalized rat S1 renal proximal tubule cells [55], similar to that observed in proximal tubules, with a reduction of about 20–40% [52, 53]. The inhibitory effect of ET, via the ET_B receptor, on Na^+-K^+ ATPase activity in human and rat renal proximal tubule cells is mediated by an increase in intracellular calcium via phosphatidyl inositol-3 kinase [55]. However, in suspensions of rabbit proximal tubule cells, ET was not found to reduce oxygen consumption, which is used as an index of Na^+-K^+ ATPase activity [56]. It remains to be determined whether or not the differential effect of ET on Na^+-K^+ ATPase activity between rats and rabbits is related to species differences.

The effect of ET on NHE3 activity in renal proximal tubule cells is still controversial. Short-term (<1 h) incubation of rat renal cortical slices with ET increases NHE3 activity that is mediated by protein kinase C and inhibition of cAMP production [57]. Short-term (<1 h) stimulation with ET of opossum kidney cells which express ET_B but not ET_A receptors, also increases NHE3 that is mediated equally by Ca^{2+}-dependent and tyrosine kinase-dependent pathways [58, 59]. However, long-term (≥6 h) stimulation of the ET_B receptor in the same opossum kidney cells leads to inhibition of NHE3 activity and expression [60]. One may conclude from these studies that in the renal proximal tubule, ET, via the ET_B receptor, inhibits sodium transport by decreasing Na^+-K^+ ATPase activity. In contrast, ET, via the ET_B receptor, acutely stimulates, but chronically inhibits NHE3 activity. However, the end-result of ET_B receptor stimulation should still be a decrease in renal proximal tubule transport because of the inhibition of Na^+-K^+ ATPase activity.

Interaction with Other G Protein-Coupled Receptors

Interaction between Endothelin A and Endothelin B Receptors
ET_A and ET_B receptors may exist as constitutive homodimers [61]. They may also exist as constitutive heterodimers [62]; therefore, there may be a 'cross-talk' between these two ET receptors. The ET_B receptor may be essential in regulating the development or expression of the ET_A receptor because ET_A receptor expression is decreased in central or peripheral tissues of ET_B receptor-deficient mice [63]. ET_A and ET_B receptor heterodimers may be responsible for a sustained Ca^{2+} signaling by delaying their internalization [64]. Both the ET_A and ET_B receptors are required for the full diuretic and natriuretic actions of ET associated with the intramedullary hyperosmotic saline infusion [65]. An increased ratio of ET_A and ET_B receptor activity is important in the development and progression of DOCA-salt-induced hypertension and organ damage [66]. Indeed, blockade of ET_B receptors in this model of hypertension increases the severity of vascular and renal proximal tubular damage [66].

Interaction with the D_3 Dopamine Receptor
Dopamine, an endogenous catecholamine, is an important regulator of sodium balance and blood pressure via renal and nonrenal mechanisms, including the regulation of appetite centers in the brain, secretion/release of hormones, and humoral agents, as well as interaction with hormones that regulate renal ion transport. Dopamine receptors are divided into D_1- and D_2-like subtypes based on their interaction with the effector enzyme adenylyl cyclase. Among D_2-like receptor subtypes, the D_3 receptor, along with the D_4 receptor, has the highest affinity for dopamine. Stimulation of the D_3 receptor increases sodium and water excretion in normotensive rats [67, 68]. The natriuresis caused by D_3 receptor stimulation is due, in part, to inhibition of Na^+-K^+ ATPase activity via a cooperative interaction with ET_B receptors [28]. The D_3 and ET_B receptors colocalize and physically interact in rat renal proximal tubules and increase each other's expression (fig. 1).

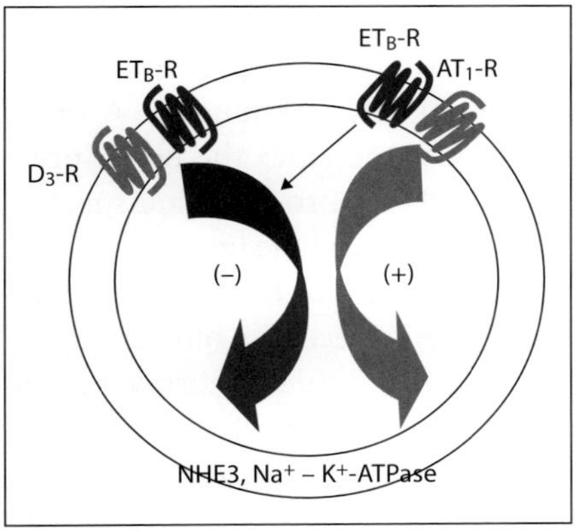

Fig. 1. Interactions among ET$_B$ (ET$_B$-R), D$_3$ dopamine (D$_3$-R), and AT$_1$ (AT$_1$-R) receptors in the regulation of sodium transport in the renal proximal tubule. Stimulation of the AT$_1$-R increases the activity of sodium transporters, sodium exchangers, and the sodium pump (e.g. $NaHCO_3$ cotransporter, Na^+H^+exchanger, and Na^+-K^+-ATPase), while their activity is decreased by the stimulation of ET$_B$-R and dopamine receptors (e.g. D$_1$-R and D$_3$-R). ET$_B$-R, D$_3$-R, and AT$_1$-R physically interact with each other. Stimulation of the ET$_B$-R increases D$_3$-R expression and function, but the opposite effect occurs with AT$_1$-R expression and function. The D$_3$-R also negatively regulates AT$_1$-R expression. In contrast, stimulation of the AT$_1$-R increases ET$_B$-R expression in the long term, and ET$_B$-R cell surface expression in the short term. Thus, the ET$_B$-R synergistically interacts with the D$_3$-R to counterbalance the stimulatory effect of the AT$_1$-R on sodium transport in the renal proximal tubule.

In immortalized renal proximal tubule cells, D$_3$ receptor stimulation increases ET$_B$ receptor expression by a calcium-mediated process which is absent in renal proximal tubule cells from spontaneously hypertensive rats [69, 70].

Interaction with the AT$_1$ Receptor

In the basal state of normal sodium intake and especially during sodium deficit, angiotensin II, via the AT$_1$ receptor, is pre-eminent in renal sodium conservation. The AT$_1$ and ET$_B$ receptors physically interact and regulate each other's expression (fig. 1). Long-term activation of the AT$_1$ receptor increases ET$_B$ receptor expression, whereas short-term activation increases cell surface ET$_B$ receptor expression in rat renal proximal tubule cells, effects that are not observed in renal proximal tubule cells from spontaneously hypertensive rats [71, 72].

Endothelin and Hypertension

Increased ET-1 plasma levels, relative to normotensive subjects, regardless of renal function, have been reported in hypertensive patients in some studies

[73–75]. Most studies, however, have found no differences in circulating ET levels between hypertensive and normotensive subjects [76]. However, the ET_B receptor may be involved in the pathogenesis of hypertension. A high-sodium diet or deoxycorticosterone-salt treatment increases the blood pressure of ET_B receptor-deficient rats. The increased renal sodium transport in these rats probably occurs at a distal tubular level because the increased blood pressure caused by an increased-sodium diet is normalized by blockade of the epithelial sodium channel with amiloride [77]. ET_B receptor blockade produces hypertension that is exaggerated by salt intake or deoxycorticosterone [78, 79]. Systemic ET_B receptor blockade also produces hypertension in mice that is maintained by the ET_A receptor [80]. These findings strongly suggest that the ET_B receptor, by itself or in conjunction with the ET_A receptor, can regulate blood pressure and decreased expression or activity of ET_B receptors increases blood pressure. However, these studies have not determined the involvement of the ET_B receptor expressed in the proximal tubule. As indicated above, the natriuretic effect of ET_B receptor agonists is impaired in spontaneously hypertensive rats. Although ET_B receptor expression is not different in renal proximal tubules between normotensive and spontaneously hypertensive rats, angiotensin II increases total ET_B receptor expression or plasma membrane expression in cells from normotensive, but not hypertensive, rats [71]. The physical interaction between ET_B and AT_1 receptors is also impaired in renal proximal tubule cells from spontaneously hypertensive rats [72]. The blunted natriuretic effect of dopamine in spontaneously hypertensive rats may also be due to an impaired physical interaction between D_3 and ET_B receptors in the proximal tubule [28], resulting in the impaired inhibition of Na^+-K^+ ATPase activity [70].

Conclusion

The renal proximal tubule in all studied species expresses the ET_B receptor. ET has a U-shaped effect on ion and fluid absorption in the renal proximal tubule. The natriuretic effect of ET_B receptor agonists and low nM concentrations of ET is, in part, due to inhibition of sodium and water transport in the proximal tubule. This is due mainly to inhibition of Na^+-K^+ ATPase activity and NHE3 activity, the latter occurring only in the long term, as acute stimulation of ET_B receptors increases NHE3 activity. The inhibition of Na^+-K^+ ATPase activity in renal proximal tubules by stimulation of the ET_B receptor that results in natriuresis is probably abetted by increased positive interaction with the D_3 receptor and negative interaction with the AT_1 receptor. The impaired natriuretic effect of ET_B receptor stimulation in spontaneously hypertensive rats may be caused by impaired ET_B receptor function that may be related to impaired interaction with D_3 and AT_1 receptors.

Acknowledgements

These studies were supported in part by grants from the National Institutes of Health, USA (HL023081, HL074940, DK039308, HL068686, HL092196), the National Natural Science Foundation of China (30470728, 30672199, 81070259), Natural Science Foundation Project of CQ CSTC (2009BA5044), and the grants for Distinguished Young Scholars of China from the National Natural Science Foundation of China (30925018).

References

1 Crowley SD, Coffman TM: In hypertension, the kidney rules. Curr Hypertens Rep 2007;9:148–153.
2 Hall JE: The kidney, hypertension, and obesity. Hypertension 2003;41:625–633.
3 Asico L, Zhang X, Jiang J, Cabrera D, Escano CS, Sibley DR, Wang X, Yang Y, Mannon R, Jones JE, Armando I, Jose PA: Lack of renal dopamine D_5 receptors promotes hypertension. J Am Soc Nephrol 2011;22:82–89.
4 Doris PA: Renal proximal tubule sodium transport and genetic mechanisms of essential hypertension. J Hypertens 2000;18:509–519.
5 Yanagisawa M, Kurihara H, Kimura S, Tomobe Y, Kobayashi M, Mitsui Y, Yazaki Y, Goto K, Masaki T: A novel potent vasoconstrictor peptide produced by vascular endothelial cells. Nature 1988;332:411–415.
6 Schneider MP, Boesen EI, Pollock DM: Contrasting actions of endothelin ET_A and ET_B receptors in cardiovascular disease. Annu Rev Pharmacol Toxicol 2007;47:731–759.
7 Romano G, Giagu P, Favret G, Bartoli E: Effect of endothelin 1 on proximal reabsorption and tubuloglomerular feedback. Kidney Blood Press Res 2000;23:360–365.
8 Kohan DE, Rossi NF, Inscho EW, Pollock DM: Regulation of blood pressure and salt homeostasis by endothelin. Physiol Rev 2011;91:1–77.
9 Licht C, Laghmani K, Yanagisawa M, Preisig PA, Alpern RJ: An autocrine role for endothelin-1 in the regulation of proximal tubule NHE3. Kidney Int 2004;65:1320–1326.
10 Matsuo G, Matsumura Y, Tadano K, Hashimoto T, Morimoto S: Effects of sarafotoxin S6c on renal haemodynamics and urine formation in anaesthetized dogs. Clin Exp Pharmacol Physiol 1997;24:487–491.
11 Hickey KA, Rubanyi G, Paul RJ, Highsmith RF: Characterization of a coronary vasoconstrictor produced by cultured endothelial cells. Am J Physiol 1985;248:C550–C556.
12 Kishi F, Minami K, Okishima N, Murakami M, Mori S, Yano M, Niwa Y, Nakaya Y, Kido H: Novel 31-amino-acid-length endothelins cause constriction of vascular smooth muscle. Biochem Biophys Res Commun 1998;248:387–390.
13 Fecteau MH, Honoré JC, Plante M, Labonté J, Rae GA, D'Orléans-Juste P: Endothelin-1 (1–31) is an intermediate in the production of endothelin-1 after big endothelin-1 administration in vivo. Hypertension 2005;46:87–92.
14 Brunner F, Brás-Silva C, Cerdeira AS, Leite-Moreira AF: Cardiovascular endothelins: essential regulators of cardiovascular homeostasis. Pharmacol Ther 2006;111:508–531.
15 Granger JP, Abram S, Stec D, Chandler D, LaMarca B: Endothelin, the kidney, and hypertension. Curr Hypertens Rep 2006;8:298–303.
16 Turner AJ, Barnes K, Schweizer A, Valdenaire O: Isoforms of endothelin-converting enzyme: why and where? Trends Pharmacol Sci 1998;19:483–486.
17 Martínez-Miguel P, Raoch V, Zaragoza C, Valdivielso JM, Rodríguez-Puyol M, Rodríguez-Puyol D, López-Ongil S: Endothelin-converting enzyme-1 increases in atherosclerotic mice: potential role of oxidized low density lipoproteins. J Lipid Res 2009;50:364–375.

18 Papachristou E, Papadimitropoulos A, Kotsantis P, Goumenos DS, Katsoris PG, Vlachojannis JG: Interaction of endothelin-1 and nitric oxide pathways in human tubular epithelial cells under the influence of cyclosporine A. Ren Fail 2010;32:727–732.
19 Pollock DM, Pollock JS: Evidence for endothelin involvement in the response to high salt. Am J Physiol Renal Physiol 2001;281:F144–F150.
20 Russell FD, Skepper JN, Davenport AP: Human endothelial cell storage granules: a novel intracellular site for isoforms of the endothelin-converting enzyme. Circ Res 1998;83:314–321.
21 Davenport AP: International Union of Pharmacology. XXIX. Update on endothelin receptor nomenclature. Pharmacol Rev 2002;54:219–226.
22 Lattmann T, Shaw S, Münter K, Vetter W, Barton M: Anatomically distinct activation of endothelin-3 and the L-arginine/nitric oxide pathway in the kidney with advanced aging. Biochem Biophys Res Commun 2005;327:234–241.
23 Goettsch W, Lattmann T, Amann K, Szibor M, Morawietz H, Münter K, Müller SP, Shaw S, Barton M: Increased expression of endothelin-1 and inducible nitric oxide synthase isoform II in aging arteries in vivo: implications for atherosclerosis. Biochem Biophys Res Commun 2001;280:908–913.
24 Yamamoto T, Suzuki H, Kubo Y, Matsumoto A, Uemura H: Endothelin A receptor-like immunoreactivity on the basal infoldings of rat renal tubules and collecting ducts. Arch Histol Cytol 2008;71:77–87.
25 Wendel M, Knels L, Kummer W, Koch T: Distribution of endothelin receptor subtypes ET_A and ET_B in the rat kidney. J Histochem Cytochem 2006;54:1193–1203.
26 Francis BN, Abassi Z, Heyman S, Winaver J, Hoffman A: Differential regulation of ET_A and ET_B in the renal tissue of rats with compensated and decompensated heart failure. J Cardiovasc Pharmacol 2004;44(Suppl 1):S362–S365.
27 Schiffrin EL: Vascular endothelin in hypertension. Vascul Pharmacol 2005;43:19–29.
28 Zeng C, Asico LD, Yu C, Villar VA, Shi W, Luo Y, Wang Z, He D, Liu Y, Huang L, Yang C, Wang X, Hopfer U, Eisner GM, Jose PA: Renal D_3 dopamine receptor stimulation induces natriuresis by endothelin B receptor interactions. Kidney Int 2008;74:750–759.
29 Lee S, Ahn D: Expression of endothelin-1 and its receptors in cisplatin-induced acute renal failure in mice. Korean J Physiol Pharmacol 2008;12:149–153.
30 Frank K, Zeier M, Gross ML, Waldherr R, Ritz E, Amann K: Comprehensive immunohistological analysis of the endothelin system in human kidney grafts. Nephrol Dial Transplant 2006;21:1365–1372.
31 Karet FE: Endothelin peptides and receptors in human kidney. Clin Sci (Lond) 1996;91:267–273.
32 Takigawa M, Sakurai T, Kasuya Y, Abe Y, Masaki T, Goto K: Molecular identification of guanine-nucleotide-binding regulatory proteins which couple to endothelin receptors. Eur J Biochem 1995;228:102–108.
33 Doi T, Sugimoto H, Arimoto I, Hiroaki Y, Fujiyoshi Y: Interactions of endothelin receptor subtypes A and B with Gi, Go, and Gq in reconstituted phospholipid vesicles. Biochemistry 1999;38:3090–3099.
34 Aramori I, Nakanishi S: Coupling of two endothelin receptor subtypes to differing signal transduction in transfected Chinese hamster ovary cells. J Biol Chem 1992;267:12468–12474.
35 Jouneaux C, Mallat A, Serradeil-Le Gal C, Goldsmith P, Hanoune J, Lotersztajn S: Coupling of endothelin B receptors to the calcium pump and phospholipase C via Gs and Gq in rat liver. J Biol Chem 1994;269:1845–1851.
36 Kitamura K, Shiraishi N, Singer WD, Handlogten ME, Tomita K, Miller RT: Endothelin-B receptors activate Gα13. Am J Physiol 1999;276:C930–C937.
37 Knotek M, Jaksić O, Selmani R, Skorić B, Banfić H: Different endothelin receptor subtypes are involved in phospholipid signalling in the proximal tubule of rat kidney. Pflugers Arch 1996;432:165–173.

38 Oakley RH, Laporte SA, Holt JA, Caron MG, Barak LS: Differential affinities of visual arrestin, beta arrestin1, and beta arrestin2 for G protein-coupled receptors delineate two major classes of receptors. J Biol Chem 2000;275:17201–10.

39 Hamdan FF, Rochdi MD, Breton B, Fessart D, Michaud DE, Charest PG, Laporte SA, Bouvier M: Unraveling G protein-coupled receptor endocytosis pathways using real-time monitoring of agonist-promoted interaction between β-arrestins and AP-2. J Biol Chem 2007;282:29089–29100.

40 Harris PJ, Zhuo J, Mendelsohn FA, Skinner SL: Haemodynamic and renal tubular effects of low doses of endothelin in anaesthetized rats. J Physiol 1991;433:25–39.

41 Perico N, Cornejo RP, Benigni A, Malanchini B, Ladny JR, Remuzzi G: Endothelin induces diuresis and natriuresis in the rat by acting on proximal tubular cells through a mechanism mediated by lipoxygenase products. J Am Soc Nephrol 1991;2:57–69.

42 Garvin JL, Herrera M, Ortiz PA: Regulation of renal NaCl transport by nitric oxide, endothelin, and ATP: clinical implications. Annu Rev Physiol 2011;73:359–376.

43 Uzuner K, Banks RO: Endothelin-induced natriuresis and diuresis are pressure-dependent events in the rat. Am J Physiol 1993;265:R90–R96.

44 Freed MI, Thompson KA, Wilson DE, Etheredge R, Jorkasky DK: Endothelin receptor antagonism does not alter renal hemodynamic responses or urinary sodium excretion in healthy humans (abstract). J Am Soc Nephrol 1996;7:1580.

45 Brooks DP, DePalma PD, Pullen M, Gellai M, Nambi P: Identification and function of putative ETB receptor subtypes in the dog kidney. J Cardiovasc Pharmacol 1995;26:S322–S325.

46 Clavell AL, Stingo AJ, Margulies KB, Brandt RR, Burnett JC Jr: Role of endothelin receptor subtypes in the in vivo regulation of renal function. Am J Physiol 1995;268:F455–F460.

47 Kamphuis C, Yates NA, McDougall JG: Differential blockade of the renal vasoconstrictor and diuretic responses to endothelin-1 by endothelin antagonist. Clin Exp Pharmacol Physiol 1994;21:329–333.

48 Bagnall AJ, Kelland NF, Gulliver-Sloan F, Davenport AP, Gray GA, Yanagisawa M, Webb DJ, Kotelevtsev YV: Deletion of endothelial cell endothelin B receptors does not affect blood pressure or sensitivity to salt. Hypertension 2006;48:286–293.

49 D'Orléans-Juste P, Labonté J, Bkaily G, Choufani S, Plante M, Honoré JC: Function of the endothelin(B) receptor in cardiovascular physiology and pathophysiology. Pharmacol Ther 2002;95:221–238.

50 Garcia NH, Garvin JL: Endothelin's biphasic effect on fluid absorption in the proximal straight tubule and its inhibitory cascade. J Clin Invest 1994;93:2572–2577.

51 Masereeuw R, Terlouw SA, van Aubel RA, Russel FG, Miller DS: Endothelin B receptor-mediated regulation of ATP-driven drug secretion in renal proximal tubule. Mol Pharmacol 2000;57:59–67.

52 Garvin J, Sanders K: Endothelin inhibits fluid and bicarbonate transport in part by reducing Na^+/K^+ATPase activity in the rat proximal straight tubule. J Am Soc Nephrol 1991;2:976–982.

53 Harris PJ, Thomas D: Effects of endothelin- 1 (ET) on proximal tubule fluid reabsorption (abstract). J Am Soc Nephrol 1990;1:416.

54 Phelps R, Guntupalli J: Effects of endothelin on Na-P1 cotransport in rat renal brush border membranes (BBM) (abstract). J Am Soc Nephrol 1990;1:580.

55 Liu Y, Yang J, Ren H, He D, Pascua A, Armando MI, Yang C, Zhou L, Felder RA, Jose PA, Zeng C: Inhibitory effect of ET_B receptor on Na^+-K^+ ATPase activity by extracellular Ca^{2+} entry and Ca^{2+} release from the endoplasmic reticulum in renal proximal tubule cells. Hypertens Res 2009;32:846–852.

56 Zeidel ML, Brady HR, Kone BC, Gullans SR, Brenner BM: Endothelin, a peptide inhibitor of Na-K-ATPase in intact renal tubular epithelial cells. Am J Physiol 1989;257:C1101–C1107.

57 Guntupalli J, DuBose TD Jr: Effects of endothelin on rat renal proximal tubule Na^+-Pi cotransport and Na^+/H^+ exchange. Am J Physiol 1994;266:F658–F666.

58 Peng Y, Moe OW, Chu T, Preisig PA, Yanagisawa M, Alpern RJ: ET_B receptor activation leads to activation and phosphorylation of NHE3. Am J Physiol 1999;276:C938–C945.
59 Chu TS, Tsuganezawa H, Peng Y, Cano A, Yanagisawa M, Alpern RJ: Role of tyrosine kinase pathways in ET_B receptor activation of NHE3. Am J Physiol 1996;271:C763–C771.
60 Chu TS, Wu KD, Wu MS, Hsieh BS: Endothelin-1 chronically inhibits Na/H exchanger-3 in ET_B-overexpressing OKP cells. Biochem Biophys Res Commun 2000;271:807–811.
61 Gregan B, Schaefer M, Rosenthal W, Oksche A: Fluorescence resonance energy transfer analysis reveals the existence of endothelin-A and endothelin-B receptor homodimers. J Cardiovasc Pharmacol 2004;44(Suppl 1):S30–S33.
62 Gregan B, Jürgensen J, Papsdorf G, Furkert J, Schaefer M, Beyermann M, Rosenthal W, Oksche A: Ligand-dependent differences in the internalization of endothelin A and endothelin B receptor heterodimers. J Biol Chem 2004;279:27679–27687.
63 Davenport AP, Kuc RE: Down-regulation of ET_A receptors in ET_B receptor-deficient mice. J Cardiovasc Pharmacol 2004;44(Suppl 1):S276–S278.
64 Evans NJ, Walker JW: Sustained Ca^{2+} signaling and delayed internalization associated with endothelin receptor heterodimers linked through a PDZ finger. Can J Physiol Pharmacol 2008;86:526–535.
65 Boesen EI, Pollock DM: Cooperative role of ET_A and ET_B receptors in mediating the diuretic response to intramedullary hyperosmotic NaCl infusion. Am J Physiol Renal Physiol 2010;299:F1424–F1432.
66 Matsumura Y, Hashimoto N, Taira S, Kuro T, Kitano R, Ohkita M, Opgenorth TJ, Takaoka M: Different contributions of endothelin-A and endothelin-B receptors in the pathogenesis of deoxycorticosterone acetate-salt-induced hypertension in rats. Hypertension 1999;33:759–765.
67 Jose PA, Asico LD, Eisner GM, Pocchiari F, Semeraro C, Felder RA: Effects of costimulation of dopamine D_1- and D_2-like receptors on renal function. Am J Physiol 1998;275:R986–R994.
68 Luippold G, Küster E, Joos TO, Mühlbauer B: Dopamine D_3 receptor activation modulates renal function in anesthetized rats. Naunyn Schmiedebergs Arch Pharmacol 1998;358:690–693.
69 Zhang Y, Fu C, Ren H, He D, Wang X, Asico LD, Jose PA, Zeng C: Impaired stimulatory effect of ET_B receptor on D_3 receptor in immortalized renal proximal tubule cells of spontaneously hypertensive rats. Kidney Blood Press Res 2011;34:75–82
70 Yu C, Yang Z, Ren H, Zhang Y, Han Y, He D, Lu Q, Wang X, Wang X, Yang C, Asico LD, Hopfer U, Eisner GM, Jose PA, Zeng C: D_3 dopamine receptor regulation of ETB receptors in renal proximal tubule cells from WKY and SHRs. Am J Hypertens 2009;22:877–883.
71 Zeng C, Hopfer U, Eisner GM, Felder RA, Jose PA: Altered AT_1 receptor regulation of ETB receptors in renal proximal tubule cells of spontaneously hypertensive rats. Hypertension 2005;46:926–931.
72 Zeng C, Wang Z, Asico LD, Hopfer U, Eisner GM, Felder RA, Jose PA: Aberrant ETB receptor regulation of AT receptors in immortalized renal proximal tubule cells of spontaneously hypertensive rats. Kidney Int 2005;68:623–631.
73 Shichiri M, Hirata Y, Ando K, Emori T, Ohta K, Kimoto S, Ogura M, Inoue A, Marumo F: Plasma endothelin levels in hypertension and chronic renal failure. Hypertension 1990;15:493–496.
74 Veglio F, Bertello P, Pinna G, Mulatero P, Rossi A, Gurioli L, Panarelli M, Chiandussi L: Plasma endothelin in essential hypertension and diabetes mellitus. J Hum Hypertens. 1993;7:321–325.
75 Schiffrin EL, Thibault G: Plasma endothelin in human essential hypertension. Am J Hypertens. 1991;4:303–308.
76 Davenport AP, Ashby MJ, Easton P, Ella S, Bedford J, Dickerson C: A sensitive radioimmunoassay measuring endothelin-like immunoreactivity in human plasma: comparison of levels in patients with essential hypertension and normotensive control subjects. Clin Sci (Lond) 1990;78:261–264.

77 Gariepy CE, Ohuchi T, Williams SC, Richardson JA, Yanagisawa M: Salt-sensitive hypertension in endothelin-B receptor-deficient rats. J Clin Invest 2000;105:925–933.
78 Williams JM, Pollock JS, Pollock DM: Arterial pressure response to the antioxidant tempol and ET_B receptor blockade in rats on a high-salt diet. Hypertension 2004;44:770–775.
79 Pollock DM, Allcock GH, Krishnan A, Dayton BD, Pollock JS: Upregulation of endothelin B receptors in kidneys of DOCA-salt hypertensive rats. Am J Physiol Renal Physiol 2000;278:F279–F286.
80 Fryer RM, Rakestraw PA, Banfor PN, Cox BF, Opgenorth TJ, Reinhart GA: Blood pressure regulation by ET_A and ET_B receptors in conscious, telemetry-instrumented mice and role of ET_A in hypertension produced by selective ET_B blockade. Am J Physiol Heart Circ Physiol 2006;290:H2554–H2559.

Chunyu Zeng, MD, PhD
Department of Cardiology, Daping Hospital, The Third Military Medical University
Chongqing 400042 (P.R. China)
Tel. +86 023 68757808, E-Mail chunyuzeng01@163.com

Barton M, Kohan DE (eds): Endothelin in Renal Physiology and Disease.
Contrib Nephrol. Basel, Karger, 2011, vol 172, pp 76–83

Role of Endothelin in Thick Ascending Limb Sodium Chloride Transport

Vanesa D. Ramseyer[a,b] · Pablo D. Cabral[a] · Jeffrey L. Garvin[a,b]

[a]Hypertension and Vascular Research Division, Henry Ford Hospital and
[b]Department of Physiology, Wayne State University, Detroit, Mich., USA

Abstract

The thick ascending limb of the loop of Henle reabsorbs 20–30% of filtered sodium chloride (NaCl) and generates the osmotic gradient necessary for water absorption in the distal nephron. It is second only to the collecting duct as a source of renal endothelin (ET)-1, which inhibits NaCl reabsorption in the thick ascending limb by reducing NaCl entry into the cell via the furosemide-sensitive $Na^+/K^+/2\ Cl^-$ cotransporter. The mechanism by which this occurs appears to be due to activation of ET_B receptors, phosphatidylinositol 3 kinase and Akt, and enhanced nitric oxide production by nitric oxide synthase 3. ET-1 may inhibit thick ascending limb NaCl absorption in either an autocrine or paracrine fashion. High-salt intake elevates ET-1 release by thick ascending limbs, although the molecular mechanism involved is unknown. Enhanced ET-1 release and inhibition of thick ascending limb NaCl absorption are thought to be among the mechanisms required to eliminate a salt load without increasing blood pressure. However, we still have much to learn about how ET-1 inhibits thick ascending limb NaCl absorption, how release and processing of ET-1 are regulated, and the receptors involved.

Role of Endothelin-1 in Thick Ascending Limb Sodium Chloride Transport

Zeidel et al. [1] first reported that endothelin (ET)-1 inhibited tubular transport in inner medullary collecting duct cells. Direct actions on nephron transport were later supported by in vivo data showing that it induced natriuresis and diuresis without affecting glomerular filtration rate or renal plasma flow [2]. Because the thick ascending limb plays an essential role in sodium (Na^+) homeostasis, reabsorbing 20–30% of filtered sodium chloride (NaCl) and creating the osmotic gradient necessary for water reabsorption [3], the effect of ET-1 on transport in this segment became the subject of subsequent research.

De Jesus Ferreira and Bailly [4] first demonstrated that 10^{-13} to 10^{-8} M ET-1 affects transport in isolated mouse cortical and medullary thick ascending limbs in a concentration-dependent manner. ET-1 applied to either the lumen or the basolateral membrane inhibited chloride transport by thick ascending limbs. ET-1 activates two different G protein-coupled receptors: ET_A and ET_B [5]. The inhibition of transport was blocked by an ET_B receptor antagonist and mimicked by the selective ET_B receptor agonist sarafotoxin 6c [4]. These data indicate that it was due to activation of the ET_B receptors. Plato et al. [6] obtained similar results in rat thick ascending limbs, demonstrating that ET-1 added to the basolateral bath reduced chloride transport.

In addition, Plato et al. [6] showed that blockade of ET_A receptors neither affected basal chloride transport nor the inhibition of chloride transport induced by ET-1 in medullary thick ascending limbs. These data indicate that ET_A receptors play no role in the regulation of thick ascending limb NaCl absorption, at least in the rat. However, systemic ET_A antagonism may enhance the ET_B-mediated effects in the kidney in general and in the thick ascending limb in particular.

NaCl is reabsorbed by the thick ascending limb via both transcellular and paracellular pathways. The first step in the transcellular pathway is entry into the cell via the luminal $Na^+/K^+/2\ Cl^-$ cotransporter. Na^+ exits via Na/K-ATPase on the basolateral membrane, while Cl^- exits via basolateral Cl^- channels or KCl cotransport. Part of the absorbed K^+ recycles across the luminal membrane via K^+ channels, generating a lumen-positive potential that drives Na^+ absorption via the paracellular pathway. Recently, Herrera et al. [7] found that ET-1 inhibits NaCl transport by blocking $Na^+/K^+/2\ Cl^-$ cotransporter-mediated Na^+ entry. Whether ET-1 affects other transporters involved in transcellular NaCl absorption or proteins that form the paracellular pathway is currently unknown. However, since it inhibits Na/K-ATPase in proximal tubules and inner medullary collecting ducts, it likely does so in the thick ascending limb as well.

In addition to NaCl, Na bicarbonate is also reabsorbed in this nephron segment [8, 9]. This process involves proton secretion by Na^+/H^+ exchange and luminal carbonic anhydrase. Although the same process occurs in the proximal tubule where ET-1 inhibits bicarbonate reabsorption [10], at present we do not know whether it alters bicarbonate transport in the thick ascending limb.

Because thick ascending limb-specific ET-1 or ET_B knockout mice have not been developed, their specific role in this nephron segment in vivo is still unknown. However, there are indirect data indicating that the thick ascending limb is important for the actions of ET-1. While we know that collecting duct-specific ET-1 and ET_B receptor knockout mice develop salt-sensitive hypertension [11, 12], Ge et al. [12] reported that blood pressure increased twice as much in mice lacking the ET-1 gene in the collecting duct as it did when ET_B receptors were absent in this nephron segment. These data suggest that part of the natriuretic effects of ET-1 is mediated by nearby cells, most likely from the thick

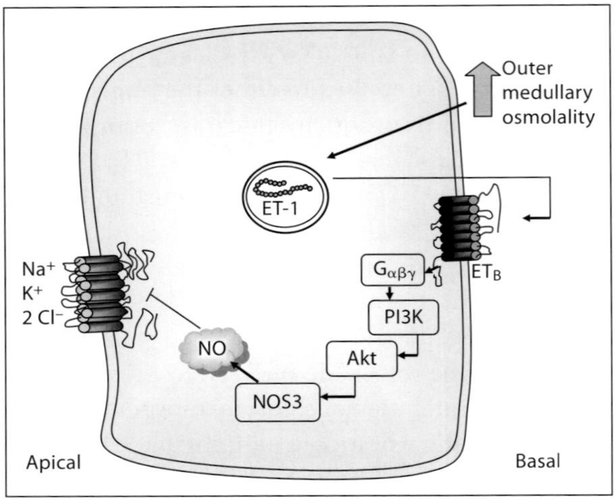

Fig. 1. Synthesis and effects of ET-1 in the thick ascending limb. High-salt intake increases renal outer medullary osmolality, which in turn enhances ET-1 release by thick ascending limbs. ET-1 acting via ET_B receptors increases NO production via phosphatidylinositol 3 kinase (PI3K), Akt and NOS3. In turn NO inhibits the $Na^+/K^+/2\ Cl^-$ cotransporter and reduces NaCl reabsorption.

ascending limb. However, we still do not know whether collecting duct-derived ET-1 inhibits transport in the thick ascending limb in vivo.

Signaling Cascades Mediating Endothelin's Effects in Thick Ascending Limbs

ET_B-induced inhibition of NaCl absorption by thick ascending limbs was initially thought to be mediated by protein kinase C (PKC). De Jesus Ferreira and Bailly [4] showed that treatment with the general PKC inhibitors staurosporine and GF109203 blocked this process in mouse thick ascending limbs. Similarly, addition of the diacylglycerol analogues 1-oleoyl-2-acetyl-sn-glycerol and 1,3-dioctanoylglycerol mimicked the effect of ET-1 on chloride transport. However, there was no increase in intracellular Ca^{++} after ET-1 or ET-3 stimulation, and thus this effect was thought to be mediated by novel PKC isoforms.

On the other hand, there is evidence showing that nitric oxide (NO) mediates ET's inhibition of transport (fig. 1). Plato et al. [6] reported that ET-1 reduced NaCl absorption via NO. Inhibition could be blocked by removing L-arginine (the substrate for NO production) from the bath and by adding Nω-Nitro-L-arginine methyl ester, a competitive NO synthase inhibitor.

ET-1-induced NO production was found to be due to stimulation of NO synthase type 3 (NOS3) because it did not stimulate NO production in NOS3 knockout mice [7]. This process appeared to be due to activation of the phosphatidylinositol 3 kinase/Akt signaling cascade. ET-1 stimulated Akt activity as measured by a fluorescence resonance energy transfer reporter and increased Akt phosphorylation at serine 473. ET-1-induced NO production was prevented by the phosphatidylinositol 3 kinase inhibitor wortmannin. Both ET-1-induced NO production and inhibition of NaCl transport were reduced by transducing thick ascending limbs with dominant-negative Akt1 [7]. The results showing that NO mediates the inhibitory effects of ET-1 on thick ascending limb NaCl reabsorption could in part explain data showing that renal infusion of ET_B agonists induces NO-mediated natriuresis by increasing cGMP [13], whereas ET_B receptor-deficient rats develop salt-sensitive hypertension [14].

The reason that one group showed that the inhibitory effect of ET-1 is mediated by PKC [4] while others have shown that it is due to NO [6, 7] is still unclear. Perhaps this discrepancy simply reflects differences in experimental design. Some PKC inhibitors, including staurosporine, inhibit other kinases. This is especially relevant since PKC activity was not measured. It is possible that activation of PKC by ET-1 increases NO production since it has been shown that PKCα can phosphorylate NOS3 at a stimulatory site and enhance NO production in endothelial cells [15]. However, whether stimulation of NOS3 by ET-1 is mediated by novel PKC isoforms in the thick ascending limb has not been explored. It is not likely that the disparate results can be explained by species differences. While Plato et al. [6] used rats and De Jesus Ferreira and Bailly [4] used mice, Herrera et al. [7] used both mice and rats.

ET-1 may increase NO not only by acutely activating NOS3 [6, 7], but also by increasing NOS3 expression. ET-1 has been shown to increase NOS3 expression in endothelial [16] and cultured cells from the inner medullary collecting duct [17]. Consistent with these findings, ET-1 stimulated NOS3 expression in thick ascending limbs [18, 19]. The increase in NOS3 expression caused by ET-1 was: (1) biphasic with a maximum effect seen at 10^{-10} M; (2) dependent on phosphatidylinositol 3 kinase activation; (3) not affected by inhibition of the ET_A receptor; and (4) blocked with an ET_B receptor antagonist [18]. Altogether, these data indicate that activation of ET_B receptors by ET-1 increases NO production and enhances NOS3 expression.

Endothelin Release by Thick Ascending Limbs

Given that ET is usually considered both a paracrine and autocrine factor, it is important to establish the source of the ET that affects thick ascending limb NaCl absorption. The thick ascending limb is the second highest ET-1 producer in the kidney after the collecting duct [20]; thus, ET's actions could be either paracrine or more likely autocrine. Of the three ET isoforms found in the kidney [21], only ET-1 and -3 are known to be expressed in the thick ascending limb [19, 20, 22, 23]. Expression of ET-1 is reportedly ten times higher than ET-3 in cultured rabbit thick ascending limbs [20], but it is not clear whether this is true for all species.

The molecular mechanisms involved in ET-1 secretion have not been studied in the thick ascending limb to our knowledge. In endothelial cells, ET-1 is released via two pathways: (1) a constitutive pathway, in which release is driven by ET-1 synthesis; and (2) a regulated pathway, in which release depends upon degranulation of vesicles storing the peptide [24]. However, whether ET-1 is stored in vesicles or whether its release follows two different pathways in the thick ascending limb is still unknown.

Increased osmolality has been shown to enhance ET-1 release from thick ascending limbs (fig. 1) [19]. High-salt intake increased renal outer medulla osmolality from 362 to 423 mOsm/kg H_2O and augmented NOS3 expression. This increase was blocked by a combined ET_A/ET_B antagonist in vivo and reproduced by raising osmolality in thick ascending limb primary cultures where it was once again blocked by an ET_B antagonist [19]. Increased ET-1 release and elevation of NOS3 expression in thick ascending limbs could help explain how a rise in blood pressure is prevented when salt intake is increased. Taken together, these data show that thick ascending limb ET-1 release is highly regulated.

In addition to release, processing of ET-1 may be regulated as well. ET-converting enzyme 1 is expressed in thick ascending limbs from humans [25] and rats [26, 27]. Its expression increases in the outer medulla of diabetic rats and in rats infused with contrast medium [26]. Similarly, in spontaneously hypertensive rats, ET-converting enzyme 1 mRNA and protein levels increase in thick ascending limbs and outer medulla, respectively [28]. On the other hand, cyclosporine injection in rats reduced both mRNA levels and protein expression in thick ascending limbs and outer medulla, respectively [27].

Although ET-1 secretion by cultured thick ascending limb cells has been reported [19, 20], it is still unclear as to whether ET-1 is released from the luminal or basolateral membrane. In rat inner medullary collecting duct cells [29], Madin-Darby canine kidney cells [30] (a model of distal nephron cells), and endothelial cells [31], ET-1 secretion is primarily abluminal, resulting in enhanced interstitial ET-1 concentrations compared to intratubular or plasma levels. These data strongly suggest that in the thick ascending limb ET-1 secretion is also abluminal.

Endothelin Receptor Expression and Regulation

In the thick ascending limb, only ET_B receptors have been demonstrated at the mRNA [32] and protein level [33]. In addition, functional studies have shown evidence of ET_B receptors but failed to demonstrate ET_A in this nephron segment [4, 6, 18]. Studies have shown that ET-1 inhibits thick ascending limb transport when added to either compartment [4, 6, 7, 34], implying that ET_B receptors are found in both membranes. However, we could find no reports directly investigating whether ET_B receptors in the thick ascending limb are predominantly found in the basolateral or luminal membrane. It is possible that when added to the lumen, ET traverses the paracellular pathway and acts on basolateral receptors, and vice versa.

Likewise there have been few studies examining regulation of ET_B receptors in the thick ascending limb. While Francis et al. [33] reported they were downregulated in heart failure, they only performed immunohistochemistry. They did not address receptor density and affinity nor did they attempt to establish the mechanism involved.

Conclusion

In summary, the thick ascending limb of the loop of Henle is an important source of ET-1 and -3 in the kidney. High-salt intake augments renal outer medullary osmolality which stimulates ET-1 release by thick ascending limbs. ET-1, in turn, activates phosphatidylinositol 3 kinase, Akt, and NOS3, elevating NO production and inhibiting $Na^+/K^+/2\ Cl^-$ cotransporter activity and chloride reabsorption by this nephron segment. These data suggest that in normotensive individuals blood pressure is maintained during high-salt intake in part by increased ET-1 release and inhibition of NaCl transport by thick ascending limbs. However, we still have much to learn about the details of ET processing and release, its regulation by other factors, and the regulation of ET_B receptor expression and activity in this nephron segment. Likewise the role of thick ascending limb-derived ET-1 or ET_B receptor activity in vivo remains to be explored. Reduced ET-1 and NO production and increased thick ascending limb transport have been shown to cause salt-sensitive hypertension. Therefore, a thorough understanding of the mechanisms that regulate these three factors is essential in the development of new strategies for the treatment of this pathology.

References

1 Zeidel ML, Brady HR, Kone BC, Gullans SR, Brenner BM: Endothelin, a peptide inhibitor of Na(+)-K(+)-ATPase in intact renaltubular epithelial cells. Am J Physiol 1989;257:C1101–C1107.

2 Perico N, Cornejo RP, Benigni A, Malanchini B, Ladny JR, Remuzzi G: Endothelin induces diuresis and natriuresis in the rat by acting on proximal tubular cells through a mechanism mediated by lipoxygenase products. J Am Soc Nephrol 1991;2:57–69.

3 Burg MB: Thick ascending limb of Henle's loop. Kidney Int 1982;22:454–464.

4 De Jesus Ferreira MC, Bailly C: Luminal and basolateral endothelin inhibit chloride reabsorption in the mouse thick ascending limb via a Ca(2+)-independent pathway. J Physiol 1997;505:749–758.

5 Kohan DE, Rossi NF, Inscho EW, Pollock DM: Regulation of blood pressure and salt homeostasis by endothelin. Physiol Rev 2011;91(1):1–77.

6 Plato CF, Pollock DM, Garvin JL: Endothelin inhibits thick ascending limb chloride flux via ET(B) receptor-mediated NO release. Am J Physiol Renal Physiol 2000;279:F326–F333.

7 Herrera M, Hong NJ, Ortiz PA, Garvin JL: Endothelin-1 inhibits thick ascending limb transport via Akt-stimulated nitric oxide production. J Biol Chem 2009;284:1454–1460.

8 Good DW: Sodium-dependent bicarbonate absorption by cortical thick ascending limb of rat kidney. Am J Physiol Renal Physiol 1985;248:F821–F829.

9 Ortiz PA, Garvin JL: Autocrine effects of nitric oxide on HCO(3)(−) transport by rat thick ascending limb. Kidney Int 2000;58:2069–2074.

10 Garvin J, Sanders K: Endothelin inhibits fluid and bicarbonate transport in part by reducing Na+/K+ ATPase activity in the rat proximal straight tubule. J Am Soc Nephrol 1991;2:976–982.

11 Ahn D, Ge Y, Stricklett PK, Gill P, Taylor D, Hughes AK, Yanagisawa M, Miller L, Nelson RD, Kohan DE: Collecting duct-specific knockout of endothelin-1 causes hypertension and sodium retention. J Clin Invest 2004;114:504–511.

12 Ge Y, Bagnall A, Stricklett PK, Strait K, Webb DJ, Kotelevtsev Y, Kohan DE: Collecting duct-specific knockout of the endothelin B receptor causes hypertension and sodium retention. Am J Physiol Renal Physiol 2006;291:F1274–F1280.

13 Nakano D, Pollock JS, Pollock DM: Renal medullary ETB receptors produce diuresis and natriuresis via NOS1. Am J Physiol Renal Physiol 2008;294:F1205–F1211.

14 Gariepy CE, Ohuchi T, Williams SC, Richardson JA, Yanagisawa M: Salt-sensitive hypertension in endothelin-B receptor-deficient rats. J Clin Invest 2000;105:925–933.

15 Partovian C, Zhuang Z, Moodie K, Lin M, Ouchi N, Sessa WC, Walsh K, Simons M: PKC alpha activates eNOS and increases arterial blood flow in vivo. Circ Res 2005;97:482–487.

16 Marsen TA, Egink G, Suckau G, Baldamus CA: Tyrosine-kinase-dependent regulation of the nitric oxide synthase gene by endothelin-1 in human endothelial cells. Pflugers Arch 1999;438:538–544.

17 Ye Q, Chen S, Gardner DG: Endothelin inhibits NPR-A and stimulates eNOS gene expression in rat IMCD cells. Hypertension 2003;41:675–681.

18 Herrera M, Garvin JL: Endothelin stimulates endothelial nitric oxide synthase expression in the thick ascending limb. Am J Physiol Renal Physiol 2004;287:F231–F235.

19 Herrera M, Garvin JL: A high-salt diet stimulates thick ascending limb eNOS expression by raising medullary osmolality and increasing release of endothelin-1. Am J Physiol Renal Physiol 2005;288:F58–F64.

20 Kohan DE: Endothelin synthesis by rabbit renal tubule cells. Am J Physiol 1991;261:F221–F226.

21 Waeber C, Hoyer D, Palacios JM: Similar distribution of [125I]sarafotoxin-6b and [125I]endothelin-1, -2, -3 binding sites in the human kidney. Eur J Pharmacol 1990;176:233–236.

22 Chen M, Todd-Turla K, Wang WH, Cao X, Smart A, Brosius FC, Killen PD, Keiser JA, Briggs JP, Schnermann J: Endothelin-1 mRNA in glomerular and epithelial cells of kidney. Am J Physiol 1993;265:F542–F550.

23 Terada Y, Tomita K, Nonoguchi H, Yang T, Marumo F: Expression of endothelin-3 mRNA along rat nephron segments using polymerase chain reaction. Kidney Int 1993;44:1273–1280.

24 Russell FD, Davenport AP: Secretory pathways in endothelin synthesis. Br J Pharmacol 1999;126:391–398.

25 Pupilli C, Romagnani P, Lasagni L, Bellini F, Misciglia N, Emoto N, Yanagisawa M, Rizzo M, Mannelli M, Serio M: Localization of endothelin-converting enzyme-1 in human kidney. Am J Physiol 1997;273:F749–F756.

26 Khamaisi M, Raz I, Shilo V, Shina A, Rosenberger C, Dahan R, Abassi Z, Meidan R, Lecht S, Heyman SN: Diabetes and radiocontrast media increase endothelin converting enzyme-1 in the kidney. Kidney Int 2008;74:91–100.
27 Nakayama Y, Nonoguchi H, Kiyama S, Ikebe M, Tashima Y, Shimada K, Tanzawa K, Tomita K: Intranephron distribution and regulation of endothelin-converting enzyme-1 in cyclosporin A-induced acute renal failure in rats. J Am Soc Nephrol 1999;10:562–571.
28 Disashi T, Nonoguchi H, Iwaoka T, Naomi S, Nakayama Y, Shimada K, Tanzawa K, Tomita K: Endothelin converting enzyme-1 gene expression in the kidney of spontaneously hypertensive rats. Hypertension 1997;30:1591–1597.
29 Kohan DE, Padilla E: Endothelin-1 is an autocrine factor in rat inner medullary collecting ducts. Am J Physiol 1992;263:F607–F612.
30 Uchida S, Horie M, Yanagisawa M, Matsushita Y, Kurokawa K, Ogata E: Polarized secretion of endothelin-1 and big ET-1 in MDCK cells is inhibited by cell Na+ flux and disrupted by NH4Cl. J Cardiovasc Pharmacol 1991;17(Suppl 7):S226–S228.
31 Wagner OF, Christ G, Wojta J, Vierhapper H, Parzer S, Nowotny PJ, Schneider B, Waldhausl W, Binder BR: Polar secretion of endothelin-1 by cultured endothelial cells. J Biol Chem 1992;267:16066–16068.
32 Terada Y, Tomita K, Nonoguchi H, Marumo F: Different localization of two types of endothelin receptor mRNA in microdissected rat nephron segments using reverse transcription and polymerase chain reaction assay. J Clin Invest 1992;90:107–112.
33 Francis BN, Abassi Z, Heyman S, Winaver J, Hoffman A: Differential regulation of ET(A) and ET(B) in the renal tissue of rats with compensated and decompensated heart failure. J Cardiovasc Pharmacol 2004;44(suppl 1):S362–S365.
34 Bailly C: Effect of luminal atrial natriuretic peptide on chloride reabsorption in mouse cortical thick ascending limb: inhibition by endothelin. J Am Soc Nephrol 2000;11:1791–1797.

Jeffrey L. Garvin, PhD
Hypertension and Vascular Research Division, Henry Ford Hospital
2799 West Grand Boulevard, Education and Research Building, 7018
Detroit, MI 48202–2689 (USA)
Tel. +1 313 916 2048, E-Mail jgarvin1@hfhs.org

Barton M, Kohan DE (eds): Endothelin in Renal Physiology and Disease.
Contrib Nephrol. Basel, Karger, 2011, vol 172, pp 84–93

Endothelins and Kidney Acidification

Donald E. Wesson

Texas A&M Health Sciences Center College of Medicine, Scott and White Healthcare,
Temple, Tex., USA

Abstract

Endothelins (ETs) were initially identified for their vasoactive properties, but they also influence kidney tubule transport including acidification. They contribute importantly to enhancing kidney acidification in response to an acid challenge to systemic acid-base balance and enhance tubule acidification in settings of chronically reduced glomerular filtration rate (GFR) as in chronic kidney disease. ETs also contribute pathophysiologically to the increased kidney acidification in some forms of chronic metabolic alkalosis. In addition, chronically increased ET activity in the kidney, as in chronic kidney disease with reduced GFR, might also mediate the progressive GFR decline observed in some nephropathies. These exciting insights have led to the need for continued investigation regarding the physiologic contribution of ETs to kidney acidification and the possible pathophysiologic consequences of the associated increased kidney ET activity.

Although data to date do not support that endothelins (ETs) mediate kidney acidification under 'control', non-acid-ingesting conditions [1], ETs contribute physiologically to kidney acidification by mediating increased kidney acid (H^+) excretion in response to an H^+ challenge to the acid-base status [1, 2] and by mediating increased per nephron acidification in the remaining nephrons of animals with chronic kidney failure induced by nephron mass reduction [3]. Because diets in industrialized societies are typically acid-inducing [4], thereby subjecting humans to a chronic H^+ challenge, and because chronic kidney disease prevalence is high and increasing in industrialized societies [5], ET-mediated kidney H^+ excretion nevertheless appears to mediate basal kidney function in humans living in industrialized societies. ETs also contribute to enhanced kidney acidification in pathophysiologic states like chronic metabolic alkalosis [6]. Furthermore, enhanced kidney ET activity in settings of increased kidney acidification contributes to kidney injury [7] with progressive glomerular filtration rate (GFR) decline [8, 9]. Hence, understanding the ET-mediated contribution to kidney acidification has important physiologic and pathophysiologic relevance to basic and clinical medicine.

Overview of the Kidney Response to H^+ Challenges

The routine H^+ challenge to systemic acid-base homeostasis for most humans is H^+ derived from liver metabolism of dietary protein [10]. Metabolism of sulfur- and phosphate-containing amino acids yields H^+, and most diets in industrialized societies include dietary protein with a high proportion of such amino acids [4]. This metabolically produced 'fixed' H^+ might increase body fluid $[H^+]$ and induce a physiologic response designed to restore optimum $[H^+]$. Two responses designed to maintain and/or restore optimal body fluid $[H^+]$ are (1) H^+ buffering [11] in which body buffers bind added H^+ to minimize the $[H^+]$ increase that would otherwise occur, and (2) 'fixed' H^+ excretion, predominantly by kidneys [10]. Because H^+-titrated buffers less effectively buffer subsequently added H^+, buffer-bound H^+ must eventually be excreted to regenerate body buffers and restore buffering capacity. Consequently, all added H^+ must be excreted to restore acid-base homeostasis; therefore, ET-mediated kidney H^+ excretion might routinely help correct the H^+ challenge induced by the diets of our industrialized society. Animal [8, 12, 13] and human [14] studies support that reduced GFR is associated with retained or incompletely excreted acid that increases plasma and kidney ET levels [8, 12–14]. These data suggest that high ET levels associated with reduced GFR contribute to the necessary increase in distal nephron acidification in the setting of reduced GFR [3], but also might contribute to progressive GFR decline observed in many such patients.

Endothelin Role in the Kidney Response to Systemic H^+ Challenges

Proximal Tubule. Unlike the large H^+ loads typically given to experimental animals to assess kidney response to a systemic acid challenge [15], the more routine H^+ challenge faced by humans is much more modest and is associated with little to no measurable changes in plasma acid-base parameters [16]. Studies using large H^+ loads show an important role for enhanced proximal tubule acidification in mediating increased kidney H^+ excretion in this setting [15]. The Na^+/H^+ exchanger type 3 (NHE3) is the major luminal H^+ transporter in the proximal tubule with more modest contributions by H^+-ATPase [17]. Large H^+ loads increased activity of NHE3 [17] and H^+-ATPase [18] in the proximal tubule. Increased NHE3 activity in this setting is mediated by ETs through ET_B receptors [19–21].

Loop of Henle. Large H^+ loads increase acidification in the thick ascending limb of the loop of Henle [22]. Because NHE is the major luminal H^+ transporter in the thick ascending limb of the loop of Henle with smaller contributions by H^+-ATPase [22], the acidification increase in response to this H^+ challenge presumably includes stimulated NHE activity. In the loop of Henle, ET stimulates local nitric oxide (NO) release [23] and NO inhibits thick ascending limb Na^+/H^+ exchange [24], suggesting that ET indirectly inhibits acidification in this segment. Because ET-induced NO action on thick ascending limb Na^+/H^+ exchange activity might more importantly inhibit NaCl reabsorption in this nephron segment [25], this indirect action of ET action in the thick ascending limb would increase distal nephron Na^+ delivery with an anticipated increase in distal nephron Na^+/H^+ exchange activity [26], enhancing distal nephron acidification. Consequently, although the direct ET effect seems to be inhibition of acidification in the loop of Henle, its net effect appears to be increased proximal and distal nephron acidification.

Distal Nephron. Large and more modest H^+ challenges to the systemic acid-base balance enhance distal nephron acidification [1, 27, 28] that is mediated by enhanced activity of NHE2 [29], H^+-ATPase [30], and H^+, K^+-ATPase [31]. ETs mediate enhanced distal nephron acidification in response to a systemic H^+ challenge [1, 2, 32, 33] by directly stimulating distal nephron Na^+/H^+ exchange [2], by stimulating H^+-ATPase through stimulation of aldosterone secretion [32], and by reducing HCO_3 secretion through increased NO activity [33]. Such studies were unable to show that ETs stimulated H^+, K^+-ATPase activity [2, 32].

Anatomic and Functional Considerations. In the kidney cortex, tubule epithelia and vascular endothelium are separated by a narrow communicating interstitial space [34], permitting possible paracrine control of tubule acidification by vascular endothelium. Human kidney microvascular endothelial cells secrete ET [35], and ET secretion by human vascular endothelium is predominantly across their basolateral surfaces [36], establishing a context whereby paracrine control of kidney epithelial acidification by ET from vascular endothelium might occur.

Factor(s) That Might Induce Increased Kidney Endothelins in Response to an H^+ Challenge to Systemic Acid-Base Status. In vivo studies show that a modest and chronic dietary H^+ challenge that increases kidney H^+ excretion and distal nephron acidification without measurable changes in plasma acid-base parameters nevertheless increases the H^+ content of the kidney interstitium [37]. Such a systemic H^+ challenge also increases kidney ET [1, 2, 8, 12–14, 32, 33] and its messenger RNA [2] through increased H^+ content [8, 12–14]. Because an acid extracelluar environment increases ET-1 secretion from human kidney microvascular cells in vitro [35], increased H^+ content of the kidney interstitium might induce ET secretion across the basolateral surfaces of the kidney microvascular endothelium that in turn enhances kidney tubule acidification through interaction with ET_B receptors located on the basolateral surface of kidney cortical epithelia [38].

General Endothelin Effects on Acidification

Cells. ET increases intracellular pH of many cell types in vitro, including skin fibroblasts [39], platelets [40], vascular smooth muscle [41], glomerular mesangial [42], cardiac myocyte [43], and kidney epithelial cells [44]. ET increases Na^+/H^+ antiporter activity in kidney epithelial membrane vesicles [45] and cortical slices [46], and is the cell membrane H^+ transporter that is most consistently influenced by ET in these indicated cell types.

Kidney Tubules. In addition to enhancing cell membrane H^+ transport, ET enhances acidification across kidney tubule epithelia. ET mediates enhanced proximal tubule acidification in chronic metabolic acidosis induced by NH_4Cl loading through stimulation of ET_B receptors, but does not appear to mediate proximal tubule acidification in non-H^+-loaded animals [20, 21]. ET mediates enhanced distal nephron acidification induced by dietary H^+ loading with NH_4^+ salts [1] and H^+-producing dietary protein [2, 32, 33], also through ET_B [1, 2]. ET also stimulates H^+ secretion in alpha intercalated cells and decreases HCO_3 secretion in beta intercalated cells of the rabbit cortical collecting duct in an in vitro model of metabolic acidosis [47]. The latter studies support that these effects of ET are direct rather than acting through other mediators. In addition, exogenous ET stimulates distal nephron acidification in vivo [48]. As observed in the proximal tubule, ET does not appear to mediate basal distal nephron acidification in control animals [1, 2, 33].

Direct Mechanisms by which H^+-Induced Endothelin Activity Increases Kidney Acidification

ET increases proximal tubule acidification in H^+-challenged animals and cells primarily, if not exclusively, through enhanced activity of the NHE3 [20, 21, 49], the major H^+ transporter in the proximal tubule [17]. ET-1-mediated stimulated NHE3 activity in opossum kidney cells (a proximal tubule cell line) appears to act through ET_B, but not ET_A, receptors [50]. The ET_B effect was mediated by the COOH-terminal tail and the second intracellular loop of ET_B [51], and involved tyrosine kinase pathways [50, 52]. ET_B stimulation of NHE3 activity is Ca^{2+}- and Rho kinase-dependent, requires phosphorylation and exocytosis of NHE3 into the apical membrane [52], and might involve Gi-mediated inhibition of cAMP accumulation [53]. Because ET increases

Na^+/H^+ exchange in the distal tubule and NHE2 appears to be the major apical NHE isoform responsible for luminal H^+ secretion in this segment [54], ET appears to increase NHE2 activity. As noted, ET stimulates H^+ secretion in alpha intercalated cells [47] whose luminal acidification is mediated predominantly by H^+-ATPase [55, 56]. These in vitro studies [47] suggest that ET can stimulate H^+-ATPase without acting through other mediators. On the other hand, ET might indirectly stimulate distal nephron H^+-ATPase in vivo [32, 33] through ET-induced stimulation of aldosterone secretion that in turn increases distal nephron H^+-ATPase activity [32]. Whether ET stimulates H^+-ATPase activity in vivo directly and/or through other mediators awaits clarification.

Indirect Mechanisms by which H^+-Induced Endothelin Activity Increases Kidney Acidification

Because many cytokines influence distal nephron acidification and ET influences levels of many of these cytokines, ET might affect kidney acidification indirectly as well as directly. As discussed earlier, ET increases distal nephron H^+-ATPase activity in vivo through stimulated aldosterone secretion [32]. Of the three major distal nephron H^+ transporters (Na^+/H^+ exchanger, H^+-ATPase, and H^+, K^+-ATPase) [57], dietary H^+ leads to ET-mediated stimulated activity of Na^+/H^+ exchange [20, 21] and H^+-ATPase [32, 33]. Dietary H^+-induced ET-mediated distal nephron acidification does not appear to be mediated by stimulated H^+, K^+-ATPase activity [32]. Although K^+ depletion increases H^+, K^+-ATPase activity [6, 58], stimulated H^+, K^+-ATPase in chronic metabolic alkalosis associated with K^+ depletion does not appear to be mediated by ET [6]. In addition, ETs might indirectly enhance kidney acidification through ET-stimulated secretion of mineralocorticoids and glucocorticoids by the adrenal cortex [59] that stimulate H^+-ATPase [57] and Na^+/H^+ exchange [60, 61], respectively, in kidney tubules as discussed earlier.

Reduced HCO_3 secretion contributes importantly to enhanced acidification induced by chronic dietary ingestion of mineral H^+ [1, 28] and H^+-producing dietary protein [2, 32, 33]. The phenomenon of reduced HCO_3 delivery to the terminal distal nephron leads to increased net H^+ excretion [62], but this phenomenon also enhances NH_4^+ secretion [57] and permits secreted H^+ to titrate non-HCO_3 buffers and thereby constitute net H^+ excretion rather than HCO_3 recovery [15]. ET decreases distal nephron HCO_3 secretion (that reduces distal nephron acidification) [2, 32, 33] and does so indirectly through stimulated NO production [33]. Other studies show that NO stimulates overall acidification in both the proximal [63] and distal [64] nephron in other settings.

Endothelin Role in the Enhanced Tubule Acidification of Chronically Reduced Glomerular Filtration Rate

ETs mediate the increased per nephron acidification in animal models of chronic kidney failure induced by nephron mass reduction [3]. These animals, which are on standard animal diets without supplemental dietary H^+, nevertheless have increased kidney interstitial levels of ET-1 [8, 9, 12–14] like those animals with intact nephron mass with supplemental dietary H^+ [1]. As described for animals with intact nephron mass given supplemental dietary H^+ [37], animals with chronically reduced nephron mass have increased interstitial H^+ [8, 9, 12–14, 37] even when nephron mass reduction and GFR are less severe, yielding plasma acid-base parameters no different from control animals with intact nephron mass [1, 13, 37]. Like the animals with intact nephron mass given supplemental dietary H^+, increased interstitial H^+ content mediates the increased interstitial levels of ET-1 [8, 9].

Endothelin Role in the Enhanced Kidney Acidification Associated with Pathophysiologic Conditions

The enhanced distal nephron H^+ secretion that 'maintains' chronic metabolic alkalosis [65–67] is mediated by kidney ETs [6]. Conversely, it is reduction of this enhanced distal nephron H^+ secretion (rather than an induced increase in distal nephron HCO_3 secretion) that 'corrects' this disturbed distal nephron acidification in this setting [67]. Because K^+ depletion is important in the maintenance of chronic metabolic alkalosis [6, 67] and because ETs mediate enhanced NHE3 activity in an autocrine fashion in proximal tubule epithelium exposed to an acid environment in vitro [68], K^+ depletion appears to be an important component of this effect. Indeed, K^+ repletion in chronic alkalosis reduces the associated increased urine ET-1 excretion [6], a surrogate of kidney ET-1 production [1].

Possible Endothelin Role in Complications of Pathophysiologic Conditions Associated with Enhanced Kidney Acidification

Animals with chronic metabolic alkalosis develop fibrosis with subsequent calcium deposition in kidney parenchyma [69]. In addition, experimental animals with reduced kidney mass develop glomerulosclerosis with tubulointerstitial fibrosis [70]. ET increases kidney matrix production and fibrosis in vitro [71], appears to mediate glomerulosclerosis in vivo [72], and its kidney levels are increased in animals with reduced GFR [6, 8, 9, 12, 13]. Consequently, increased ET activity in these and possibly other settings might contribute to associated progressive kidney injury through ET-mediated glomerulosclerosis and fibrosis. Indeed, ETs mediate tubulointerstitial injury in animals with intact nephron mass that ingest an H^+-inducing diet [7], GFR decline induced by metabolic acidosis in animals with reduced nephron mass [8, 9], and GFR decline in animals with less severely reduced nephron mass without metabolic acidosis but with H^+ retention [9]. Consistent with these animal studies, amelioration of the metabolic acidosis of humans with low GFR using dietary alkali reduced parameters of kidney injury, slowed GFR decline, and reduced kidney ET production [73]. Furthermore, humans with moderately reduced GFR and plasma acid-base parameters not different from those with normal GFR have increased urine excretion of ET-1, consistent with increased kidney ET-1 levels, that were reduced with dietary alkali [74]. These studies support that, like the animal studies described previously, humans with reduced GFR have increased kidney ET levels that physiologically contribute to increase kidney acidification but pathophysiologically contribute to kidney injury and progressive GFR decline. This hypothesis must be tested with further studies.

References

1 Wesson DE: Endogenous endothelins mediate increased distal tubule acidification induced by dietary acid in rats. J Clin Invest 1997;99:2203–2211.
2 Khanna A, Simoni J, Hacker C, Duran M-J, Wesson DE: Increased endothelin activity mediates augmented distal nephron acidification induced by dietary protein. J Am Soc Nephrol 2004;15:2266–2275.
3 Wesson DE: Endogenous endothelins mediate augmented acidification in remnant kidneys. J Am Soc Nephrol 2001;12:1826–1835.
4 Remer T: Influence of nutrition on acid-base balance-metabolic aspects. Eur J Nutr 2001;40:214–220.
5 US Renal Data System: USRDS 2007 Annual Data Report. Bethesda, The National Institutes of Health, National Institute of Diabetes and Digestive and Kidney Diseases, 2008.
6 Wesson DE: Na^+/H^+ exchange mediates increased distal tubule acidification in chronic alkalosis. Kidney Int 1998;53:945–951.
7 Wesson DE, Nathan T, Rose T, Simoni J, Tran RM: Dietary protein induces endothelin-mediated kidney injury through enhanced intrinsic acid production. Kidney Int 2007;71:210–217.
8 Phisitkul S, Hacker C, Simoni J, Tran RM, Wesson DE: Dietary protein causes a decline in the glomerular filtration rate of the remnant kidney mediated by metabolic acidosis and endothelin receptors. Kidney Int 2008;73:192–199.
9 Wesson DE, Simoni J: Acid retention during kidney failure induces endothelin and aldosterone production which lead to progressive GFR decline, a situation ameliorated by alkali diet. Kidney Int 2010;78:1128–1135.
10 Lennon EJ, Lemann J Jr, Litzow JR: The effects of diet and stool composition on the net external acid balance of normal subjects. J Clin Invest 1966;45:1601–1607.
11 Lemann J Jr, Litzow JR, Lennon EF: The effects of chronic acid loads in normal man: further evidence for the participation of bone mineral in the defense against chronic metabolic acidosis. J Clin Invest 1966;45:1608–1614.
12 Wesson DE, Simoni J: Increased tissue acid mediates progressive GFR decline in animals with reduced nephron mass. Kidney Int 2009;75:929–935.

13 Wesson DE, Simoni J: Acid retention during renal failure induces endothelin and aldosterone production which lead to progressive decline of the GFR, a situation ameliorated by alkali diet. Kidney Int 2010;78:1128–1135.
14 Wesson DE, Simoni J, Broglio K, Sheather S: Acid retention accompanies reduced GFR in humans and increases plasma levels of endothelin and aldosterone. Am J Physiol Renal Physiol 2011;300:F830–F837.
15 Buerkert J, Martin D, Trigg D: Segmental analysis of the renal tubule in buffer production and net acid formation. Am J Physiol 1983;244:F442–F454.
16 Kurtz I, Maher T, Hulter HN, Schambelan M, Sebastian A: Effect of diet on plasma acid-base composition in normal humans. Kidney Int 1983;24:670–680.
17 Preisig PA, Ives HE, Cragoe EJ Jr, et al: Role of the Na^+/H^+ antiporter in rat proximal tubule bicarbonate absorption. J Clin Invest 1987;80:970–978.
18 Chambrey R, Paillard M, Podevin RA: Enzymatic and functional evidence for adaptation of the vacuolar H(+)-ATPase in proximal tubule apical membranes from rats with chronic metabolic acidosis. J Biol Chem 1994;269:3243–3250.
19 Licht C, Laghmani K, Yanagisawa M, et al: An autocrine role for endothelin-1 in the regulation of proximal tubule NHE3. Kidney Int 2004;65:1320–1326.
20 Laghmani K, Preisig PA, Moe OW, et al: Endothelin-1/endothelin-B receptor-mediated increases in NHE3 activity in chronic metabolic acidosis. J Clin Invest 2001;107:1563–1569.
21 Laghmani K, Preisig PA, Alpern RJ: The role of endothelin in proximal tubule proton secretion and the adaptation to a chronic metabolic acidosis. J Nephrol 2002;15(suppl 5):S75–S87.
22 Capasso G, Unwin R, Rizzo M, Pica A, Giebisch G: Bicarbonate transport along the loop of Henle: molecular mechanisms and regulation. J Nephrol 2002;15(suppl 5):S88–S96.
23 Plato CF, Garvin JL: Nitric oxide, endothelin and nephron transport: potential interactions. Clin Exp Pharmacol Physiol 1999;26:2626–268.
24 Garvin JL, Hong NJ: Nitric oxide inhibits sodium/hydrogen exchange activity in the thick ascending limb. Am J Physiol 1999;277:F377–F382.
25 Plato CF, Pollock DM, Garvin JL: Endothelin inhibits thick ascending limb chloride flux via ET(B) receptor-mediated NO release. Am J Physiol 2000;279:F326–F333.
26 Weinstein AM: A mathematical model of rat distal convoluted tubule. I. Cotransporter function in early DCT. Am J Physiol 2005;289:F699–F720.
27 Levine DZ: An in vivo microperfusion study of distal tubule bicarbonate reabsorption in normal and ammonium chloride rats. J Clin Invest 1985;75:588–595.
28 Wesson DE: Reduced HCO_3 secretion mediates increased distal nephron acidification induced by dietary acid. Am J Physiol 1996;271:F670–F678.
29 Malnic G, Fernandez R, Cassola AC, et al: Mechanisms and regulation of H^+ transport in distal tubule epithelial cells. Wien Klin Wochenschr 1997;109:429–434.
30 Malnic G, Fernandez R, Lopes MJ: The role of the distal nephron in the regulation of acid-base equilibrium by the kidney. Braz J Med Biol Res 1994;27:831–850.
31 Silver RB, Mennitt PA, Satlin LM: Stimulation of apical H-K-ATPase in intercalated cells of cortical collecting duct with chronic metabolic acidosis. Am J Physiol 1996;270:F539–F547.
32 Khanna A, Simoni J, Wesson DE: Endothelin-induced increased aldosterone activity mediates augmented distal nephron acidification as a result of dietary protein. J Am Soc Nephrol 2005;16:1929–1935.
33 Wesson DE, Simoni J, Prabhakar S: Endothelin-induced increased nitric oxide activity mediates augmented distal nephron acidification as a result of dietary protein. J Am Soc Nephrol 2006;17:406–413.
34 Kriz W, Kaissling B: Structural organization of the mammalian kidney; in Seldin D, Giebisch G (eds): The Kidney. Physiology and Pathophysiology. New York, Raven Press, 1992, p 207.
35 Wesson DE, Simoni J, Green DF: Reduced extracellular pH increases endothelin-1 secretion by human renal microvascular cells. J Clin Invest 1998;101:578–583.

36 Wagner OF, Christi G, Wojta J, et al: Polar secretion of endothelin-1 by cultured endothelial cells. J Biol Chem 1992;267:16066–16068.
37 Wesson DE: Dietary acid increases blood and renal cortical acid content in rats. Am J Physiol 1998;274:F97–F103.
38 Dean R, Zhuo J. Alcorn D, et al: Cellular distribution of ^{125}-endothelin-binding in rat kidney following in vivo labeling. Am J Physiol 1994;267:F845–F852.
39 Gardner JP, Maher E, Aviv A: Calcium mobilization and Na^+/H^+ antiport activation by endothelin in human skin fibroblasts. FEBS Lett 1989;256:38–42.
40 Touyz RM, Lariviere R, Schiffrin EL: Endothelin influences pHi of human platelets through protein kinase C mediated pathways. Thrombosis Res 1995;78:55–65.
41 Hubel CA, Highsmith RF. Endothelin-induced changes in intracellular pH and Ca^{2+} in coronary smooth muscle: role of Na(+)-H(+) exchange. Biochemical J 1995;310:1013–1020.
42 Simonson MS, Wann S, Mene P, et al. Endothelin stimulates phospholipase C, Na^+/H^+ exchange, c-fos expression, and mitogenesis in rat mesangial cells. J Clin Invest 1989;83:708–712.
43 Wang H, Sakuai, Endoh M: Pharmacological analysis by HOE642 and KB-R9032 of the role of Na(+)/H(+) exchange in the endothelin-1 induced Ca(2+) signaling in rabbit ventricular mycoytes. Brit J Pharmacol 2000;131:638–644.
44 Walter R, Helmle-Kolb C, Forgo J, et al: Stimulation of Na^+/H^+ exchange activity by endothelin in opossum kidney cells. Pflugers Archiv 1995;430:137–144.
45 Eiam-Ong S, Hilden SA, King AJ, et al: Endothelin-1 stimulates the Na^+/H^+ and Na^+/HCO_3- transporters in rabbit renal cortex. Kid Int 1992;42:18–24.
46 Guntupalli J, Dubose TD Jr: Effects of endothelin on rat renal proximal tubule Na(+)-Pi cotransport in Na^+/H^+ exchange. Am J Physiol 1994;266:F658–F666.
47 Tsuruoka S, Watanabe S, Purkerson JM, Fujimura A, Schwartz GJ: Endothelin and nitric oxide mediate the adaptation of the cortical collecting duct to metabolic acidosis. Am J Physiol Renal Physiol 2006;291:F866–F873.
48 Wesson DE, Dolson GM: Endothelin-1 increases rat distal tubule acidification in vivo. Am J Physiol 1997;273:F586–F594.
49 Chu T-S, Peng Y, Cano A, et al: $Endothelin_B$ receptor activates NHE-3 by a Ca^{2+}-dependent pathway in OKP cells. J Clin Invest 1996;97:1454–1462.
50 Chu TS, Tsuganezawa H, Cano A, Yanagisawa M, Alpern RJ: Role of tyrosine kinase pathways in ET_B activation of NHE3. Am J Physiol 1996;271:C763–C771.
51 Laghmani K, Sakamoto A, Yanagisawa M, Preisig AP, Alpern RJ: A consensus sequence in endothelin B receptor loop is required for NHE3 activation by endothelin-1. Am J Physiol 2005;288:F32–F39.
52 Preisig PA: The acid-activated signaling pathway: starting with Pyk2 and ending with increased NHE3 activity. Kid Int 2007;72:1324–1329.
53 Cano A, Miller RT, Alpern RJ, Preisig PA: Angiotensin II stimulation of Na-H antiporter activity is cAMP independent in OKP cells. Am J Physiol 1994;266:C1603–C1608.
54 Wang T, Hropot M, Aronson PS, Giebisch G: Role of NHE isoforms in mediating bicarbonate reabsorption along the nephron. Am J Physiol 2001;281:F1117–F1122.
55 Tsuruoka S, Schwartz GJ: Adatation of rabbit cortical collecting duct HCO_3-transport to metabolic acidosis in vitro. J Clin Invest 1996;97:1076–1084.
56 Weiner ID, Frank AE, Wingo CS: Apical proton secretion by the inner stripe of the outer medullary collecting duct. Am J Physiol 1999;276:F606–F613.
57 Fernandez R, Lopes MJ, de Lira RF, et al: Mechanism of acidification along cortical distal tubule of the rat. Am J Physiol 1994;266:F218–F226.
58 Eiam-Ong S, Kurtzman NA, Sabatini S: Regulation of collecting tubule adenosine triphosphatases by aldosterone and potassium. J Clin Invest 1993;91:2385–2392.
59 Delarue C, Conlon JM, Remy-Jouet I, et al: Endothelins as local activators of adrenocortical cells. J Mol Endocrinol 2004;32:1–7.
60 Baum M, Moe OW, Gentry DL, Aplern RJ: Effect of glucocorticoids on renal cortical NHE-3 and NHE-1 mRNA. Am J Physiol 1994;267:F437–F442.

61 Ambuhl PM, Yang X, Peng Y, et al: Glucocorticoids enhance acid activation of the Na^+/H^+ exchanger 3 (NHE3). J Clin Invest 1999;103:429–435.
62 Knepper, MA, Packer R, Good DW: Ammonium transport in the kidney. Physiol Rev 1989;69:179–248.
63 Wang T: Nitric oxide regulates HCO_3^- and Na^+ transport by a cGMP-mediated mechanism in the kidney proximal tubule. Am J Physiol 1997;272:F242–F248.
64 Tsuruoka S, Schwartz GJ, Wakaumi M, et al. Nitric oxide production modulates cyclosporine A-induced distal renal tubular acidosis in the rat. J Pharmacol Exp Ther 2003;305:840–845.
65 Wesson DE: Augmented bicarbonate reabsorption by both the proximal and distal nephron contributes to the maintenance of chloride-deplete metabolic alkalosis in rats. J Clin Invest 1989;84:1460–1469.
66 Wesson DE, Dolson GM: Augmented bidirectional HCO_3 transport by rat distal tubule in chronic alkalosis. Am J Physiol 1991;261:F308–F331.
67 Wesson DE: Combined potassium and chloride repletion corrects augmented H^+ secretion by distal tubules in chronic alkalosis. Am J Physiol 1994;265:F592–F603.
68 Amemiya M, Mori H, Imamura S, et al: Stimulation of NHE3 in OKP cells by an autocrine mechanism. Nephron: Exp Nephrol 2004;96:e23–e30.
69 Levine DZ: Single-nephron studies: implications for acid-base regulation. Kidney Int 1990;38:744–761.
70 Grond J, Beukers JY, Schilthuis MS, et al: Analysis of renal structural and functional features in two rat strains with a different susceptibility to glomerular sclerosis. Lab Invest 1986;54:77–83.
71 Ruiz-Ortega M, Gomez-Garre D, Alcazar R, et al: Involvement of angiotensin II and endothelin in matrix production and renal sclerosis. J Hypertension 1994;(supplement 12):S51–S58.
72 Remuzzi G: Role of endothelin in the development of glomerulosclerosis. Kidney Blood Press Res 1996;19:182–183.
73 Phisitkul S, Khanna A, Simoni J, Broglio K, Sheather S, Rajab H, Wesson DE: Amelioration of metabolic acidosis in subjects with low GFR reduces kidney endothelin production, reduces kidney injury, and better preserves GFR. Kidney Int 2010;77:617–623.
74 Mahajan A, Simoni J, Sheather S, Broglio K, Rajab MH, Wesson DE: Daily oral sodium bicarbonate preserves glomerular filtration rate by slowing its decline in early hypertensive nephropathy. Kidney Int 2010;78:303–309.

Donald E. Wesson, MD
Texas A&M Health Sciences Center College of Medicine
Scott and White Healthcare
2401 South 31st Street, Office 407H
Temple, TX 76508 (USA)
Tel. +1 254 724 2368, E-Mail DWESSON@swmail.sw.org

Barton M, Kohan DE (eds): Endothelin in Renal Physiology and Disease.
Contrib Nephrol. Basel, Karger, 2011, vol 172, pp 94–106

Endothelin and Collecting Duct Sodium and Water Transport

Donald E. Kohan

Division of Nephrology, University of Utah Health Sciences Center, Salt Lake City, Utah, USA

Abstract

The renal collecting duct (CD) produces and binds more endothelin (ET)-1 than any other region of the kidney. ET-1 has the potential to act as an autocrine regulator of CD function since ET-1 is secreted abluminally and ET receptors are located primarily on the basolateral side of the CD cell. A large number of in vitro studies have supported this notion of an autocrine function for ET-1, demonstrating that the peptide, largely through activation of the ET_B receptor, inhibits both sodium (Na) and water reabsorption in the CD. The physiologic relevance of these findings has been confirmed in vivo wherein mice with CD-specific knockout of ET-1 are hypertensive on a normal Na diet and develop worsened hypertension associated with Na retention when placed on a high-Na diet. Similarly, CD ET-1-deficient mice demonstrate enhanced responsiveness to vasopressin and an impaired ability to excrete an acute water load. CD-specific knockout of both ET_A and ET_B receptors together causes a similar hypertensive and Na-retaining phenotype. The mechanisms by which ET-1 exerts its effects on CD Na and water reabsorption are being increasingly understood. ET-1 inhibits epithelial Na channel (ENaC) activity through src- and mitogen-activated protein kinase-dependent pathways; channel number is also reduced by a β1Pix-dependent mechanism. In addition, nitric oxide is an important modulator of ET-1 actions on the ENaC, although the mechanism by which this occurs remains to be determined. ET-1 reduces CD water reabsorption by inhibition of vasopressin-stimulated adenylyl cyclase activity through G_i and protein kinase C-dependent pathways, leading to a reduction in cellular cAMP levels. Taken together, these findings indicate that the CD ET system is an important physiologic regulator of systemic blood pressure and volume homeostasis.

Endothelin (ET)-1, though initially described as an endothelial cell-derived vasoconstrictor, has now been shown to be a collecting duct (CD)-derived autocrine factor that plays an important role in controlling blood pressure (BP) and renal sodium (Na) and water excretion. This review will describe the role of

ET-1 in the physiologic regulation of renal Na water transport and discuss the mechanisms by which such regulation occurs. As will become apparent, the CD ET system provides a vital link between extracellular fluid volume (ECFV) and renal fluid and electrolyte excretion.

Endothelin Production by the Collecting Duct

The CD is the major cellular source of ET-1 in the kidney, with progressively greater ET-1 synthesis moving from the cortex to the inner medulla [1–3]. The CD also expresses relatively high ET-3, although much less than ET-1 [2]. Since the inner medullary CD (IMCD) is comprised almost exclusively of principal cells, it is likely that this cell type is the main source of ET-1 in the CD. The vast majority of ET-1 is secreted abluminally [4, 5].

Indirect evidence suggests that CD ET synthesis parallels Na intake. In general, Na loading increases urinary ET-1 excretion [6–8]. Since urinary ET-1 excretion is markedly reduced during normal or high Na intake in the absence of CD ET-1 [6], these findings suggest that CD ET-1 production is enhanced by expanding ECFV (although ECFV has not been strictly measured). How salt loading increases CD ET-1 synthesis is uncertain, but it appears unrelated to circulating agents such as vasopressin (AVP), bradykinin, or norepinephrine [5, 9, 10]. Aldosterone paradoxically (i.e. during ECFV contraction) increases IMCD ET-1 production [11, 12], in part through derepression of gene expression by inhibition of Dot1a-Af9-induced histone hypomethylation (fig. 1) [13].

Another possibility is that local renal factors modulate CD ET-1 production in response to Na loading; such factors include renal interstitial pressure, extracellular tonicity, tubule fluid flow rate, and activity of paracrine substances derived from neighboring cells. There is no information on the effect of interstitial pressure or paracrine factors on CD ET-1 synthesis. Increased renal interstitial NaCl concentration occurs during high Na intake [14] and can increase urinary ET-1 excretion [15] and ET-1 production by cultured cells [9, 16]. However, hypertonic NaCl has also been reported to inhibit ET-1 secretion by MDCK or rat IMCD cells [17, 18]. Intuitively, it seems unlikely that interstitial hypertonicity would be the major stimulator of CD ET-1 synthesis during Na loading and potentially ECFV expansion since extracellular tonicity is primarily increased during ECFV contraction.

Another possibility for local control of CD ET-1 production is tubule fluid flow. Conditions that increase tubule flow, including salt or water loading, increase urinary ET-1 excretion [19–21]. Recent studies found that shear stress augments ET-1 mRNA in mpk-CCD cells, a cortical CD (CCD) cell line [22]. Similarly, shear stress stimulates ET-release by endothelial cells [23]. Thus, Na loading expansion may increase CD ET-1 production, at least in part, through increased tubule fluid flow.

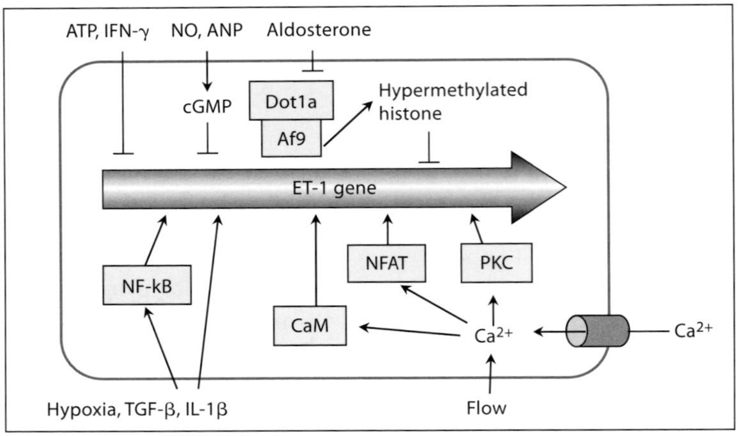

Fig. 1. Regulation of ET-1 production in the CD. Stimulators of ET-1 biosynthesis include aldosterone, hypoxia, transforming growth factor-β (TGF-β), interleukin-1β (IL-1β), and flow. These enhance ET-1 production by activation PKC, nuclear factor of activated T cells (NFAT), calmodulin (CaM) and nuclear factor (NF)-κB. In addition, aldosterone represses disrupter of telomeric silencing alternative splice variant (Dot1a), reducing interaction with AF9 and derepressing ET-1 gene transcription. Inhibitors of ET-1 production include ATP, interferon-γ (IFN-γ), nitric oxide (NO), and atrial natriuretic peptide (ANP), the latter two acting through cGMP.

CD ET-1 synthesis is under control by a variety of signaling pathways (fig. 1). Shear stress-stimulated CD ET-1 mRNA accumulation is dependent upon extracellular and intracellular Ca^{2+}, protein kinase C (PKC) and phospholipase C (PLC) [22]. In cultured rat IMCD cells, inhibition of calmodulin (CaM) or CaM kinase markedly reduces IMCD ET-1 secretion and mRNA; this response is largely mediated at the transcriptional level [24]. ET-1 promoter-reporter constructs yielded maximal activity within 1.2- to 1.7-kb 5′ to the transcription start site in IMCD cells, unlike maximal ET-1 promoter activity in the proximal 0.36-kb region in aortic endothelial cells, suggesting novel regulation of ET-1 synthesis in IMCD cells. Further studies revealed that two regions in the 500-bp region identified above were nuclear factor of activated T cells (NFAT) consensus-binding sites and that mutation of these regions largely abolished ET-1 promoter activity [25]. These two sites selectively bound NFATc1 or NFATc3, suggesting a regulatory pathway for ET-1 biosynthesis in IMCD involving Ca^{2+}, CaM, CaM kinase, and NFAT.

Endothelin Receptor Expression by the Collecting Duct

The medulla and the CD are the major sites of renal ET binding in the kidney (215, 428, 492, 515). Numerous studies indicate that the CD, and particularly the IMCD, expresses predominantly ET_B receptors (ET_B-R) [26–30]. The CD also expresses ET_A-R, although to a lesser degree than ET_B-R [29, 31–33]. The large majority of CD ET-R appear to be located on the basolateral side [4, 34]. Given the predominant basolateral ET-1 secretion [4, 10, 17], these findings suggest that ET-1 has the potential to act in an autocrine manner in the CD.

CD ET-R expression may be regulated. AVP can reduce CD ET-R number, while AVP, forskolin, or dibutyryl cAMP decreases ET-1 binding to rat CCD and IMCD [35, 36]. Protein kinase A inhibition prevents the cAMP response, most likely via the ET_B-R, since only this ET-R isoform contains a protein kinase A-induced phosphorylation site. CD ET-R may be subject to heterologous desensitization in that angiotensin II decreases ET_A and ET_B-R binding and mRNA in rat IMCD [37]. As will be described, since ET-1 is a natriuretic factor in the CD, AVP and angiotensin II downregulation of CD ET-R function may promote the Na-retaining effects of these peptides.

Overview of Endothelin Regulation of Collecting Duct Sodium Transport

ET-1 directly reduces Na transport in the CD. Part of this effect may be due to decreased Na/K ATPase activity [38]; however, the bulk of evidence supports an inhibitory effect on the epithelial Na channel (ENaC). Studies in cultured CD [39], distal nephron A6 cells [40], and 3T3 cells stably expressing ENaC [41] indicate that ET-1 can reduce both channel open probability (Po) and cell surface number. ET-1 decreases ENaC Po in isolated split-open rat CCD [34]. Mice with ET-1 knocked out specifically in principal cells of the CD (CD ET-1 KO) have elevated systolic BP by approximately 15 mm Hg during normal salt intake [6]. On a high-Na diet, these mice develop marked hypertension (by approx. 35 mm Hg systolic BP) associated with weight gain and decreased urinary Na excretion relative to genetic controls. Progressive elevations in renal perfusion pressure are accompanied by a diminished natriuretic response in CD ET-1 KO mice as compared to control animals [42]. ENaC is likely involved in the hypertension in CD ET-1 KO mice since amiloride abrogates the hypertension and Na retention [6]. Hence, CD-derived ET-1 functions as an autocrine inhibitor of Na reabsorption and this effect is substantially mediated by suppression of ENaC activity.

Endothelin Receptors and Collecting Duct Sodium Transport

ET-1 inhibition of CD Na transport is primarily mediated by ET_B-R (fig. 2). ET_B-R activation reduces ENaC Po in A6 cells [40] and blocks ET-1 inhibition of ENaC Po in rat CCD [34]. Intramedullary infusion of an ET_B-R antagonist

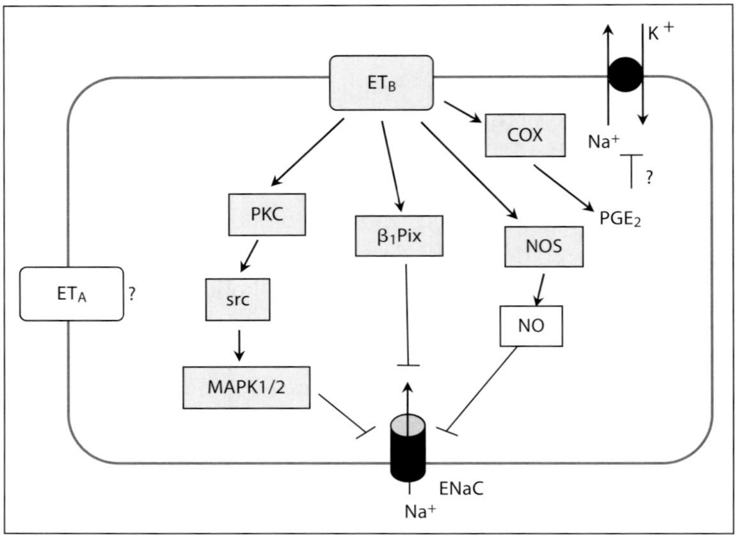

Fig. 2. ET-1 regulation of CD Na transport. ET-1, primarily through activation of ET_B receptors, may inhibit Na/K ATPase activity via cyclooxygenase (COX)-mediated production of prostaglandin E_2 (PGE_2). The major means of inhibition of Na transport is through decreased ENaC activity which is mediated by nitric oxide synthase (NOS)-stimulated production of NO, PAK-interacting exchange factor (β1Pix), and serial activation of protein kinase C, src tyrosine kinase (src), and mitogen-activated protein kinase (MAPK). The role of ET_A receptors is unclear.

reduces volume expansion-induced natriuresis [43], while intramedullary infusion of an ET_B-R agonist causes natriuresis [44]. Rats with a naturally occurring null mutation in ET_B-R that are rescued from lethality by a transgene conferring gut-specific ET_B-R expression (to prevent intestinal aganglionosis) develop aggravated hypertension in response to a high-salt intake and this is abrogated by amiloride [45]. Of greatest relevance, mice with CD-specific knockout of the ET_B-R gene (CD ET_B KO) have a modest increase in systolic BP during normal salt intake, and develop worsened hypertension and Na retention (by approx. 20 mm Hg) during high-salt intake [46]. Notably, the degree of hypertension in CD ET_B KO mice is about 50% of that seen in CD ET-1 KO animals, suggesting that CD ET_B does not fully mediate the antihypertensive and natriuretic effects of CD-derived ET-1.

The role of ET_A-R in regulating CD Na transport remains uncertain. ET-1, in the presence of ET_B-R antagonism, stimulates Na channel Po in A6 cells [40], suggesting an antinatriuretic effect of ET_A-R activation. However, intramedullary ET-1 infusion increased Na excretion in ET_B-R-deficient rats and this was prevented by ET_A-R blockade [47] (this response was observed only in female animals, possibly due to ET-1-induced decreased medullary blood flow only in males). Mice with CD-specific knockout of ET_A-R (CD ET_A KO) have no change in arterial BP or urinary Na excretion, regardless of the amount of Na intake [48]. Thus, these studies suggest that ET_A-R in the CD does not mediate ET-1 inhibition of Na reabsorption.

Unfortunately, viewing CD ET_A-R and ET_B-R as independently functioning receptors may be too simplistic. In a follow-up study, the ET_A-R and ET_B-R genes were simultaneously disrupted in the CD (CD $ET_{A/B}$ KO). These animals had hypertension and Na retention similar to CD ET-1 KO mice and greater than CD ET_B KO animals [42]. The reasons for these findings are unknown, but indicate that new paradigms must be considered. Conceivably, both ET_A-R and ET_B-R can exert natriuretic effects, but ET_B-R is more effective (since ET_B-R are much more prevalent in CD). Thus, absence of CD ET_A-R could be obscured by abundant ET_B-R, while absence of ET_B-R may be partially compensated for by ET_A-R. Another consideration is that ET-R may exist as homo- and heterodimers with functional consequences. For example, ET_B-R homodimers and $ET_{A/B}$ heterodimers may normally exist since the ET_B-R isoform predominates. In CD ET_B KO, however, ET_A-R homodimers may form and exert a natriuretic effect, while ET_B-R homodimers may fully compensate for any loss of ET_A-R in CD ET_A KO. Although these are difficult issues to examine in vivo, these considerations highlight the importance of viewing CD ET-R isoform effects on Na transport as potentially interdependent.

Mechanisms of Endothelin Regulation of Collecting Duct Sodium Transport

A number of kinases mediate, at least in part, ET-1 inhibition of CD NaCl transport (fig. 2). Although PKC may be involved in ET-1 inhibition of AVP-stimulated chloride reabsorption in the rat CCD [49], PKC does not appear to mediate ET-1 actions on ENaC in the CCD [34]. src kinase is likely a mediator of ET-1 inhibition of ENaC; ET-1 suppression of ENaC activity was src kinase-dependent in 3T3 cells transfected with ENaC [41]. src kinase was required for ET_B-R-mediated inhibition of ENaC Po in isolated split-open CCD [34]. Mitogen-activated protein kinase (MAPK) is also required for ET-1 signaling; ET-1 increases MAPK activity in rat outer medullary CD and IMCD [50], while ET_B-R inhibition of ENaC was MAPK1/2-dependent in the isolated CCD preparation [34]. Notably, MAPK-mediated ENaC phosphorylation decreases both ENaC Po and number [51, 52]. Finally, long-term exposure of cultured CD cells to ET-1 reduces ENaC channel number through increased β1Pix binding to 14–3–3β, preventing 14–3–3β interaction with the ubiquitin ligase Nedd4–2, thereby augmenting ENaC internalization and degradation [39]. Taken together, the above studies indicate that several kinases and/or signaling molecules are effectors of ET-1 inhibition of ENaC in the CD, including src and MAPK and β1Pix.

Another mediator of ET-1 inhibition of CD Na reabsorption is nitric oxide (NO; fig. 2). ET-1 can stimulate NO production by the CD. ET-1 activation of ET_B-R enhances NO synthase (NOS) 1-dependent NO production and cGMP production in rat IMCD cells [53]. ET-1, possibly via ET_A-R, increases NOS1 protein expression in IMCD3 cells [54], while antagonism of either ET-R isoform reduces NOS3 mRNA and protein expression in rat IMCD cells [55]. In addition, IMCD from ET_B-R-deficient rats have impaired NOS activity [56]. Such ET-1 stimulation of NO production has the potential to inhibit CD Na reabsorption. NO can inhibit Cl transport in the thick ascending limb [57], decrease Na transport in the CCD [58], and reduce ENaC activity in cultured CD [59]. Most notably, studies in CD ET-1 KO mice strongly implicate NO in mediating the natriuretic effect of ET-1 in the CD [42]. CD ET-1 KO mice demonstrated reduced inner medullary NOS1 and NOS3 activities that were associated with decreased urinary NO metabolite excretion during normal or high-Na intake as well as during any given magnitude of renal perfusion pressure. In these same mice, NOS inhibition had a very modest effect on BP, while it markedly increased BP in control animals. In fact, NOS inhibition resulted in no difference in BP between CD ET-1 KO and control mice. Finally, NOS1 blockade abrogated the natriuresis induced by intramedullary infusion of an ET_B-R agonist [44]. Taken together, these studies implicate NO as an important mediator of the inhibitory effect of ET-1 on Na reabsorption by the CD.

There is some data to suggest that prostaglandins play a role in ET-1 inhibition of CD Na transport. ET-1 inhibition of Na/K ATPase activity in rabbit IMCD is prevented by cyclooxygenase (COX) inhibition [38], and ET-1 increases prostaglandin E_2 production by IMCD (through activation of ET_B-R and COX-2) (fig. 2) [60]. However, nonspecific COX or COX-2 antagonism had no effect on BP or urinary Na excretion in CD ET-1 KO mice [61]. Furthermore, COX may even be upregulated in IMCD of CD ET-1 KO since IMCD prostaglandin E_2 levels are increased in these animals [61]. Thus, there is no convincing evidence that prostaglandins are involved in the natriuretic effect of ET-1 in the CD.

Endothelin Regulation of Collecting Duct Water Transport

Similar to Na loading, high water intake increases urinary ET-1 excretion [18, 21, 62]. ET-1, in turn, can exert a potent diuretic effect. Exogenous ET-1 or ET_B-R agonists promote renal water excretion in the presence or absence of renal hemodynamic changes [63, 64]. ET-1 reduces AVP-stimulated water transport in rat CD [49, 65]. CD ET-1 KO mice have impaired diuresis in response to an acute water load; reduced plasma AVP levels during normal water intake; and increased urine osmolality, aquaporin-2 mRNA, and aquaporin-2 phosphorylation following systemically administered AVP [66]. Hence, CD-derived ET-1

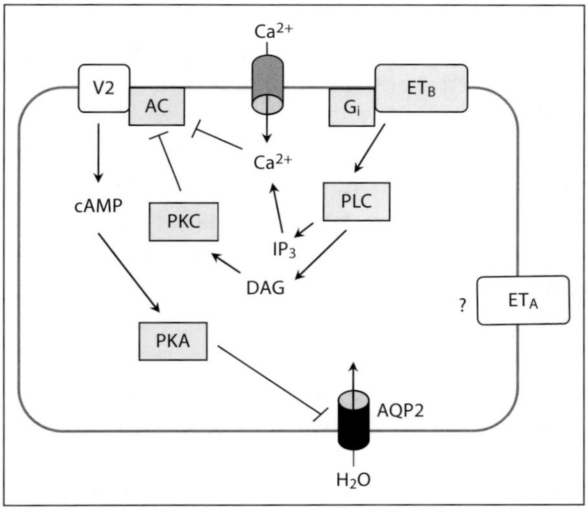

Fig. 3. ET-1 regulation of CD water transport. ET-1, primarily through activation of ET_B receptors, inhibits water transport by decreasing adenylyl cyclase (AC) activity, leading to reduced cAMP, and decreased protein kinase A (PKA)-mediated phosphorylation of the aquaporin-2 (AQP2) water channel. ET_B receptors exerts their effect on AC through a G protein (G_i)-dependent pathway, phospholipase C-dependent production of inositol triphosphate (IP_3) and diacylglycerol (DAG), protein kinase C (PKC, and increases in intracellular Ca^{2+} concentration. The role of ET_A receptors is unclear.

appears to tonically inhibit AVP responsiveness in the CD, thereby functioning as an autocrine diuretic factor.

The inhibitory effect of ET-1 on CD water transport is largely due to modulation of AVP-stimulated adenylyl cyclase activity (fig. 3). ET decreases AVP-stimulated adenylyl cyclase-dependent cAMP accumulation in all regions of the CD in rats and pigs [9, 60, 67]. This effect of ET-1 on cAMP levels can occur in an autocrine manner because exposure to anti-ET-1 antibodies augments AVP-stimulated cAMP accumulation in cultured IMCD cells [4]. ET-1 inhibition of AVP-stimulated water reabsorption is fully dependent upon suppression of adenylyl cyclase activity since ET-1 does not reduce dibutyryl cAMP-stimulated water transport in the IMCD [68]. Finally, acutely isolated IMCD segments from CD ET-1 KO mice have increased responsiveness to AVP or forskolin (assessed by cAMP accumulation) [66] that is paralleled by increased adenylyl cyclase 5 and/or 6 expression [69]. In total, these findings indicate that CD-derived ET-1 inhibits AVP responsiveness in the CD by reducing adenylyl cyclase activity.

ET_B-R appears to be the main ET-R isoform involved in suppression of AVP-responsiveness in the CD (fig. 3). The inhibitory effect of ET-1 on AVP-induced cAMP accumulation in rat IMCD is prevented by ET_B-R antagonism (ET_A-R blockers were ineffective), while ET_B-R agonists reduce AVP-stimulated cAMP content in this same cell type [60, 70]. Mice with ET_A-R disruption specifically in the CD (CD ET_A KO) manifest increased ability to eliminate a water load, elevated plasma AVP levels, and decreased AVP- and forskolin-stimulated cAMP accumulation in the IMCD [48]. Restated, CD ET_A KO mice have reduced AVP responsiveness, suggesting that ET_B-R expression facilitates a diuresis perhaps due to loss of counter-regulatory ET_A-R. This issue of which ET-R isoform modulates AVP actions in the CD and the nature of such modulation is highly complex; ET-R homodimerization or heterodimerization, constitutive activation (in the absence of ligands), and differential responsiveness of ET-R isoforms are possibilities that are largely unexplored.

ET-1 reduces cAMP accumulation in CD through a variety of signaling pathways (fig. 3). The G protein G_i is involved since pertussis toxin prevents ET-1 inhibition of AVP-stimulated cAMP and water transport in the IMCD [60, 65]. PKC blockade largely abolishes ET-1 inhibition of AVP-dependent cAMP accumulation and water reabsorption in the CD; ET-1 also increases inositol phosphates in the CD [60, 71, 72]. Alterations in intracellular Ca^{2+} concentration ($[Ca^{2+}]_i$) also appear to play a role in ET-1 regulation of AVP action. While the data is somewhat variable, ET-1 appears to cause an initial rapid rise, followed by a sustained elevation, in CD $[Ca^{2+}]_i$ [9, 73]. The initial response is likely due to release of Ca^{2+} from cell stores [65, 74]. The sustained ET-1-induced increase in $[Ca^{2+}]_i$ likely requires extracellular Ca^{2+} entry through, at least partially, dihydropyridine-sensitive Ca^{2+} channels [74, 75]. While the role of $[Ca^{2+}]_i$ in mediating ET-1 inhibition of water transport has not been directly assessed, increased $[Ca^{2+}]_i$ per se can reduce AVP-stimulated adenylyl cyclase activity in rat IMCD (through phospholipase C activation of PKC) [76]. Finally, although prostaglandin and NO production are enhanced by ET-1, studies have failed to show a role of these substances in mediating ET-1 inhibition of AVP-induced cAMP accumulation in the CD [53, 60, 67].

Conclusion

CD-derived ET-1 is an important regulator of arterial BP and ECFV. ET-1 exerts these effects largely through autocrine inhibition of Na and water reabsorption in the CD. A variety of signaling mechanisms mediate ET-1 actions once bound to CD ET_B-R, including multiple protein kinases, NO, Ca^{2+}, phospholipase C, and others. CD ET_A-R may also exert an inhibitory effect on Na transport, but this remains speculative. Given that the CD contains relatively large amounts of ET-1 and its cognate receptors, and that knockout of ET-1 or both of its receptors in the CD causes marked hypertension, it is apparent that this system plays an important role in normal physiology and has the potential to be involved in deranged BP and volume balance during pathophysiologic processes affecting the CD. Consideration of the importance of this system in control of volume and pressure homeostasis should also be given when ET-R antagonists are used in the clinical arena.

References

1 Chen M, Todd-Turla K, Wang W-H, Cao X, Smart A, Brosius FC, Killen PD, Keiser JA, Briggs JP, Schnermann J: Endothlein-1 mRNA in glomerular and epithelial cells of kidney. Am J Physiol 1993;265:F542–F550.

2 Kohan DE: Endothelin synthesis by rabbit renal tubule cells. Am J Physiol 1991;261:F221–F226.

3 Ujiie K, Terada Y, Nonoguchi H, Shinohara M, Tomita K, Marumo F: Messenger RNA expression and synthesis of endothelin-1 along rat nephron segments. J Clin Invest 1992;90:1043–1048.

4 Kohan DE, Padilla E: Endothelin-1 is an autocrine factor in rat inner medullary collecting ducts. Am J Physiol;1992;263:F607–F612.

5 Todd-Turla KM, Zhu XL, Shu X, Chen M, Yu T, Smart A, Killen PD, Fejes-Toth G, Briggs JP, Schnermann JB: Synthesis and secretion of endothelin in a cortical collecting duct cell line. Am J Physiol 1996;271:F330–F339.

6 Ahn D, Ge Y, Stricklett PK, Gill P, Taylor D, Hughes AK, Yanagisawa M, Miller L, Nelson RD, Kohan DE: Collecting duct-specific knockout of endothelin-1 causes hypertension and sodium retention. J Clin Invest 2004;114:504–511.

7 Cuzzola F, Mallamaci F, Tripepi G, et al: Urinary adrenomedullin is related to ET-1 and salt intake in patients with mild essential hypertension. Am J Hypertens 2001;14:224–230.

8 Malatino LS, Bellanuova I, Cataliotti A, et al: Renal endothelin-1 is linked to changes in urinary salt and volume in essential hypertension. J Nephrol 2000;13:178–184.

9 Migas I, Bäcker A, Meyer-Lehnert H, Kramer HJ: Endothelin synthesis by porcine inner medullary collecting duct cells. Am J Hyperten 1995;8:748–752.

10 Todd-Turla KM, Chen M, Kretzler M, Rose D, Smart A, Yu T, Zhu X, Fehes-Toth G, Briggs JP, Schnermann J: Synthesis and polarized secretion of endothelin in a cortical collecting duct cell line (M1). Hypertension 1994;24:408.

11 Hsieh TJ, Lin SR, Lee YJ, Shin SJ, Lai YH, Hsu CH, Tsai JH: Increased renal medullary endothelin-1 synthesis in prehypertensive DOCA- and salt-treated rats. Am J Physiol Renal Physiol 2000;279:F112–F121.

12 Gumz ML, Popp MP, Wingo CS, Cain BD: Early transcriptional effects of aldosterone in a mouse inner medullary collecting duct cell line. Am J Physiol Renal Physiol 2003;285:F664–F673.

13 Zhang W, Xia X, Reisenauer MR, Rieg T, Lang F, Kuhl D, Vallon V, Kone BC: Aldosterone-induced Sgk1 relieves Dot1a-Af9-mediated transcriptional repression of epithelial Na+ channel alpha. J Clin Invest 2007;117:773–783.

14 Herrera M, Garvin J: A high-salt diet stimulates thick ascending limb eNOS expression by raising medullary osmolality and increasing release of endothelin-1. Am J Physiol 2005;288:F58–F64.

15 Boesen EI, Pollock DM: Acute increases of renal medullary osmolality stimulate endothelin release from the kidney. Am J Physiol Renal Physiol 2007;292:F185–F191.

16 Yang T, Terada Y, Nonoguchi H, Ujiie K, Tomita K, Marumo F: Effect of hyperosmolality on production and mRNA expression of endothelin-1 in inner medullary collecting duct. Am J Physiol 1993;264:F684–F689.

17 Schramek H, Gstraunthaler G, Willinger CC, Pfaller W: Hyperosmolality regulates endothelin release by Madin-Darby canine kidney cells. J Am Soc Nephrol 1993;4:206–213.

18 Kohan DE, Padilla E: Osmolar regulation of endothelin-1 production by rat inner medullary collecting duct. J Clin Invest 1993;91:1235–1240.

19 Modesti P, Cecioni I, Migliorini A, Naldoni A, Costoli A, Vanni S, Serneri G: Increased renal endothelin formation is associated with sodium retention and increased free water clearance. Am J Physiol 1998;275:H1070–H1077.
20 Worgall S, Manz F, Kleschin K, Feth F, Rascher W: Elevated urinary excretion of endothelin-like immunoreactivity in children with renal disease is related to urine flow rate. Clin Nephrol 1994;41:331–337.
21 Zeiler M, Löffler B-M, Bock HA, Thiel G, Basel K: Water diuresis increases endothelin-1 excretion in humans. J Am Soc Nephrol 1995;6:751.
22 Lyon-Roberts B, Strait K, van Perusem E, Kittikulsuth W, Pollock JS, Pollock DM, Kohan DE: Flow regulation of collecting duct endothelin-1 production. Am J Physiol Renal Physiol 2011;300:F650–F656.
23 da Silva RF, Chambaz C, Stergiopulos N, Hayoz D, Silacci P: Transcriptional and post-transcriptional regulation of preproendothelin-1 by plaque-prone hemodynamics. Atherosclerosis 2007;194:383–390.
24 Strait KA, Stricklett PK, Kohan JL, Miller MB, Kohan DE: Calcium regulation of endothelin-1 synthesis in rat inner medullary collecting duct. Am J Physiol Renal Physiol 2007;293:F601–F606.
25 Strait KA, Stricklett PK, Kohan RM, Kohan DE: Identification of two nuclear factor of activated T-cells (NFAT)-response elements in the 5′-upstream regulatory region of the ET-1 promoter. J Biol Chem 2010;285:28520–28528.
26 Takemoto F, Uchida S, Ogata E, Kurokawa K: Endothelin-1 and endothelin-3 binding to rat nephrons. Am J Physiol 1993;264:F827–F832.
27 Karet FE, Kuc RE, Davenport AP: Novel ligands BQ123 and BQ3020 characterize endothelin receptor subtypes ET_A and ET_B in human kidney. Kidney Int 1993;44:36–42.
28 Terada Y, Tomita K, Nonoguchi H, Yang T, Marumo F: Different localization and regulation of two types of vasopressin receptor messenger RNA in microdissected rat nephron segments using reverse transcription polymerase chain reaction. J Clin Invest 1993;92:2339–2345.
29 Wendel M, Knels L, Kummer W, Koch T: Distribution of endothelin receptor subtypes ET_A and ET_B in the rat kidney. J Histochem Cytochem 2006;54:1193–1203.
30 Kohan DE, Hughes AK, Perkins SP: Characterization of endothelin receptors in the inner medullary collecting duct of the rat. J Biol Chem 1992;267:12336–12340.
31 Cernacek P, Legault L, Stewart DJ, Levy M: Specific endothelin binding sites in renal medullary collecting duct cells: lack of interaction with ANP binding and cGMP signalling. Can J Physiol Pharmacol 1992;70:1167–1174.
32 Omatsu S, Tomoyoshi T: Immunohistochemical study on endothelin in rat kidney. Hinyokika Kiyo 1997;43:109–114.
33 Yamamoto T, Suzuki H, Kubo Y, Matsumoto A, Uemura H: Endothelin A receptor-like immunoreactivity on the basal infoldings of rat renal tubules and collecting ducts. Arch Histol Cytol 2008;71:77–87.
34 Bugaj V, Pochynyuk O, Mironova E, Vandewalle A, Medina JL, Stockand JD: Regulation of the epithelial Na+ channel by endothelin-1 in rat collecting duct. Am J Physiol Renal Physiol 2008;295:F1063–F1070.
35 Takemoto F, Uchida S, Katagirl H, Oka Y, Nakao A, Kurokawa K: Desensitization of endothelin-1 binding by vasopressin via a cAMP-mediated pathway in rat CCD. Am J Physiol 1995;268:F385–F390.
36 Wong NL, Wong BP, Tsui JK: Vasopressin regulates endothelin-B receptor in rat inner medullary collecting duct. Am J Physiol Renal Physiol 2000;278:F369–F374.
37 Wong NL, Tsui JK: Angiotensin regulates endothelin-B receptor in rat inner medullary collecting duct. Metabolism 2001;50:661–666.
38 Zeidel ML, Brady HR, Kone BC, Gullans SR, Brenner BM: Endothelin, a peptide inhibitor of Na^+-K^+-ATPase in intact tubular epithelial cells. Am J Physiol 1989;257:C1101–C1107.
39 Pavlov TS, Chahdi A, Ilatovskaya DV, Levchenko V, Vandewalle A, Pochynyuk O, Sorokin A, Staruschenko A: Endothelin-1 inhibits the epithelial Na+ channel through betaPix/14-3-3/Nedd4-2. J Am Soc Nephrol 2010;21:833–843.

40 Gallego MS, Ling BN: Regulation of amiloride-sensitive sodium channels by endothelin-1 in distal nephron cells. Am J Physiol 1996;271:F451–F460.
41 Gilmore ES, Stutts MJ, Milgram SL: SRC family kinases mediate epithelial Na+ channel inhibition by endothelin. J Biol Chem 2001;276:42610–42617.
42 Schneider MP, Ge Y, Pollock DM, Pollock JS, Kohan DE: Collecting duct-derived endothelin regulates arterial pressure and Na excretion via nitric oxide. Hypertension 2008;51:1605–1610.
43 Guo X, Yang T: Endothelin B receptor antagonism in the rat renal medulla reduces urine flow rate and sodium excretion. Exp Biol Med (Maywood) 2006;231:1001–1005.
44 Nakano D, Pollock JS, Pollock DM: Renal medullary ETB receptors produce diuresis and natriuresis via NOS1. Am J Physiol Renal Physiol 2008;294:F1205–F1211.
45 Gariepy CE, Ohuchi T, Williams SC, Richardson JA, Yanagisawa M: Salt-sensitive hypertension in endothelin-B receptor-deficient rats. J Clin Invest 2000;105:925–933.
46 Ge Y, Bagnall A, Stricklett PK, Strait K, Webb DJ, Kotelevtsev Y, Kohan DE: Collecting duct-specific knockout of the endothelin B receptor causes hypertension and sodium retention. Am J Physiol Renal Physiol 2006;291:F1274–F1280.
47 Nakano D, Pollock DM: Contribution of endothelin A receptors in endothelin 1-dependent natriuresis in female rats. Hypertension 2009;53:324–330.
48 Ge Y, Stricklett PK, Hughes AK, Yanagisawa M, Kohan DE: Collecting duct-specific knockout of the endothelin A receptor alters renal vasopressin responsiveness, but not sodium excretion or blood pressure. Am J Physiol Renal Physiol 2005;289:F692–F698.
49 Tomita K, Nonguchi H, Terada Y, Marumo F: Effects of ET-1 on water and chloride transport in cortical collecting ducts of the rat. Am J Physiol 1993;264:F690–F696.
50 Terada Y, Yamada T, Takayama M, Nonoguchi H, Sasaki S, Tomita K, Marumo F: Presence and regulation of Raf-1-K (kinase), MAPK-K, MAP-K, and S6-K in rat nephron segments. J Am Soc Nephrol 1995;6:1565–1577.
51 Falin RA, Cotton CU: Acute downregulation of ENaC by EGF involves the PY motif and putative ERK phosphorylation site. J Gen Physiol 2007;130:313–328.
52 Shi H, Asher C, Chigaev A, Yung Y, Reuveny E, Seger R, Garty H: Interactions of beta and gamma ENaC with Nedd4 can be facilitated by an ERK-mediated phosphorylation. J Biol Chem 2002;277:13539–13547.
53 Stricklett PK, Hughes AK, Kohan DE: Endothelin-1 stimulates NO production and inhibits cAMP accumulation in rat inner medullary collecting duct through independent pathways. Am J Physiol Renal Physiol 2006;290:F1315–F1319.
54 Sullivan JC, Goodchild TT, Cai Z, Pollock DM, Pollock JS: Endothelin(A) (ET(A)) and ET(B) receptor-mediated regulation of nitric oxide synthase 1 (NOS1) and NOS3 isoforms in the renal inner medulla. Acta Physiol (Oxf) 2007;191:329–336.
55 Ye Q, Chen S, Gardner DG: Endothelin inhibits NPR-A and stimulates eNOS gene expression in rat IMCD cells. Hypertension 2003;41:675–681.
56 Taylor TA, Gariepy CE, Pollock DM, Pollock JS: Gender differences in ET and NOS systems in ET_B receptor-deficient rats: effect of a high salt diet. Hypertension 2003;41:657–662.
57 Plato C, Pollock D, Garvin J: Endothelin inhibits thick ascending limb chloride flux via ET(B) receptor-mediated NO release. Am J Physiol Renal Physiol 2000;279:F326–F333.
58 Ortiz PA, Garvin JL: Role of nitric oxide in the regulation of nephron transport. Am J Physiol Renal Physiol 2002;282:F777–F784.
59 Helms MN, Yu L, Malik B, Kleinhenz DJ, Hart CM, Eaton DC: Role of SGK1 in nitric oxide inhibition of ENaC in Na+-transporting epithelia. Am J Physiol Cell Physiol 2005;289:C717–C726.
60 Kohan DE, Padilla E, Hughes AK: Endothelin B receptor mediates ET-1 effects on cAMP and PGE2 accumulation in rat IMCD. Am J Physiol 1993;265:F670–F676.
61 Ge Y, Strait KA, Stricklett PK, Yang T, Kohan DE: Role of prostaglandins in collecting duct-derived endothelin-1 regulation of blood pressure and water excretion. Am J Physiol Renal Physiol 2007;293:F1805–F1810.

62 Mattyus I, Zimmerhack LB, Schwarz A, Hentschel M, Brandis M, Milteny M, Tulassay T: Renal excretion of endothelin in children is influenced by age and diuresis. Acta Pædiatr 1994;83:468–472.
63 Kamphuis C, Yates NA, McDaugall JG: Differential blockade of the renal vasoconstrictor and diuretic responses to endothelin-1 by endothelin antagonist. Clin Exp Pharmacol Physiol 1994;21:329–333.
64 Matsuo G, Matsummura Y, Tadano K, Hashimoto T, Morimoto S: Effects of sarafotoxin S6c on renal haemodynamics and urine formation in anaesthetized dogs. Clin Exp Pharmacol Physiol 1997;24:487–491.
65 Nadler SP, Zimplemann JA, Hebert RL: Endothelin inhibits vasopressin-stimulated water permeability in rat terminal inner medullary collecting duct. J Clin Invest 1992;90:1458–1466.
66 Ge Y, Ahn D, Stricklett PK, Hughes AK, Yanagisawa M, Verbalis JG, Kohan DE: Collecting duct-specific knockout of endothelin-1 alters vasopressin regulation of urine osmolality. Am J Physiol Renal Physiol 2005;288:F912–F920.
67 Tomita K, Nonoguchi H, Marumo F: Effects of endothelin on peptide-dependent cyclic adenosine monophosphate accumulation along the nephron segments of the rat. J Clin Invest 1990;85:2014–2018.
68 Oishi R, Nonoguchi H, Tomita K, Marumo F: Endothelin-1 inhibits AVP-stimulated osmotic water permeability in rat medullary collecting duct. Am J Physiol 1991;261:F951–F956.
69 Strait KA, Stricklett PK, Kohan DE: Altered collecting duct adenylyl cyclase content in collecting duct endothelin-1 knockout mice. BMC Nephrol 2007;8:8.
70 Woodcock EA, Land SL: Functional endothelin ET_B receptors on renal papillary tubules. Eur J Pharmacol 1993;247:93–95.
71 Tomita K, Nonoguchi H, Marumo F: Inhibition of Fluid Transport by Endothelin through Protein Kinase C in Collecting Duct of Rats. Basel, Karger, 1991.
72 Woodcock E, Land S: Endothelin receptors in rat renal papilla with a high affinity for endothelin-3. Eur J Pharmacol 1991;208:255–260.
73 Naruse M, Uchida S, Ogata E, Kurokawa K: Endothelin-1 increases cell calcium in mouse collecting tubule cells. Am J Physiol 1991;261:F720–F725.
74 Korbmacher C, Boulpaep EL, Giebisch G, Geibel J: Endothelin increases $[Ca^{2+}]_i$ in M-1 mouse cortical collecting duct cells by a dual mechanism. Am J Physiol 1993;265:C349–C357.
75 Kurokawa K, Yoshitomi K, Ikeda M, Uchida S, Naruse M, Imai M: Regulation of cortical collecting duct function: effect of endothelin. Am Heart J 1993;125:582–588.
76 Teitelbaum I, Berl T: Increased cytosolic Ca^{2+} inhibits AVP-stimulated adenylyl cyclase activity in rat IMCT cells by activation of PKC. Am J Physiol 1994;266:F486–F490.

Donald E. Kohan, MD, PhD
Division of Nephrology, University of Utah Health Sciences Center
1900 East 30 North
Salt Lake City, UT 84132 (USA)
Tel. +1 801 581 2726, E-Mail donald.kohan@hsc.utah.edu

Barton M, Kohan DE (eds): Endothelin in Renal Physiology and Disease.
Contrib Nephrol. Basel, Karger, 2011, vol 172, pp 107–119

Endothelin in the Control of Renal Sympathetic Nerve Activity

Ulla C. Kopp

Departments of Internal Medicine and Pharmacology, University of Iowa Carver College of Medicine and Department of Veterans Affairs Medical Center, Iowa City, Iowa, USA

Abstract

The kidney is densely innervated by sympathetic nerves. Increases in renal sympathetic nerve activity (RSNA) decrease urinary sodium excretion. The kidney also has abundant afferent sensory innervation, located primarily in the renal pelvic wall. Sympathetic nerve fibers and afferent nerve fibers often run separately but intertwined in the same nerve bundles in the renal pelvic wall, providing anatomic support for a functional interaction between RSNA and afferent renal nerve activity (ARNA). Activation of RSNA increases ARNA, which in turn decreases RSNA by activation of the renorenal reflexes. Thus, RSNA-induced increases in ARNA exert a powerful negative feedback control of RSNA via activation of the renorenal reflexes in the overall goal of maintaining low RSNA to facilitate urinary sodium excretion. A high-sodium diet enhances and a low-sodium diet reduces the RSNA-induced increases in ARNA. The physiologic importance of the dietary-induced changes in the RSNA-mediated increases in ARNA is underlined by salt-sensitive hypertension in rats lacking afferent renal innervation. Endothelin (ET), ET_A receptors (R), and ET_B-R are present in the renal pelvic wall. ET plays a modulatory role in the activation of the afferent renal nerves that is dependent on dietary sodium intake. In a high-sodium diet, increased activation of ET_B-R facilitates the interaction between RSNA and ARNA resulting in suppression of RSNA, via activation of the renorenal reflexes, to limit sodium retention. In a low-sodium diet, increased activation of renal pelvic ET_A-R suppresses the interaction between RSNA and ARNA which increases RSNA via impairment of the renorenal reflex mechanism, eventually leading to sodium retention. These findings suggest that the increased renal sympathetic nerve activity and salt-sensitive hypertension in ET-1/ET_B-R-deficient subjects is, at least in part, related to suppressed interaction between RSNA and ARNA.

Renal Innervation: Sympathetic Nerves

The kidney has a rich supply of sympathetic nerves that innervate all segments of the renal vasculature, with the highest innervation along the afferent glomerular arterioles [1]. Also, the sympathetic nerves innervate the renin-containing granular cells in the afferent arterioles. All parts of the nephron are innervated by sympathetic nerves, with the greatest number of neuroeffector junctions in the proximal tubule. Sympathetic nerves have also been localized in the renal pelvic wall [2]. Increases in renal sympathetic nerve activity (RSNA) result in decreases in renal blood flow, decreases in urinary sodium excretion, and increases in renin secretion rate. The activation threshold of RSNA for reducing renal hemodynamics is higher than that for altering renal excretory function and renin secretion. Whereas renin secretion and urinary sodium excretion are modulated by low frequencies of RSNA, high RSNA frequencies are required to reduce renal blood flow.

Modulatory Role of Endothelin on Renal Sympathetic Nerve Activity and Neurotransmission

In addition to its well-known effects on renal blood flow and urinary sodium excretion, there is considerable evidence for ET modulating renal neurotransmitter release. Renal administration of endothelin (ET-1) inhibits the increase in renal norepinephrine (NE) release produced by renal nerve stimulation at low-level frequencies that decrease urinary sodium excretion but have little or no effects on renal blood flow [3–5]. ET-3 and an ET_B receptor (R) agonist produce similar effects as ET-1, suggesting that activation of ET_B-R modulates the release of NE by a presynaptic mechanism [4, 5]. Activation of ET_B-R reduces the antidiuretic and antinatriuretic responses to renal nerve stimulation without altering the small renal vasoconstrictor responses to low-level renal nerve stimulation in dogs [3–5]. However, this has not been a uniform finding. Studies in rabbits failed to show an effect of a nonselective ET-R antagonist on changes in renal hemodynamics or urinary sodium excretion produced by renal nerve stimulation [6]. Whether the increase in renal NE release produced by renal nerve stimulation was altered by ET-R antagonists in rabbits has not been reported. It has also been reported that ET-1 enhances the postsynaptic effects of NE on the vascular responses to nerve simulation in various in vitro preparations, including isolated perfused rat kidney [7].

Taken together, available data strongly suggest that ET-1-mediated activation of ET_B-R inhibits stimulated renal NE release by a prejunctional mechanism that may reduce the antinatriuretic effects of increased RSNA.

In addition to its effect on presynaptic NE release, there is considerable evidence for ET modulating sympathetic nerve activity (SNA) by central and/or peripheral mechanisms [8–10]. Interestingly, the modulatory role of ET-1 on RSNA is dependent on dietary sodium. ET_B-R-deficient rats fed a high-sodium diet are characterized by increased arterial pressure [e.g. 9, 11]. Although, it is important to keep in mind that measurements of multifiber nerve activity need to be interpreted with caution, the studies by Ling et al. [8] demonstrating increased basal RSNA in ET-1-deficient mice fed a high-sodium diet by two different methods together with impairment of the baroreflex control of RSNA in these mice provide evidence for increased RSNA in ET-1-deficient mice fed a high-sodium diet. Inappropriately increased RSNA in high-sodium diets most likely contributes to the development of salt-sensitive hypertension in these mice.

Among the mechanisms contributing to salt-sensitive hypertension in ET_B-R-deficient and ET_B-R-blocked rats is increased ET-1-mediated activation of ET_A-R [9, 11] resulting in impaired renal function [9]. In addition, studies demonstrating a greater depressor response in ET_B-R-deficient rats versus wild-type rats to a ganglionic blocker suggest that salt-sensitive hypertension in the ET_B-R-deficient rats is associated with increased SNA [9].

The increased SNA and RSNA in ET_B-R-deficient rats [9] and ET-1-deficient mice [8] fed a high-sodium diet together with the inhibitory effects of ET-1-mediated activation of ET_B-R on renal presynaptic NE release [3–5] suggest that activation of ET_B-R plays an important role in reducing sympathetic neural control of renal function to facilitate urinary sodium excretion during high-sodium dietary conditions in healthy individuals.

Mechanisms Contributing to Endothelin-1-Mediated Control of Renal Sympathetic Nerve Activity: Role of the Afferent Renal Nerves

The demonstration of ET-1 and ET-1 mRNA in dorsal root ganglia (DRG) [12] suggests the presence of ET-1 in peripheral sensory nerves. Although much attention has been placed on ET as a mediator of pain [e.g. 13], there is also evidence for ET in the maintenance of water and sodium homeostasis, as discussed above [9, 11, 14]. A role for ET-1 in the regulation of blood pressure via the sensory nerves was suggested by studies in rats treated neonatally with capsaicin to destroy all sensory nerves [15]. These rats develop salt-sensitive hypertension, which is reduced by ET_A-R antagonists in a similar fashion as in ET_B-R-deficient and ET_B-R-blocked rats [9, 11].

ET-1 modulates the carotid baroreceptor reflexes. Central administration of ET-1 and ET-3, the latter having higher affinity for ET_B-R than ET_A-R [16], increases baroreflex sensitivity [17] which eventually leads to reduced SNA. Local administration of ET-1 into an isolated baroreceptor preparation suppresses the baroreceptor activity, possibly via activation of ET_A-R [18]. A role for ET-1 in the activation of afferent renal nerves was hypothesized by our findings that rats with selective afferent renal denervation are characterized by a similar salt-sensitive hypertension [19] as ET-1-deficient mice and ET_B-R-deficient rats [8, 9, 11].

Afferent Renal Innervation

The kidney has abundant afferent innervation – the fibers proceed from the kidney to the neuraxis and contain substance P and calcitonin gene-related peptide as neurotransmitters in the peripheral nerve terminals. The sensory nerve endings are located primarily in the renal pelvic wall [2, 10, 20, 21]. Sympathetic nerve fibers and afferent nerve fibers often run separately but intertwined in the same nerve bundles in the renal pelvic smooth muscle layer [2], providing anatomical support for a possible functional interaction between RSNA and afferent renal nerve activity (ARNA). Indeed, stretching the renal pelvic wall by increasing renal pelvic pressure activates the afferent renal nerves and leads to reflex decreases in RSNA and increased natriuresis, i.e. a renorenal reflex response [1]. The magnitude of increases in renal pelvic pressure required to increase ARNA is in the range of that produced by modest volume expansion, i.e. ≥5 mm Hg [22, 23], suggesting an important physiological role for the renal pelvic mechanosensory nerves in the renal control of water and sodium homeostasis.

The responsiveness of the renal mechanosensory nerves is also modulated by changes in RSNA. Reflex increases and decreases in RSNA increase and decrease ARNA, respectively [2, 10, 20]. Taken together, there is strong evidence for a negative feedback system where increases in RSNA increase ARNA, which in turn decrease RSNA via activation of the renorenal reflex in normotensive rats. The importance of this renorenal reflex mechanism in the control of RSNA and sodium excretion is highlighted by the impaired responsiveness of the renal mechanosensory nerves both to increases in renal pelvic pressure and increases in RSNA in spontaneously hypertensive rats, which are characterized by increased RSNA and sodium retention [1, 24, 25].

Among the mechanisms activated by stretching the renal pelvic wall is bradykinin activating protein kinase C leading to the induction of cyclooxygenase-2 (COX-2) and increased renal pelvic synthesis of prostaglandin E_2 (PGE_2) [26]. PGE_2 activates EP4 receptors on renal sensory nerves [27] which leads to activation of the cAMP-protein kinase A transduction pathway and release of substance P from the afferent renal nerves by a Ca^{++}-dependent mechanism. Substance P, in turn, increases ARNA (fig. 1). The presence of bradykinin-binding sites and COX-2 mRNA in the subepithelial layer of the pelvic wall [28, 29] suggests that bradykinin-induced induction of COX-2 leads to increased PGE_2 synthesis in the tissue surrounding the afferent renal nerves [2, 10, 20, 21]. The notion of activation of adenylyl cyclase/cAMP/protein kinase A leading to increases in Ca_i^{++} by activation of N-type Ca^{++} channels and substance P release in peripheral sensory nerves [30] is supported

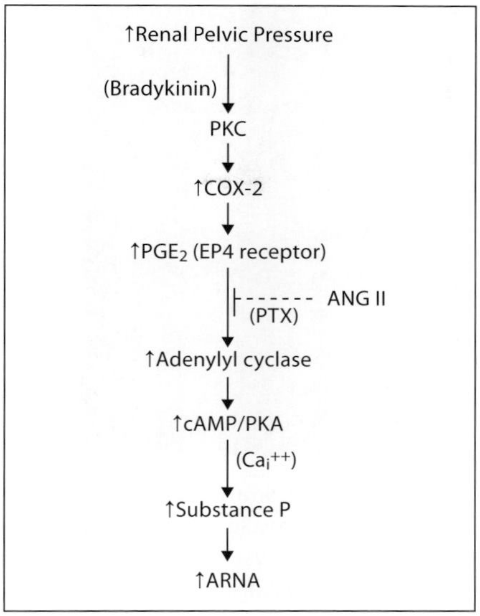

Fig. 1. Stretching the renal pelvic wall by increases in renal pelvic pressure leads to bradykinin-mediated activation of B2 receptors resulting in activation of protein kinase C (PKC), which leads to induction of COX-2 and renal pelvic release of PGE_2. PGE_2 stimulates EP4 receptors on the afferent renal nerves leading to activation of the adenylyl cyclase/cAMP/protein kinase A (PKA) transduction pathway which in turn results in increases in substance P release from the afferent renal nerves involving activation of N-type Ca_i^{++} channels. Substance P increases ARNA. PTX = Petussis toxin.

by the presence of the EP4 receptor on the renal pelvic sensory nerves and studies in cultured DRG neurons [31]. The latter studies showed that activation of adenylyl cyclase results in increases in Ca_i^{++} and release of calcitonin gene-related peptide which are blocked by inhibiting N-type Ca^{++} channels or protein kinase A.

The responsiveness of the renal mechanosensory nerves is enhanced by high and suppressed by low-sodium diets because of an interaction between PGE_2 and angiotensin (ANG) II at the peripheral sensory nerve endings (fig. 1) [32–34]. In conditions of low-sodium dietary intake, high levels of endogenous ANG II in renal pelvic tissue reduces the PGE_2-mediated activation of adenylyl cyclase via a pertussis toxin-sensitive mechanism [32], leading to impairment of the renorenal reflexes. Conversely in conditions of high-sodium dietary intake, characterized by low levels of endogenous ANG II in renal pelvic tissue [32], there is little or no inhibition of the PGE_2-mediated activation of adenylyl cyclase. Importantly, in a high-sodium diet, the activation threshold for the mechanosensory nerves is <3 mm Hg, suggesting that the afferent renal nerves are tonically active in high-sodium dietary conditions [33].

The ARNA responses to increases in RSNA are also enhanced by high-sodium and suppressed by low-sodium diets [10, 20]. These findings highlight the physiological importance of the renorenal reflex mechanisms. Namely, in high-sodium dietary conditions, enhancement of the RSNA-induced increases in ARNA increases the inhibitory renorenal reflex control of RSNA, resulting in reduction of RSNA to facilitate sodium excretion. Conversely, in low-sodium dietary conditions, suppression of the RSNA-induced increases in ARNA results

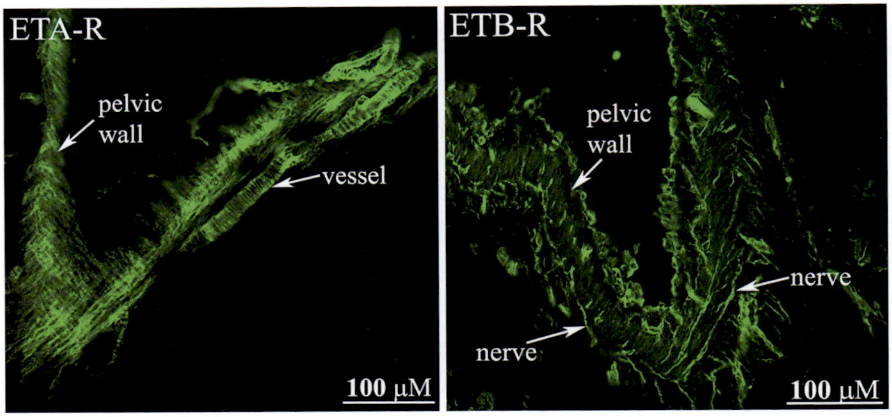

Fig. 2. Left panel, immunofluorescence shows the presence of ET_A-R-like immunoreactivity on smooth muscle cells in the renal pelvic wall and adjacent blood vessels in renal pelvic tissue (arrows). Right panel, ET_B-immunoreactivity is present on nerve fiber-like structures (arrows) among smooth muscle cells (arrows) in the renal pelvic wall (adapted from [10]).

in increased RSNA by reducing the renorenal reflex inhibition of RSNA, eventually leading to sodium retention.

Endothelin-1 Modulates the Altered Responsiveness of Renal Mechanosensory Nerves Produced by Various Dietary Sodium Intakes

Anatomical support for ET-1 and its receptors playing a role in the renorenal reflexes is derived from the presence of ET-1 in the renal pelvic wall and DRG at T_9-L_1 [35], which contain the majority of the cell bodies of the afferent renal nerves [36]. Both ET_A-R and ET_B-R are synthesized in the renal pelvic wall and T_9-L_1 DRG [10]. However, the location of the two ET-R differs. Whereas, ET_B-R are located on or close to unmyelinated Schwann cells surrounding the sensory nerves in the renal pelvic wall, the ET_A-R are located on smooth muscle cells in close conjunction to the renal sensory nerves (fig. 2) [10]. The differential location of ET_A-R and ET_B-R in renal pelvic tissue suggests that activation of the two ET-R modulates the responsiveness of the renal sensory nerves by different mechanisms.

In rats fed a normal sodium diet, endogenous renal ET-1 has little or no effect on the responsiveness of the renal sensory nerves. However, in rats fed a high-sodium diet, the enhanced responsiveness of the renal sensory nerves to increases in renal pelvic pressure is reduced by renal pelvic administration of BQ788, an ET_B-R antagonist. Conversely in a low-sodium diet, the suppressed responsiveness of the renal sensory nerves to increases in renal pelvic pressure is enhanced by BQ123, an ET_A-R antagonist (fig. 3) [35]. In the presence of BQ788

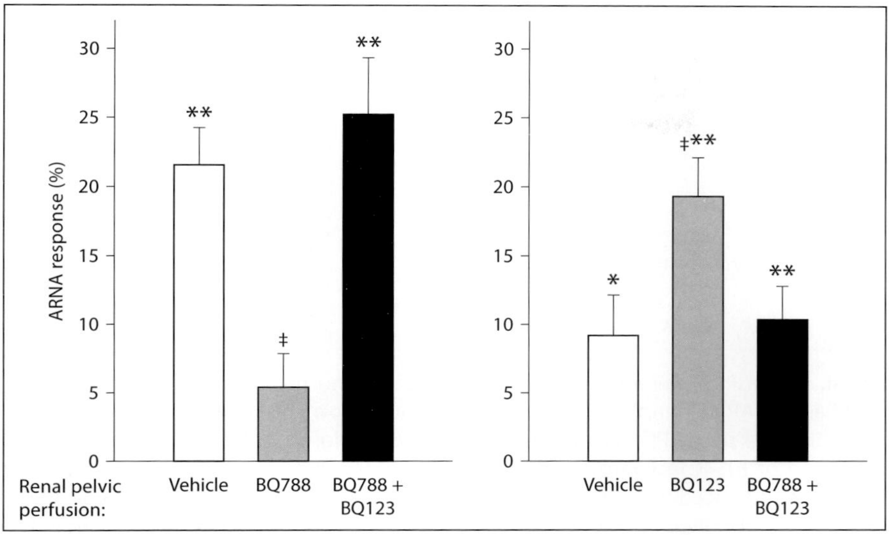

Fig. 3. Effects of ET_B-R (left panel) and ET_A-R (right panel) antagonists and their combination on the ARNA responses to increased renal pelvic pressure. In rats fed a high-sodium diet (left panel), renal pelvic perfusion with the ET_B-R antagonist BQ788 reduced the ARNA responses to increasing renal pelvic pressure 7.5 mm Hg. Adding the ET_A-R antagonist BQ123 to the renal pelvic perfusate containing BQ788 restored the ARNA responses to increased renal pelvic pressure. In rats fed a low-sodium diet (right panel), renal pelvic perfusion with BQ123 enhanced the ARNA responses to increasing renal pelvic pressure. Adding BQ788 to the renal pelvic perfusate containing BQ123 reduced the ARNA responses to increased renal pelvic pressure towards that seen in the presence of the vehicle. * $p < 0.05$; ** $p < 0.01$ vs. baseline; ‡ $p < 0.01$ vs. ARNA responses during pelvic perfusion with vehicle or BQ788 + BQ123 (adapted from [35]).

in high-sodium diet rats, the suppressed responsiveness of renal sensory nerves is related to increased activation of ET_A-R. Conversely in the presence of BQ123 in rats on a low-sodium diet, the increased responsiveness of the renal sensory nerves is due to increased activation of ET_B-R.

The in vivo data were reproduced in vitro using an isolated renal pelvic wall preparation. PGE_2 was used to activate renal sensory nerves, as PGE_2 is one of the mechanisms involved in the activation of the afferent renal nerves following increases in renal pelvic pressure (fig. 1). The findings that incubation of the renal pelvic wall with BQ788 suppressed and BQ123 enhanced the PGE_2-mediated release of substance P from pelvises derived from rats fed high- and low-sodium diets, respectively, suggest that the modulatory effects of ET-1 on ARNA (fig. 3, 4) were due to ET-1 activating mechanisms at the peripheral sensory nerve endings in the renal pelvic wall. Renal ET-1 concentrations are not modulated by dietary sodium [9, 35] and the similar nature of the responses to ET_B-R and ET_A-R antagonists in vivo and in vitro argues against urinary ET-1

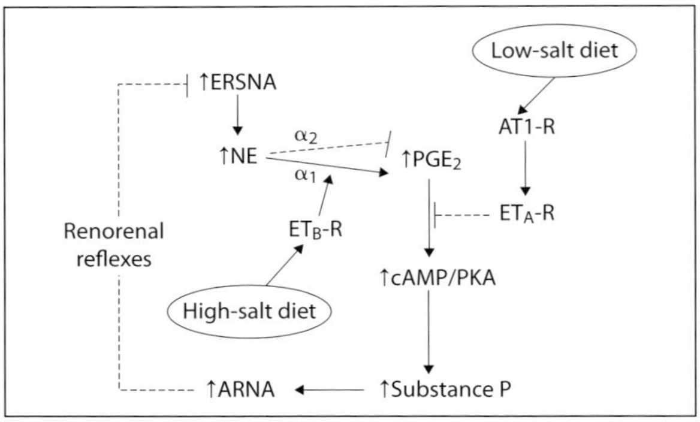

Fig. 4. Increasing efferent RSNA (ERSNA) increases the release of NE which activates α_1-AR and α_2-AR on peripheral sensory nerves leading to increases and decreases, respectively, in renal pelvic PGE_2 synthesis/release. In conditions of high-sodium (salt) intake, activation of ET_B-R facilitates/enhances the α_1-AR-mediated synthesis/release of PGE_2, leading to enhanced release of substance P and increases in ARNA. The increased ARNA exerts a negative feedback control of RSNA via activation of the renorenal reflexes in the overall goal of maintaining a low level of RSNA during high-sodium dietary intake. In conditions of low-sodium intake, increased endogenous levels of ANG II and expression of ET_A-R on smooth muscle fibers close to the sensory nerves in the renal pelvic wall leads to ET-1-induced suppression of NE-induced release of substance P and ARNA by a mechanism(s) downstream of PGE_2 synthesis. Suppression of ARNA leads to disinhibition of the renorenal reflex control of RSNA, which contributes to compensatory sodium retention during low-sodium dietary intake.

concentrations modulating the responsiveness of renal mechanosensory nerves. Instead, the increased expression of ET_A-R in the renal pelvic wall in rats fed a low-sodium diet [10] may contribute to the differential effects of ET-1 on the responsiveness of the renal mechanosensory nerves in high- and low-sodium diets.

Further studies have shown that activation of renal pelvic ET_A-R contributes to the ANG II-mediated suppression of the responsiveness of the renal mechanosensory nerves (fig. 4) [37], a similar interaction between ET_A-R and ANG II type I receptors as reported in the cardiovascular system [e.g. 38, 39].

Endothelin-1 Modulates the Altered Interaction between Renal Sympathetic Nerve Activity and Afferent Renal Nerve Activity Produced by Various Dietary Sodium Intakes

Increases in RSNA increase the release of the neurotransmitter NE resulting in increases and decreases in ARNA by stimulation of α_1-adrenoceptors (AR) and α_2-AR, respectively, on the renal pelvic sensory nerve fibers (fig. 4). The RSNA-induced increases in ARNA is critically dependent on intact renal pelvic synthesis/release of PGE_2. NE-mediated activation of α_1-AR and α_2-AR increases and decreases, respectively, renal pelvic release of PGE_2 in association with parallel changes in substance P release (fig. 4) [20]. As discussed above, the interaction between RSNA and ARNA is modulated by dietary sodium. Increased activation of renal pelvic α_2-AR contributes to the reduced interaction between RSNA and ARNA during low-sodium intake, whereas minimal activation of α_2-AR contributes to the enhanced RSNA-ARNA interaction under conditions of high-sodium intake [20].

Importantly, ET-1 also plays a modulatory role in the differential ARNA responses to increases in RSNA during various sodium intakes. Whereas blocking either ET_A-R or ET_B-R has no effect on the increases in ARNA produced by reflex increases in RSNA in normal sodium diets, activation of renal pelvic ET_B-R and ET_A-R contributes importantly to the enhanced and reduced activation, respectively, of the afferent renal nerves by reflex increases in RSNA in high- and low-sodium diets [10].

In vitro studies using an isolated renal pelvic wall preparation showed that incubating renal pelvises from rats on a high-sodium diet with BQ788 reduced the NE-induced increases in renal pelvic release of PGE_2 and substance P, suggesting an interesting interaction between NE-mediated stimulation of α_1-AR and ET-1-mediated stimulation of ET_B-R in the activation of renal sensory nerves. The data suggest that activation of ET_B-R increases the responsiveness of the renal sensory nerves by increasing renal PGE_2 synthesis (fig. 4). The role of ET_A-R in the suppressed RSNA-ARNA interaction in a low-sodium diet appears to involve mechanisms downstream of PGE_2 synthesis. BQ123 had only marginal effects on the NE-induced release of substance P from renal pelvises derived from low-sodium diet rats. However, in the presence of subthreshold concentrations of PGE_2 for substance P release, BQ123 markedly enhanced the responsiveness of the renal sensory nerves to NE in rats on a low-sodium diet.

Possible Mechanisms Involved in the Activation of Endothelin B and A Receptors Involved in the Activation of the Renal Sensory Nerves

Numerous studies link activation of ET_B-R to increased PGE_2 synthesis. In view of the ET_B-R located on Schwann cells enveloping sensory nerves in the renal pelvic wall [10], studies in glial cells showing increased intracellular Ca^{++}, COX-2, and PGE_2 synthesis in response to ET-1 and ET-3 [40, 41] are of interest. Because activation of protein kinase C in the renal pelvic wall leads to induction of COX-2, increased PGE_2 synthesis, and substance P release (fig. 1) [26], we hypothesize that activation of ET_B-R contributes to the NE-mediated release of PGE_2, possibly involving activation of phospholipase C/protein kinase C, resulting in the release of substance P and activation of the renal sensory nerves (fig. 4).

The presence of ET_A-R on smooth muscle cells in small vessels in and adjacent to the renal pelvic wall may suggest that the ET_A-R-induced modulation of renal sensory nerve activation is related to local vasoconstriction with ischemia leading to an increase in oxygen free radicals [42], which have been shown to impair carotid baroreceptor activity [43]. Further, the findings that BQ123 only enhances the responsiveness to NE in the presence of PGE_2 together with studies showing activation of ET_A-R contributes to the ANG II-induced suppression of the activation of renal mechanosensory nerves [37] may suggest that the ET_A-R-induced suppression of the renal sensory nerves involves inhibition of the PGE_2-mediated activation of adenylyl cyclase by a pertussis toxin-sensitive mechanism (fig. 4) [32].

Role of Endothelin A Receptor in Pathological Conditions of Increased Renal Sympathetic Nerve Activity

Congestive heart failure (CHF) is characterized by increased RSNA and sodium retention [1]. Activation of the carotid and arterial baroreflexes are impaired in heart failure. In addition, the renorenal reflexes are impaired in CHF rats [44]. Whereas suppression of the renorenal reflexes in physiological conditions of sodium retention produced by, for example, a low-sodium diet is an appropriate response, an impairment of the renorenal reflexes would aggravate the sodium retention in pathological conditions of sodium retention. A role for ET-1 in the pathogenesis of CHF was suggested by the findings of elevated cardiac and renal ET-1 levels [45–47] and increased expression of renal ET_A-R in CHF, the latter being correlated with worsening of cardiac function [45]. A role for ET_A-R in the increased RSNA in CHF was suggested by the fall in RSNA by chronic administration of BQ123 in rabbits with CHF, most likely by a mechanism(s) involving improvement of the carotid baroreceptor reflexes [48]. Also, BQ123 enhanced the renorenal reflex responses to activation of renal mechanosensory nerves [44]. Because BQ123 also enhanced the PGE_2-mediated release of substance P from isolated renal pelvises derived from CHF rats, these data suggest that ET-1-mediated activation of ET_A-R in the renal pelvic wall contributes to the suppressed responsiveness of renal mechanosensory nerves in heart failure.

Conclusion

ET-1 contributes importantly to the renal mechanisms involved in the control of RSNA and maintenance of sodium balance by at least two mechanisms. Available data in dogs suggest that ET-1-mediated activation of ET_B-R on presynaptic renal sympathetic nerve terminals leads to decreases in NE release and reduced antinatriuretic responses to increases in RSNA during normal sodium dietary conditions. Whether the ET_B-R-mediated inhibition of presynaptic release of NE would be enhanced in high-sodium dietary conditions is currently unknown. The effects of ET-1 on the interaction between RSNA and ARNA is highly dependent on dietary sodium intake. In a high-sodium diet, increased activation of ET_B-R on renal pelvic sensory nerves facilitates the interaction between RSNA and ARNA resulting in suppression of RSNA, via activation of the renorenal reflexes, to limit sodium retention. In a low-sodium diet, increased activation of renal pelvic ET_A-R suppresses the interaction between RSNA and ARNA which increases RSNA via impairment of the renorenal reflex mechanism, eventually leading to sodium retention. These findings suggest that the increased RSNA/SNA and salt-sensitive hypertension in ET-1/ET_B-R-deficient subjects is, at least in part, related to suppressed interaction between RSNA and ARNA.

Acknowledgement

The work performed in the author's laboratory was supported by the National Institutes of Health, Heart, Lung and Blood Institute, RO1 HL66068 and by research grants from the Department of Veterans Affairs and the American Heart Association, Heartland Affiliate.

References

1 Dibona GF, Kopp UC: Neural control of renal function. Physiol Rev 1997;77:75–197.
2 Kopp UC, Cicha MZ., Smith LA, Mulder J, Hokfelt T: Renal sympathetic nerve activity modulates afferent renal nerve activity by PGE_2-dependent activation of α_1- and α_2-adrenoceptors on renal sensory nerve fibers. Am J Physiol 2007;293:R1561–R1572.
3 Takagi H, Hisa H, Satoh S: Effects of endothelin on adrenergic neurotransmission in the dog kidney. Eur J Pharmacol 1991;203:291–294.
4 Matsuo G, Matsumura Y, Tadano K, Hashimoto T, Morimoto S: Involvement of nitric oxide in endothelin ET_B receptor-mediated inhibitory actions on antidiuresis and norepinephrine overflow induced by stimulation of renal nerves in anesthetized dogs. J Cardiovasc Pharmacol 1997;30:325–331.
5 Matsumura Y, Matsuo G, Egi Y, Tadano K, Morimoto S: Inhibitory effects of endothelin-3 on antidiuresis and norepinephrine overflow induced by stimulation of renal nerves in anesthetized dogs. J Cardiovasc Pharmacol 1996;28:278–284.
6 Boesen EI, Anderson WP, Evans RG, Kett MM: Endogenous endothelins and the response to electrical renal nerve stimulation in anaesthetized rabbits. Autonom Neurosci 2007;132:8–15.
7 Reid JJ: Endothelin-1 may be a physiological modulator of vasoconstriction in rat kidney. J Cardiovasc Pharmacol 1993;22(suppl 8):s67–s70.
8 Ling G-Y, Cao W-H, Onodera M, Ju K-H, Kurihara H, Kurihara Y, Yazaki Y, Kumada M, Fukuda Y, Kuwaki T: Renal sympathetic nerve activity in mice: comparison between mice and rats and between normal and endothelin-1 deficient mice. Brain Res 1998;808:238–249.
9 Ohkita M, Wang Y, Nguyen NDT, Tsai Y-H, Williams SC, Wiseman RC, Killen PD, Li S, Yanagisawa M, Gariepy CE: Extrarenal ET_B plays a significant role in controlling cardiovascular responses to high dietary sodium in rats. Hypertension 2005;45:940–946.
10 Kopp UC, Grisk O, Cicha MZ, Smith LA, Steinbach A, Schlüter T, Mähler N, Hökfelt: Dietary sodium modulates the interaction between efferent renal sympathetic nerve activity and afferent renal nerve activity: role of endothelin. Am J Physiol 2009;297:R337–R351.

11 Pollock DM, Pollock JS: Evidence for endothelin involvement in the response to high salt. Am J Physiol 2001;281:F144–F150.
12 Giaid A, Gibson SJ, Ibrahim NBN, Legon S, Bloom SR, Yanagisawa M, Masaki T, Varndell IM, Polak JM: Endothelin 1, an endothelium-derived peptide, is expressed in neurons of the human spinal cord and dorsal root ganglia. Proc Natl Acad Sci USA 1989;86:7634–7638.
13 Klass M, Hord A, Wilcox M, Denson D, Csete M: A role for endothelin in neuropathic pain after constriction injury of the sciatic nerve. Anesth Analg 2005;101:1757–1762.
14 Ahn D, Ge Y, Stricklett PK, Gill P, Taylor D, Hughes AK, Yanagisawa M, Miller L, Nelson RD, Kohan DE: Collecting duct-specific knockout of endothelin-1 causes hypertension and sodium retention. J Clin Invest 2004;114:504–511.
15 Ye DZ, Wang DH: Function and regulation of endothelin-1 and its receptors in salt sensitive hypertension induced by sensory nerve degeneration. Hypertension 2002;39:673–678.
16 Rubanyi GM, Polokoff MA: Endothelins: molecular biology, biochemistry, pharmacology, physiology and pathophysiology. Pharmacol Rev 1994;46:325–415.
17 Itoh S, van den Buuse M: Sensitization of baroreceptor reflex by central endothelin in conscious rats. Am J Physiol 1991;260:H1106–H1112.
18 Li DP, Fan MZ, He RR: Effects of endothelin on carotid baroreceptor activity in anesthetized rats. Sheng Li Xu Bao 1998;50:532–538.
19 Kopp UC, Jones SY, DiBona GF: Afferent renal denervation impairs baroreflex control of efferent renal sympathetic nerve activity. Am J Physiol 2008;295:R1882–R1890.
20 Kopp UC, Cicha MZ, Smith LA, Ruohonen S, Scheinin M, Fritz N, Hökfelt T: Dietary sodium modulates the interaction between efferent and afferent renal nerve activity by altering activation of α_2-adrenoceptors on renal sensory nerves. Am J Physiol 2011;300:R298–R310.
21 Liu L, Barajas L: The rat renal nerves during development. Anat Embryol 1993;188:345–361.
22 Kopp UC, Smith LA, Pence A: Na^+-K^+-ATPase inhibition sensitizes renal mechanoreceptors activated by physiological increases in renal pelvic pressure. Am J Physiol 1994;267:R1109–R1117.
23 Genovesi S, Pieruzzi F, Wijnmaalen P, Centonza L, Golin R, Zanchetti A, Stella A: Renal afferents signaling diuretic activity in the cat. Circ Res 1993;73:906–913.
24 Kopp UC, Smith LA, DiBona GF: Impaired renorenal reflexes in spontaneously hypertensive rats. Hypertension 1987;9:69–75.
25 Kopp UC, Cicha MZ, Smith LA: Impaired interaction between efferent and afferent renal nerve activity in SHR involves increased activation of α_2-adrenoceptors. Hypertension 2011;57:640–647.
26 Kopp UC, Farley DM, Cicha MZ, Smith LA: Activation of renal mechanosensitive neurons involves bradykinin, protein kinase C, PGE_2 and substance P. Am J Physiol 2000;278:R937–R946.
27 Kopp UC, Cicha MZ, Nakamura K, Nüsing RM, Smith LA, Hökfelt T: Activation of EP4 receptors contributes to prostaglandin E2 mediated stimulation of renal sensory nerves. Am J Physiol 2004;287:F1269–F1282.
28 Manning DC, Snyder SH: Bradykinin receptors localized by quantitative autoradiography in kidney, ureter, and bladder. Am J Physiol 1989;256:F909–F915.
29 Kopp UC, Cicha MZ, Smith LA, Haeggström JZ, Samuelsson B, Hökfelt T: Cyclooxygenase-2 involved in stimulation of renal mechanosensitive neurons. Hypertension 2000;35:373–378.
30 Kopp UC, Cicha MZ: PGE_2 increases substance P release from renal pelvic sensory nerves via activation of N-type calcium channels. Am J Physiol 1999;276:R1241–R1248.
31 Hou L, Wang X: PKC and PKA, but not PKG mediate LPS induced CGRP release and $[Ca^2]_i$ elevation in DRG neurons of neonatal rats. J Neurosci Res 2001;66:592–600.
32 Kopp UC, Cicha MZ, Smith LA: Angiotensin blocks substance P release from renal sensory nerves by inhibiting PGE_2-mediated activation of cAMP. Am J Physiol Renal Physiol 2003;285:F472–F483.

33 Kopp UC, Cicha MZ, Smith LA: Endogenous angiotensin modulates PGE_2-mediated release of substance P from renal mechanosensory nerve fibers. Am J Physiol 2002;282:R19–R30.
34 Kopp UC, Cicha MZ, Smith LA: PGE_2 increases release of substance P from renal sensory nerves by activating the cAMP-PKA transduction cascade. Am J Physiol 2002;282:R1618–R1627.
35 Kopp UC, Cicha MZ, Smith LA: Differential effects of endothelin on the activation of renal mechanosensory nerves: stimulatory in high and inhibitory in low sodium diet. Am J Physiol Regul Integr Comp Physiol 2006;291:R1545–R1556.
36 Donovan MK, Wyss JM, Winternitz SR: Localization of renal sensory neurons using fluorescent dye technique. Brain Res 1983;259:119–122.
37 Kopp UC, Cicha MZ, Smith LA: Activation of endothelin-A receptors contributes to angiotensin-induced suppression of renal sensory nerve activation. Hypertension 2007;49:141–147.
38 Ito H, Hirata Y, Adachi S, Tanaka M, Tsujino M, Koike A, Nogami A, Marumo F, Hiroe M: Endothelin-1 is an autocrine/paracrine factor in the mechanism of angiotensin II-induced hypertrophy in cultured rat cardiomyocytes. J Clin Invest 1993;93:398–403.
39 Rajagopalan S, Laursen JB, Borthayre A, Kurz S, Keiser J, Haleen S, Giaid A, Harrison DG: Role for endothelin-1 in angiotensin II-mediated hypertension. Hypertension 1997;30:29–34.
40 Koyama Y, Mizobata T, Yamamoto N, Hashimoto H, Matsuda T, Baba A: Endothelins stimulate expression of cyclooxygenase 2 in rat cultured astrocytes. J Neurochem 1999;73:1004–1011.
41 Verkhratsky A, Orkand RA, Kettenmann H: Glial calcium: homeostasis and signalling function. Physiol Rev 1998;78:99–141.
42 Loomis ED, Sullivan JC, Osmond DA, Pollock DM, Pollock JS: Endothelin mediates superoxide production and vasoconstriction through activation of NADPH oxidase and uncoupled nitric-oxide synthase in the rat aorta. J Pharmacol Exp Ther 2005;315:1058–1064.
43 Li Z, Mao HZ, Abboud FM, Chapleau MW: Oxygen-derived free radical contribute to baroreceptor dysfunction in atherosclerotic rabbits. Cir Res 1996;79:802–811.
44 Kopp UC, Cicha MZ, Jones SY: Activation of endothelin A receptors contributes to impaired responsiveness of renal mechanosensory nerves in congestive heart failure. Can J Physiol Pharmacol 2010;88:622–629.
45 Motte S, van Beneden R, Mottet J, Rondolet B, Mathieu M, Havaux X, Lause P, Clercx C, Ketelslegers JM, Naeije R, McEntee K: Early activation of cardiac and renal endothelin systems in experimental heart failure. Am J Physiol 2003;285:H2482–H2491.
46 Luchner A, Jougasaki M, Friedrich E, Borgeson DD, Stevens TL, Redfield MM, Riegger GAJ, Burnett JC: Activation of cardiorenal and pulmonary tissue endothelin-1 in experimental heart failure. Am J Physiol 2000;279:R974–R979.
47 Kobayashi T, Miyauchi T, Sakai S, Kobayashi M, Yamaguchi I, Goto K, Sugishita Y: Expression of endothelin-1, ET_A and ET_B receptors and ECE and distribution of endothelin-1 in failing rat heart. Am J Physiol 1999;276:H1197–H1206.
48 Liu J-L, Pliquett RU, Brewer E, Cornish KG, Shen Y-T, Zucker IH: Chronic endothelin-1 blockade reduces sympathetic nerve activity in rabbits with heart failure. Am J Physiol 2001;280:R1960–1913.

Ulla C. Kopp, PhD
Department of Internal Medicine
VA Medical Center, Bldg. 41, Highway 6W
Iowa City, IA 52246 (USA)
Tel. +1 319 338 0581, Ext. 7608, E-Mail ulla-kopp@uiowa.edu

Barton M, Kohan DE (eds): Endothelin in Renal Physiology and Disease.
Contrib Nephrol. Basel, Karger, 2011, vol 172, pp 120–138

Endothelin and Podocyte Injury in Chronic Kidney Disease

Cécile Fligny[a] · Matthias Barton[b] · Pierre-Louis Tharaux[a]

[a]INSERM and Université Paris-Descartes, Paris Cardiovascular Research Centre – PARCC, Paris, France; [b]Molecular Internal Medicine, University of Zürich, Zürich, Switzerland

Abstract

The identification of patients at increased risk for chronic kidney disease offers the potential to prevent or delay end-stage renal disease and the associated cardiovascular events. Data from recently completed controlled clinical trials of endothelin (ET) receptor blockers confirmed their potent antiproteinuric effect after a number of preclinical studies. A spectrum of proteinuric glomerular diseases results from podocyte abnormalities and, in return, impact podocyte structure and function. Because podocytes are cells in the glomerulus that form a crucial component of the glomerular filtration barrier, contributing to size selectivity and maintaining a large filtration surface, we focus on evidence that suggest ET-1 may promote podocyte injury thereby aggravating albumin urinary loss and alteration of the glomerular microvasculature. Systematic confrontation of animal models and studies in human subjects should help decipher pathophysiological mechanisms whereby the local renal ET system promotes podocyte injury and chronic kidney disease in specific pathophysiological contexts. Current evidence suggests that more experimental and clinical attention should be paid to conditions with increased vascular or endocapillary ET-1 production on the one hand, and in diseases with altered podocyte phenotype and survival such as focal segmental glomerulosclerosis and crescentic glomerulonephritis on the other. These conclusions may assist clinicians in creating optimal clinical trials for patients at increased risk for or with overt chronic kidney disease.

Proteinuria is a hallmark of glomerular dysfunction and an indicator of the severity of chronic renal disease [1]. In the intact kidney, the glomerular filtration barrier prevents urinary protein leaking into the urine. The amount of albumin excreted into the urine, a protein that is specific to assess the functional integrity of the glomerular filtration barrier [2–4], is not only a good indicator of overall glomerular structural integrity, but also serves as a good predictor of cardiovascular events related to renal disease [5, 6]. Indeed, patients with advanced renal disease have a greatly increased risk of dying from cardiovascular complications [7–13].

The glomerular filtration barrier is located on the glomerular basement membrane and consists of podocytes with the opposing slit diaphragms and glomerular endothelial cells (fig. 1). Podocytes (also known as glomerular or visceral epithelial cells) are terminally differentiated [4] and highly specialized cells with a complex cytoarchitecture [14]. They are attached to the glomerular capillary through so-called 'foot processes', which explains the terminology. Foot process detachment or 'effacement' is a key feature of podocyte injury [4, 15]. Recently, podocyte loss [16] and podocyturia [17] have been recognized as potential diagnostic markers of the severity of glomerular disease. Characteristic features of podocytes are interdigitated foot processes with filtration slits in between. These are bridged by the slit diaphragm, which plays a major role in establishing the selective permeability of the glomerular filtration barrier [4]. Injury to podocytes leads to proteinuria, a hallmark of most glomerular diseases [4, 18]. Podocytes express numerous G protein-coupled receptors implicated in the progression of glomerular diseases, including receptors for angiotensin II, thromboxane A_2, E-series prostaglandins, platelet activation factor, and cysteinyl leukotrienes. Podocytes also express endothelin A (ET_A) and ET_B receptors [19–22], which are therapeutic targets and sensitive to pharmacological blockade [19].

In the present review, we assembled experimental and clinical data with a pathophysiological rationale focusing largely on podocytes. When possible, we compared data for the expression of ET-1 and its receptors with the results of intervention studies for various proliferative and nonproliferative podocyte diseases, of various causes, with a view to identifying the situations in which this system may play an important role. Indeed, although the effects of ET-1 on hemodynamics and glomerular remodeling [23–28] and on mesangial phenotype [29–32] are well documented, evidence from previous preclinical [19] and recent clinical studies [33–35] indicate direct effects on the components of the filtration barrier of podocyte function [19, 36], as evidenced by blood pressure-independent reductions of pre-existing proteinuria. The antiproteinuric effect of blocking ET_A receptors is currently a much debated topic in efforts to understand the relationships between the hemodynamic and glomerular structural modifications affected by ET-1.

We will review the experimental evidence for a regulatory role of ET-1 in podocyte function and hypothesize a nonexclusive pathophysiological involvement of the ET-1 system in different glomerular diseases with podocyte injury. First, ET-1 produced by injured glomerular or preglomerular endothelium would be freely filtered through the filtration barrier and act on podocyte ET receptors. This is expected to occur in severe hypertension [23, 37, 38], hemolytic uremic syndrome [39] and thrombotic microangiopathy [40, 41], sickle cell disease [42], connective tissue diseases [43], and vasculitis [44–46]. Due to space limitation, we will not review these conditions or diabetic nephropathy, but refer the reader instead to excellent reviews of these topics [47–51], Gagliardini et al., this vol., pp. 171–184; Pollock and Pollock, this vol., pp. 149–159; Benz and Amann, this vol., pp. 139–148; Tharaux, this vol., pp.185–199. We will briefly review evidence for increased ET-1 activity in the vasculature in IgA nephropathy where vascular lesions are frequent and of prognostic value. Second, ET-1 may be produced de novo by mesangial cells. We will also review the current evidence for ET-1 production by mesangial cells, in particular IgA nephropathy, and by infiltrating leukocytes. Third, scarce evidence suggest that podocytes (visceral epithelial cells) and glomerular parietal epithelial cells produce ET-1 that affects the podocyte phenotype in a paracrine and autocrine fashion in vitro [19, 20]. In this regard we will review the current evidence in focal segmental glomerulosclerosis (FSGS) and crescentic glomerulonephritis.

Regulatory Role of Endothelin for Podocyte Function: Role of Endothelin A Receptors

Binding of ET-1 to rat podocytes has been reported by radioligand-binding studies [52], and Rebibou et al. [53] found ET-binding sites on human podocytes and increases in calcium flux in response to ET-1. It has been previously proposed that the neighboring cells in close proximity of podocytes (fig. 1) – which also produce ET – may also affect podocyte function and structure through ET-mediated interactions [22, 54]. In cultured podocytes, injury induced by puromycin aminonucleoside (which causes FSGS in vivo) [55, 56] causes apoptosis and MMP-9 upregulation, which is completely prevented by ET_A receptor blockade (fig. 2c) or ET_A receptor RNA silencing (fig. 2d). Puromycin aminonucleoside also causes complete disruption of the actin cytoskeleton, which can be rescued by blocking ET_A (but not ET_B) receptors (fig. 2e, 3) [19]. These data are supported by similar observations from Benigni's group [20] who extended these findings by showing that ET-1 occurs in response to exogenous protein upregulation of podocytes and that it is also associated with cytoskeleton disruption. Additionally, exogenous ET-1 induces cytoskeleton rearrangement that can be prevented in an ET_A receptor-sensitive manner [21]. Cultured podocytes synthesize ET-1 [57, 58], which inhibits podocyte proliferation because ET_A receptor gene silencing stimulates podocyte growth (fig. 2e).

Preclinical studies suggest that ET_A antagonists can in part restore podocyte morphology and function under normotensive [19] and hypertensive conditions [36] as well as prevent glomerular podocyte loss in experimental diabetes [59]. Glomerular expression of MMP-9, a podocyte injury marker, is downregulated in rats with FSGS after ET_A receptor blockade in vivo (fig. 2a), suggesting that endogenous ET also activates this pathway [19]. In addition, ET_A blockade in rats with FSGS markedly downregulates renal p21waf/cip protein [19],

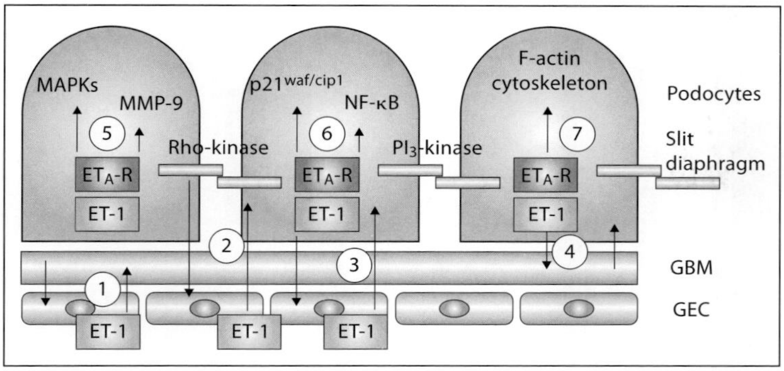

Fig. 1. Effects of ET and its interactions with the glomerular capillary basement membrane (GBM), its endothelial cells (GEC), and podocytes/slit diaphragm. Also shown are the proposed signaling pathways by which ET-1 contributes to podocyte and glomerular injury. ET-1 is produced by cells on both sides of the GBM, namely glomerular endothelial cells (GEC) and podocytes. With regard to interacting with the different cellular components of the glomerular capillary, ET released from GEC may directly act on the GBM (1), with the slit diaphragm (2), or with the podocyte (3). ET-1 release and signaling may also work in the reverse direction. Finally, ET-1 released from podocytes may interact with the GBM and vice versa (4). In the podocyte, ET-1 activates ET_A receptors (ET_A-R) and promotes glomerular injury and sclerosis through MAPKs p38 and p44/p42 (5), the growth promoter and cdk-inhibitor $p21^{waf/cip1}$ and NF-κB (6), and causes disruption of the F-actin cytoskeleton (7). In addition, ET-1 promotes slit diaphragm dysfunction via activation of Rho-kinase and PI_3-kinase. Figure reproduced from Barton [47] with permission of the publisher.

a promoter of glomerulosclerosis [60] and cell growth [61]. The mechanism underlying the partial healing of glomerular podocytes by ET_A antagonists is currently unclear, particularly since the cells are terminally differentiated; however, recent evidence suggests that visceral epithelial cells (podocytes) can be replaced by parietal epithelial cells [62–64]. It has been previously proposed that this may be one of the underlying mechanisms contributing to the restoration of podocyte structure and function after ET receptor blockade [47, 51].

Glomerulonephritis with Endocapillary Proliferation or Expansion

Experimental Mesangioproliferative Diseases, Glomerulonephritis Following Streptococcal Infection, Lupus-Associated Nephropathy, and Kidney Transplantation

The mitogenic or comitogenic effect of ET-1 on mesangial cells was described above. In the model of glomerulonephritis with mesangial proliferation induced by the injection of antithymocyte antibodies, an increase is observed in the glomerular expression of ET-1 and ET_B receptors [65]. Jesus Egido and colleagues

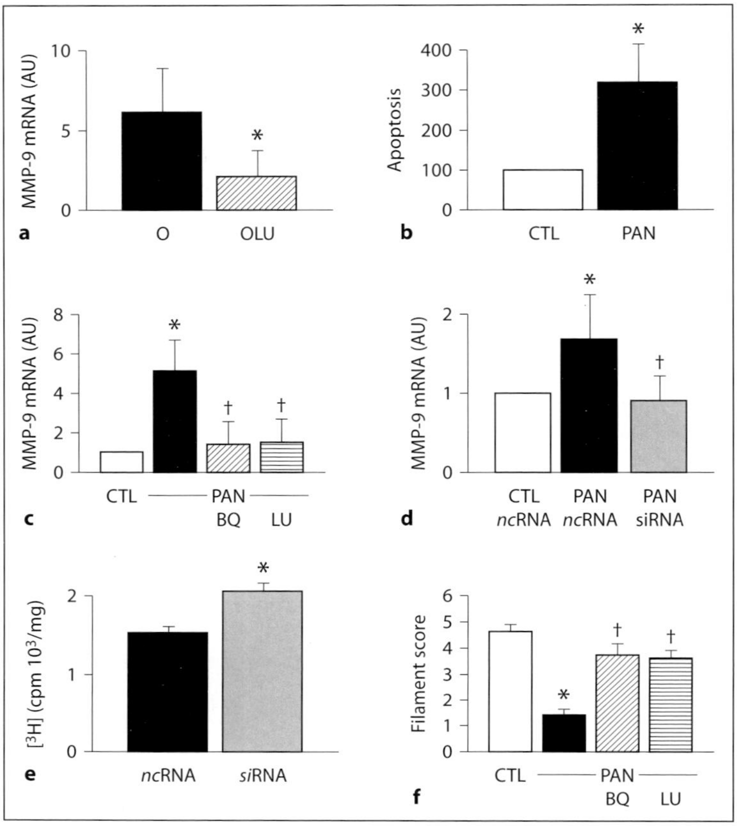

Fig. 2. **a** Inhibition of MMP-9 gene expression in vivo in laser-dissected glomeruli of aged rats with established FSGS treated with the ET_A receptor antagonist darusentan/LU135252 (OLU) for 4 weeks compared with placebo (O). Gene expression levels are given as arbitrary units (AU) using the $\Delta\Delta_{CT}$ method. **b-f** Effect of ET_A receptor blockade on human podocyte function in vitro: treatment with puromycin aminonucleoside (50 µg/ml, PAN), which causes FSGS in rats in vivo, resulted in podocyte apoptosis (**b**, $p < 0.05$ vs. control). Similarly, puromycin aminonucleoside (PAN)-induced podocyte injury marker MMP-9, which was equally prevented by peptide (BQ) or nonpeptide (LU, darusentan) ET_A receptor-selective receptor antagonists (**c**), or ET_A receptor RNA interference (siRNA); ncRNA, noncoding RNA control (**d**). Silencing of ET_A receptors through *si*RNA increased podocyte growth as determined de novo DNA synthesis in human podocytes measured by thymidine incorporation (**e**). PAN treatment of mouse podocytes caused severe disruption of the actin cytoskeleton (see also fig. 3 of this article), an effect that was largely prevented by ET_A antagonists BQ-123 (peptide, 10 nmol/l) or LU135252 (darusentan, nonpeptide, 30 nmol/l) on podocyte cytoskeleton disruption. O = old; OLU = old, darusentan; CTL = control. * $p < 0.05$ versus CTL; † $p < 0.05$ versus PAN alone/old. Figure reproduced from Ortmann et al. [19] with permission of the publisher.

[66] also reported that an ET_A antagonist prevents proteinuria and histological lesions in another model of mesangial proliferation, induced by the repeated systemic injection of ovalbumin in rats. A role for podocytes in this process has been suggested as Benigni's group [20] recently reported that albumin causes podocyte cytoskeleton disruption and upregulation of podocyte ET-1 expression. In humans with acute glomerulonephritis following streptococcal infection in which intense endocapillary mesangial proliferation is observed, together with an extensive leukocyte infiltrate, the local activation of ET-1 has not been investigated or reported. However, a sample of 18 children with glomerulonephritis was found to have higher plasma ET concentrations than 14 healthy children [67]. The investigators also observed a positive correlation between plasma ET-1 concentration and arterial blood pressure, consistent with a link between endothelial activation and plasma ET concentration in the context of acute poststreptococcal glomerulonephritis. Although elevated ET-1 plasma concentrations may reflect systemic and renal endothelial stress leading to activation of ET-1 synthesis with potential influence on glomeruli, no unequivocal role can be asserted given the paracrine actions of this peptide [68].

Only two studies have reported the mesangial expression of ET-1 in humans: lupus-associated nephropathy with moderate mesangial proliferation [69] and kidney transplantation with cyclosporine treatment [70]. Remarkably, even after discontinuation of the immunosuppressant, ET_A receptor blockade as monotherapy is capable of preventing much of the allograft rejection-associated mesangial proliferation and glomerulosclerosis [71].

IgA Nephropathies

Lehrke et al. [72] determined preproET-1 mRNA levels by quantitative RT-PCR and evaluated ET-1, ET_A, and ET_B protein levels semiquantitatively by immunohistochemical methods in kidney biopsy specimens from a sample of 26 patients with atypical glomerular nephropathy including mostly IgA nephropathy with proteinuria. These authors found that preproET-1 mRNA levels were significantly higher in patients with high proteinuria (>2 g/day) and moderately altered creatinine clearance (<80 ml/min) than in other subjects. No change was found in ET_A mRNA levels, but ET_B mRNA levels were found to be higher in the patients with the highest proteinuria. A complementary immunohistochemical study on a sample of kidney biopsy specimens from patients with mesangial IgA deposition demonstrated marked immunoreactivity for ET-1 in the proximal tubule and in the glomerular endothelium, and, in some cases, also in parietal epithelial cells [72].

Although culture studies have demonstrated that cultured mesangial cells and podocytes can produce ET-1, only a faint expression of glomerular immunoreactive ET-1 was seen by Rastaldi et al. [73] in remnant glomeruli of 25 biopsies with advanced glomerular and interstitial damage. This pattern contrasted with the marked ET-1 expression in many small-sized vessels, as well as some tubule sections and interstitial mononuclear cells that correlated with histopathological damage and alteration of GFR [73]. Indeed, monocytes are strongly suspected of playing a role in the physiopathology of Berger's disease and a positive correlation has indeed been found between the preproET-1 mRNA levels in monocytes on the one hand, and creatininemia, the severity of histological kidney lesions, and proteinuria on the other [74]. The stimulation of ET-1 production by mesangial cells cultured in the presence of polynuclear neutrophils from patients with IgA nephropathies has suggested a role for reactive oxygen species in this activation [75].

In these two cases, the relationship between the activation of monocytes and polynuclear neutrophils and progression of the renal disease is only suggestive. However, the chronic inhibition of ET_A in the ddY mouse model has demonstrated that this approach protects against proteinuria and glomerular lesions [76], although the underlying mechanisms remain unclear. There has, as yet, been no description of activation of the ET system or of clinical trials of ET receptor antagonists for the most serious forms of IgA nephropathy, for which new treatments would be welcome.

Noncrescentic Lupus-Associated Nephropathies

ET has the activity of a proinflammatory cytokine, and inhibition of ET_A receptors is generally associated with a reduction of inflammation in the kidney and other tissues [77–79]. Meanwhile, the cellular targets that secrete cytokines in an ET receptor-dependent fashion in vivo have yet to be identified. On the other hand, IL-1β increases renal ET-1 production [80]. Only a few studies have been carried out on these diseases in animals, but their findings nonetheless implicate ET-1 in the physiopathology of lupus-associated nephropathy. In NZB/W F1 lupus mice, ET-1 and ET receptor levels increase very significantly with progression of the nephropathy and are decreased by methylprednisolone treatment [44]. Long-term preventive effects of ET receptor antagonists have been demonstrated in this mouse model of lupus (NZB/W F1 mice), with a decrease in proteinuria and histological lesions, an improvement in glomerular filtration, and a smaller increase in arterial blood pressure [45, 81]. In humans, bosentan has been shown to limit both the severity of secondary pulmonary arterial hypertension in some subjects [82, 83] as well as skin ulcerations [84], indicating a probable pathophysiological role in this vascular disease apart from the kidney as well.

Plasma ET-1 concentrations may be high in lupus patients [85] and a pioneering study by Herman et al. [69] reported the presence of immunoreactivity for ET-1 in mesangial cells in the renal biopsy specimens of subjects with lupus, which was then confirmed by Lehrke et al. [72]; however, whereas ET-1 is normally restricted to the endothelium in the human kidney [69, 86], the factors stimulating ET-1 synthesis in lupus have not yet been clearly identified. Serum factors may be involved [46], and there is evidence that inflammatory factors play also a role. Thus, in subjects with lupus nephritis and chronic renal insufficiency, the plasma ET-1 concentration increases linearly with declining renal function, whereas the fraction of ET-1 excreted in urine increases exponentially over a period of 6 months. In the absence of renal insufficiency, the fraction of ET-1 excreted and the ET-1/creatinine ratio are also higher in subjects with lupus than in controls and are correlated with C-reactive protein concentrations, compatible with systemic inflammatory activation [87]. Of note, because most of the plasma ET-1 filtered through glomeruli is subject to degradation by neutral endopeptidase [88, 89] in the proximal tubule and by an acidic highly specific ET-1-inactivating metalloendopeptidase, ET-1 in the urine is of tubular rather than glomerular origin [90]. Thus, urinary ET-1 likely reflects the production of this peptide by tubulointerstitial diseases of the kidney and the link with podocyte damage remains to be shown.

The Endothelin System in Glomerular Epithelial Diseases (Podocytopathies)

Minimal Change Disease: A Podocyte Disease without Inflammatory Infiltrate
A semiquantitative and morphological immunohistochemical study compared renal ET-1 expression in 25 adult subjects with minimal change disease and acute renal insufficiency with that of 28 subjects with the same nephropathy in the absence of renal insufficiency [91]. The patients with renal insufficiency were found to have higher ET-1 levels in the endothelium of the glomeruli and tubules. The precise cell types responsible for producing ET-1 in this context remain to be identified. There was no clear correlation between ET-1 levels and the rate of response to steroids or the stability of remission, which did not differ significantly between the two groups [91]. Thus, an increase in the renal production of ET-1 is associated with an insufficiency of renal function with strong proteinuria in this context. Angiotensin II treatment or proteinuria may stimulate ET-1 synthesis, but the possible pathophysiological role of this stimulation remains to be demonstrated.

Glomerulosclerosis due to Nephron Reduction
In rats with chronic renal insufficiency induced by surgical reduction of the number of nephrons, renal preproET-1 mRNA production and the urinary excretion of ET-1 increase proteinuria and the deterioration of renal function in a correlated manner [92]. The pioneering work of Benigni et al. [93] demonstrated that the blockade of ET receptors prevented the decline of renal function, but not proteinuria, in a rat model of severe nephron reduction.

In a model of focal and segmental hyalinosis and glomerulosclerosis (FSGS) obtained by the administration of a puromycin aminonucleoside, Nakamura et al. [94] reported an increase in glomerular ET-1 synthesis during the progression

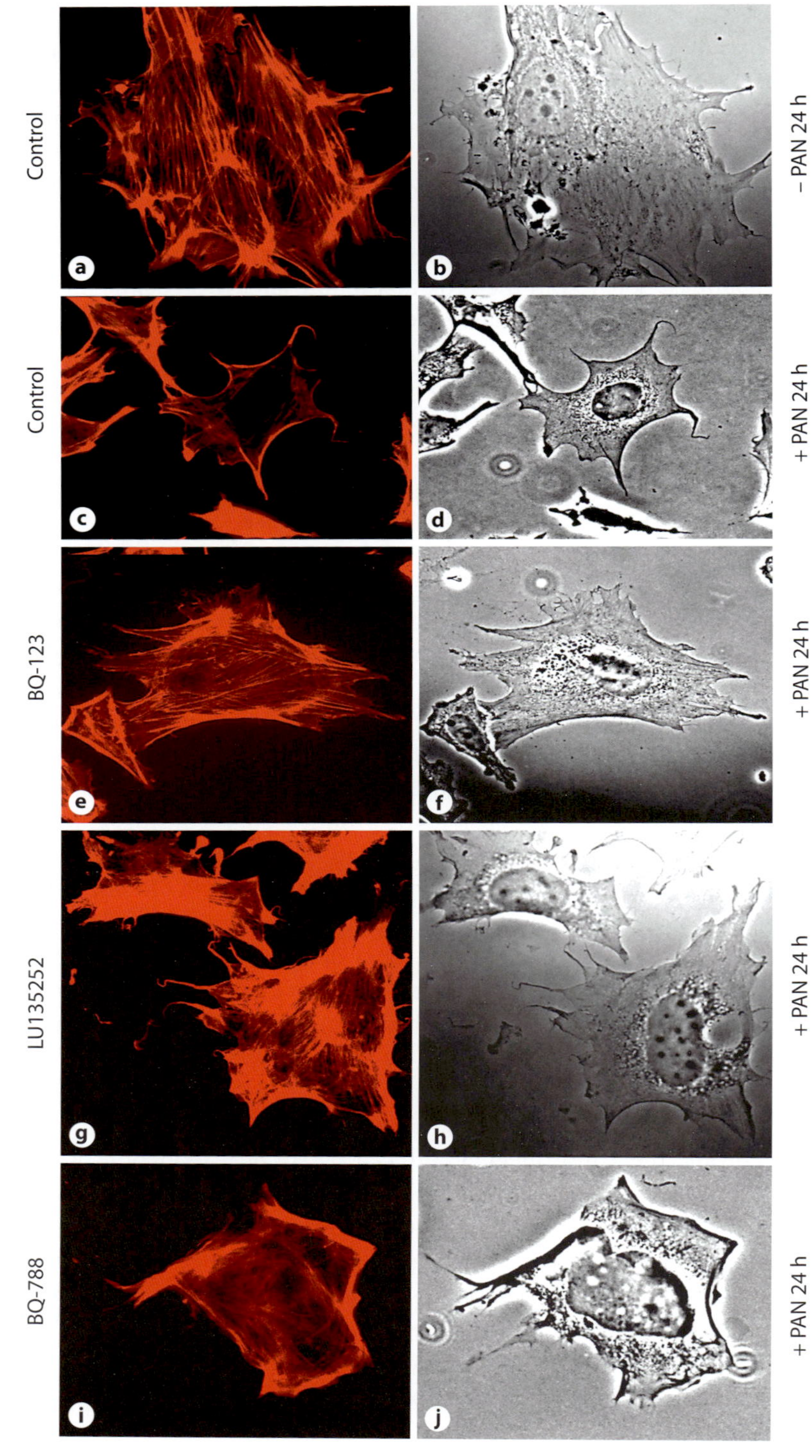
Control
– PAN 24 h
Control
+ PAN 24 h
BQ-123
+ PAN 24 h
LU135252
+ PAN 24 h
BQ-788
+ PAN 24 h
a
b
c
d
e
f
g
h
i
j

of FSGS. Glomerular cell type(s) involved in ET-1 synthesis were not identified. When the progression of the nephropathy is slowed by feeding the rats a low-protein diet, less ET-1 is synthesized locally, suggesting a role for hyperfiltration as an inductor of local ET-1 synthesis. This has not yet been tested in mechanistic studies. However, these findings can be compared with those suggesting a role for vimentin in the mechanically transduced stimulation of ET-1 activity in a mouse nephron reduction model [95]. The increase in renal (mostly endothelial) ET-1 synthesis observed in a mouse model of sickle cell disease [42] with glomerulomegaly and hyperfiltration complicated by FSGS may also be due to mechanically transduced stimulation. ET_A or ET_A and ET_B combined blockade displayed contrasted effectiveness in blunting albuminuria and injury in experimental diabetic models with initial hyperfiltration [79, 96–98]. The cellular source of ET-1 and the mechanisms underlying this stimulation also remain to be determined in these cases.

Nevertheless, hemodynamics may be just part of the pathophysiological cause of glomerular damage. Indeed, FSGS with proteinuria and podocyte damage due to aging in aged Wistar rats could be reduced by greater than 50% after only 4 weeks of darusentan treatment, with no change in blood pressure or glomerular filtration rate, and without affecting tubulointerstitial renal injury [19]. Also, in malignant hypertension, beneficial effects of ET_A blockade on podocyte structure was independent of sustained hypertension [36, 99], again making the argument for a potential structural effect of the treatment. Thus, ET_A-mediated apoptosis and loss of podocytes in experimental FSGS may in part be due to the direct cytopathogenic effect of ET_A-signals (fig. 4).

There have been few clinical studies, but those published to date suggest that in humans renal ET-1 levels increase during nephropathies complicated by renal insufficiency. In a controlled study comparing patients with chronic renal insufficiency and healthy subjects, urinary ET-1 excretion levels were significantly higher in the patients than the controls [100]. High urinary ET-1 output has also been reported in subjects with sickle cell disease and microalbuminuria with normal or 'strong' renal function (i.e. with glomerular hyperfiltration). In these

Fig. 3. Effects of puromycin aminonucleoside (PAN) on morphology and actin cytoskeleton fiber organization of cultured podocytes in vitro as visualized by actin-phalloidin immunofluorescence (left panels) and corresponding phase-contrast microphotographs (right panels). Representative examples of podocyte morphology in the absence (**a**, **b**) or presence of puromycin aminonucleoside (10 μg/ml) alone (**c**, **d**) or after pretreatment with ET_A receptor antagonists BQ-123 (10 nmol/l; **e**, **f**) and darusentan (LU, 30 nmol/l; **g**, **h**), or the ET_B receptor antagonist BQ-788 (10 nmol/l; **i**, **j**). PAN treatment resulted in cell shrinkage, foot process effacement, and cytoskeleton disruption with actin fibers completely disappearing (**c**). PAN-induced podocyte injury was largely prevented by the ET_A receptor antagonists BQ-123 or LU135252. The ET_B receptor selective antagonist BQ-788 had no effect (**i**, **j**). Figure reproduced from Ortmann et al. [19] with permission of the publisher.

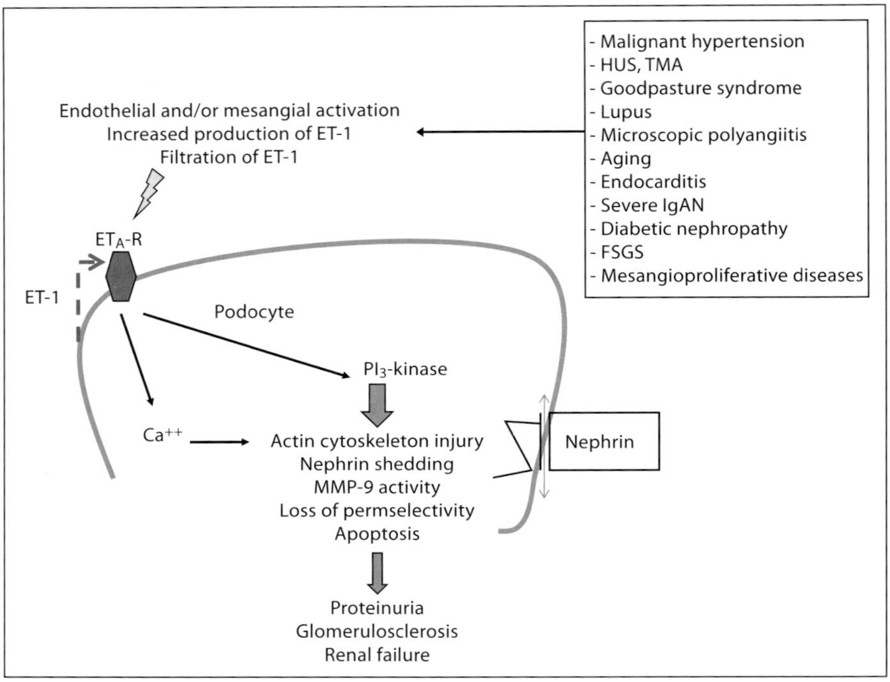

Fig. 4. Pathophysiological model: ET-1 produced by injured glomerular or preglomerular endothelium, with potential additional contribution of mesangial synthesis, would be freely filtered through the filtration barrier and act on podocyte ET receptors. ET_A receptor (ET_A-R) activity has been characterized in cultured podocytes where it is coupled to calcium signaling and promotes actin cytoskeleton rearrangement. In addition to promotion of nephron shedding, these cascades of ET-1-mediated events would lead to alteration of the filtration barrier with loss of its permselectivity. Additionally, ET-1 induces synthesis of MMP-9 thereby amplifying its pathophysiological alteration of the filtration barrier and proinflammatory activity. ET_A activation was also shown to contribute to proapoptotic pathways leading to loss of podocytes in vivo, again contributing to albuminuria and destruction of the glomerular capillary. The dashed arrow represents possible autocrine and paracrine synthesis of ET-1 by podocytes, and thus source of a potential aggravating loop. HUS = Hemolytic uremic syndrome; TMA = thrombotic microangiopathy; IgAN = IgA nephropathy or Berger's disease; MMP = matrix metalloproteinase; PI_3-kinase = Phosphatidylinositol 3-kinase.

subjects, the elevated urinary ET-1 excretion, which was four times higher than that in carefully matched healthy controls, was correlated with microalbuminuria, suggesting a pathological link between the loss of glomerular permselectivity and the renal synthesis of ET-1 in this context preceding progression to FSGS [101]. To our knowledge, there have been no studies of the interactions between the ET system and polymorphisms of the ApoL1 gene strongly associated with the risk of FSGS in subjects of African descent [102].

Finally, proteinuria has been shown to be rapidly and substantially decreased (–26±4%) by the administration of an ET_A antagonist in subjects with chronic renal insufficiency and proteinuria, but not diabetes, already treated with an RAS inhibitor [103]. This effect on proteinuria is accompanied by a modest decrease in arterial blood pressure and a significant increase in renal blood flow with no change in glomerular filtration rate. The authors of this study reported this decrease in proteinuria to be persistent, whereas arterial blood pressure returned to previous levels when the treatment was stopped, consistent with a local rather than systemic effect of ET_A on the physiopathology of proteinuria [34]. Dhaun et al. [104] also observed a synergy between this approach and RAS blockade.

Glomerulonephritis with Extracapillary Proliferation
There is a critical absence of systematic studies of the elements of the ET-1 system as a function of the type of lupus-associated nephropathy, although clinical and experimental evidence suggest its implication as reviewed above. Such studies are required to obtain information about the physiopathology of this heterogeneous group of nephropathies and, possibly, to determine which class of lupus-associated nephropathy would be the most suitable for testing ET receptor antagonists. In particular, no study has yet demonstrated activation of the ET-1 system in the most inflammatory and destructive forms of type III and IV lupus-associated glomerulonephritis, such as rapidly progressing glomerulonephritis with endo- and extracapillary proliferation. This is also true for glomerulonephritis with extracapillary proliferation of other causes, for which few data are currently available. Of important note, immunoreactive ET-1 was not found in glomeruli from 18 patients with necrotizing/crescentic IgA glomerulonephritis [73]. We also found no characterization of the ET system in collapsing FSGS.

Conclusion

Podocyte damage is frequently observed in conditions with local activation of the ET-1 system. Increases in ET-1 synthesis by the renal endothelium with glomerular damage have been reported principally in experimental and clinical conditions in which endothelial activation occurs, potentially causing abnormal glomerular hemodynamics and inflammation (fig. 4) [23, 37–46]. This endothelial synthesis of ET-1 can increase the capillary pressure of glomerular filtration, leukocyte and platelet recruitment, proteinuria, and glomerular inflammation, with downstream ischemic consequences.

The addition of local nonendothelial sources of ET-1 would likely accentuate kidney remodeling and failure. Preclinical studies and the all too rare anatomical and preclinical studies have shown that the synthesis of ET-1 may be decompartmentalized, either extra-endothelially in the renal arteriolar smooth muscle cells in salt-sensitive or NO-deficient hypertension [23, 105], or in the mesangial cells in nephropathies with mesangial IgA deposition and glomerulonephritis with endocapillary proliferation (see above).

A number of pathophysiological questions remain unanswered. The expression of ET_A and ET_B on podocytes has been confirmed [19–22], but the exact physiological and pathophysiological roles on podocytes of local ET-1 synthesis by endothelial cells, vascular smooth muscle cells, and/or mesangial cells are still unclear. Indeed, this peptide is filtered and should therefore display effects on podocytes [19]. Furthermore, although podocytes can synthesize ET-1 in a dedifferentiated state in vitro, a physiological or pathophysiological role of the possible autocrine secretion of this peptide remains to be demonstrated in vivo, particularly in crescentic glomerulonephritis and FSGS. These conclusions and further studies may assist clinicians in optimally designing clinical trials for patients at increased risk for chronic kidney disease.

Acknowledgements

This work was supported by the Swiss National Science Foundation grants No. 3200-108258 and K33KO-122504 (to Dr. Barton) and the INSERM Contrat d'Interface (to Dr. Tharaux).

References

1 Dimitrov BD, Perna A, Ruggenenti P, Remuzzi G: Predicting end-stage renal disease: Bayesian perspective of information transfer in the clinical decision-making process at the individual level. Kidney Int 2003;63:1924–1933.
2 Lemley KV, Blouch K, Abdullah I, Boothroyd DB, Bennett PH, Myers BD, Nelson RG: Glomerular permselectivity at the onset of nephropathy in type 2 diabetes mellitus. J Am Soc Nephrol 2000;11:2095–2105.
3 Andersen S, Blouch K, Bialek J, Deckert M, Parving HH, Myers BD: Glomerular permselectivity in early stages of overt diabetic nephropathy. Kidney Int 2000;58:2129–2137.
4 Pavenstadt H, Kriz W, Kretzler M: Cell biology of the glomerular podocyte. Physiol Rev 2003;83(1):253–307.
5 Lambers Heerspink HJ, Brantsma AH, de Zeeuw D, Bakker SJ, de Jong PE, Gansevoort RT: Albuminuria assessed from first-morning-void urine samples versus 24-hour urine collections as a predictor of cardiovascular morbidity and mortality. Am J Epidemiol 2008;168:897–905.
6 Reffelmann T, Dorr M, Volzke H, Friedrich N, Krebs A, Ittermann T, Felix SB: Urinary albumin excretion, even within the normal range, predicts an increase in left ventricular mass over the following 5 years. Kidney Int 2010;77:1115–1122.
7 Solomon SD, Lin J, Solomon CG, Jablonski KA, Rice MM, Steffes M, Domanski M, Hsia J, Gersh BJ, Arnold JM, Rouleau J, Braunwald E, Pfeffer MA: Influence of albuminuria on cardiovascular risk in patients with stable coronary artery disease. Circulation 2007;116:2687–2693.

8 Nobakhthaghighi N, Kamgar M, Bekheirnia MR, McFann K, Estacio R, Schrier RW: Relationship between urinary albumin excretion and left ventricular mass with mortality in patients with type 2 diabetes. Clin J Am Soc Nephrol 2006;1:1187–1190.
9 Perkovic V, Verdon C, Ninomiya T, Barzi F, Cass A, Patel A, Jardine M, Gallagher M, Turnbull F, Chalmers J, Craig J, Huxley R: The relationship between proteinuria and coronary risk: a systematic review and meta-analysis. PLoS Med 2008;5:e207.
10 Jackson CE, Solomon SD, Gerstein HC, Zetterstrand S, Olofsson B, Michelson EL, Granger CB, Swedberg K, Pfeffer MA, Yusuf S, McMurray JJ: Albuminuria in chronic heart failure: prevalence and prognostic importance. Lancet 2009;374:543–550.
11 van der Velde M, Matsushita K, Coresh J, Astor BC, Woodward M, Levey A, de Jong P, Gansevoort RT, Levey AS, de Jong PE, El-Nahas M, Eckardt KU, Kasiske BL, Ninomiya T, Chalmers J, Macmahon S, Tonelli M, Hemmelgarn B, Sacks F, Curhan G, Collins AJ, Li S, Chen SC, Hawaii Cohort KP, Lee BJ, Ishani A, Neaton J, Svendsen K, Mann JF, Yusuf S, Teo KK, Gao P, Nelson RG, Knowler WC, Bilo HJ, Joosten H, Kleefstra N, Groenier KH, Auguste P, Veldhuis K, Wang Y, Camarata L, Thomas B, Manley T: Lower estimated glomerular filtration rate and higher albuminuria are associated with all-cause and cardiovascular mortality. A collaborative meta-analysis of high-risk population cohorts. Kidney Int 2011;79:1341–1352.
12 Warnock DG, Muntner P, McCullough PA, Zhang X, McClure LA, Zakai N, Cushman M, Newsome BB, Kewalramani R, Steffes MW, Howard G, McClellan WM: Kidney function, albuminuria, and all-cause mortality in the REGARDS (Reasons for Geographic and Racial Differences in Stroke) study. Am J Kidney Dis 2010;56:861–871.
13 Matsushita K, van der Velde M, Astor BC, Woodward M, Levey AS, de Jong PE, Coresh J, Gansevoort RT: Association of estimated glomerular filtration rate and albuminuria with all-cause and cardiovascular mortality in general population cohorts: a collaborative meta-analysis. Lancet 2010;375:2073–2081.
14 Faul C, Asanuma K, Yanagida-Asanuma E, Kim K, Mundel P: Actin up: regulation of podocyte structure and function by components of the actin cytoskeleton. Trends Cell Biol 2007;17:428–437.
15 Shankland SJ: The podocyte's response to injury: role in proteinuria and glomerulosclerosis. Kidney Int 2006;69:2131–2147.
16 Kriz W: The role of the podocyte in the degeneration of a renal glomerulus. Adv Nephrol Necker Hosp 1997;27:1–13.
17 Petermann A, Floege J: Podocyte damage resulting in podocyturia: a potential diagnostic marker to assess glomerular disease activity. Nephron Clin Pract 2007;106:c61–66.
18 Meyer TW, Bennett PH, Nelson RG: Podocyte number predicts long-term urinary albumin excretion in Pima Indians with type II diabetes and microalbuminuria. Diabetologia 1999;42:1341–1344.
19 Ortmann J, Amann K, Brandes RP, Kretzler M, Munter K, Parekh N, Traupe T, Lange M, Lattmann T, Barton M: Role of podocytes for reversal of glomerulosclerosis and proteinuria in the aging kidney after endothelin inhibition. Hypertension 2004;44:974–981.
20 Morigi M, Buelli S, Angioletti S, Zanchi C, Longaretti L, Zoja C, Galbusera M, Gastoldi S, Mundel P, Remuzzi G, Benigni A: In response to protein load podocytes reorganize cytoskeleton and modulate endothelin-1 gene: implication for permselective dysfunction of chronic nephropathies. Am J Pathol 2005;166:1309–1320.
21 Morigi M, Buelli S, Zanchi C, Longaretti L, Macconi D, Benigni A, Moioli D, Remuzzi G, Zoja C: Shigatoxin-induced endothelin-1 expression in cultured podocytes autocrinally mediates actin remodeling. Am J Pathol 2006;169:1965–1975.
22 Collino F, Bussolati B, Gerbaudo E, Marozio L, Pelissetto S, Benedetto C, Camussi G: Preeclamptic sera induce nephrin shedding from podocytes through endothelin-1 release by endothelial glomerular cells. Am J Physiol Renal Physiol 2008;294:F1185–1194.
23 Tharaux PL, Chatziantoniou C, Casellas D, Fouassier L, Ardaillou R, Dussaule JC: Vascular endothelin-1 gene expression and synthesis and effect on renal type I collagen synthesis and nephroangiosclerosis during nitric oxide synthase inhibition in rats. Circulation 1999;99:2185–2191.

24 Dussaule JC, Tharaux PL, Boffa JJ, Fakhouri F, Ardaillou R, Chatziantoniou C: Mechanisms mediating the renal profibrotic actions of vasoactive peptides in transgenic mice. J Am Soc Nephrol 2000;11(Suppl 16):S124–S128.
25 Berthold H, Munter K, Just A, Kirchheim HR, Ehmke H: Contribution of endothelin to renal vascular tone and autoregulation in the conscious dog. Am J Physiol 1999;276:F417–F424.
26 Qiu C, Samsell L, Baylis C: Actions of endogenous endothelin on glomerular hemodynamics in the rat. Am J Physiol 1995;269:R469–R473.
27 Tsuchiya K, Naruse M, Sanaka T, Naruse K, Zeng ZP, Nitta K, Demura H, Shizume K, Sugino N: Renal and hemodynamic effects of endothelin in anesthetized dogs. Am J Hypertens 1990;3:792–795.
28 Kon V, Yoshioka T, Fogo A, Ichikawa I: Glomerular actions of endothelin in vivo. J Clin Invest 1989;83:1762–1767.
29 Bakris GL, Fairbanks R, Traish AM: Arginine vasopressin stimulates human mesangial cell production of endothelin. J Clin Invest 1991;87:1158–1164.
30 Simonson MS, Dunn MJ: Endothelin-1 stimulates contraction of rat glomerular mesangial cells and potentiates beta-adrenergic-mediated cyclic adenosine monophosphate accumulation. J Clin Invest 1990;85:790–797.
31 Simonson MS, Wann S, Mene P, Dubyak GR, Kester M, Dunn MJ: Endothelin-1 activates the phosphoinositide cascade in cultured glomerular mesangial cells. J Cardiovasc Pharmacol 1989;13(Suppl 5):S80–S83, discussion S84.
32 Badr KF, Murray JJ, Breyer MD, Takahashi K, Inagami T, Harris RC: Mesangial cell, glomerular and renal vascular responses to endothelin in the rat kidney. Elucidation of signal transduction pathways. J Clin Invest 1989;83:336–342.
33 Mann JF, Green D, Jamerson K, Ruilope LM, Kuranoff SJ, Littke T, Viberti G: Avosentan for overt diabetic nephropathy. J Am Soc Nephrol 2010;21:527–535.
34 Wenzel RR, Littke T, Kuranoff S, Jurgens C, Bruck H, Ritz E, Philipp T, Mitchell A: Avosentan reduces albumin excretion in diabetics with macroalbuminuria. J Am Soc Nephrol 2009;20:655–664.
35 Kohan DE, Pritchett Y, Molitch M, Wen S, Garimella T, Audhya P, Andress DL: Addition of atrasentan to renin-angiotensin system blockade reduces albuminuria in diabetic nephropathy. J Am Soc Nephrol 2011;22:763–772.
36 Opocensky M, Kramer HJ, Backer A, Vernerova Z, Eis V, Cervenka L, Certikova Chabova V, Tesar V, Vaneckova I: Late-onset endothelin-A receptor blockade reduces podocyte injury in homozygous Ren-2 rats despite severe hypertension. Hypertension 2006;48:965–971.
37 Whitworth CE, Veniant MM, Firth JD, Cumming AD, Mullins JJ: Endothelin in the kidney in malignant phase hypertension. Hypertension 1995;26:925–931.
38 Sventek P, Li JS, Grove K, Deschepper CF, Schiffrin EL: Vascular structure and expression of endothelin-1 gene in L-NAME-treated spontaneously hypertensive rats. Hypertension 1996;27:49–55.
39 Petruzziello TN, Mawji IA, Khan M, Marsden PA: Verotoxin biology: molecular events in vascular endothelial injury. Kidney Int Suppl 2009:S17–S19.
40 Bitzan MM, Wang Y, Lin J, Marsden PA: Verotoxin and ricin have novel effects on preproendothelin-1 expression but fail to modify nitric oxide synthase (ecNOS) expression and NO production in vascular endothelium. J Clin Invest 1998;101:372–382.
41 Nangaku M, Yamada K, Gariepy CE, Miyata T, Inagi R, Kurokawa K, Yanagisawa M, Fujita T, Johnson RJ: ET(B) receptor protects the tubulointerstitium in experimental thrombotic microangiopathy. Kidney Int 2002;62:922–928.
42 Sabaa N, de Franceschi L, Bonnin P, Castier Y, Malpeli G, Debbabi H, Galaup A, Maier-Redelsperger M, Vandermeersch S, Scarpa A, Janin A, Levy B, Girot R, Beuzard Y, Leboeuf C, Henri A, Germain S, Dussaule JC, Tharaux PL: Endothelin receptor antagonism prevents hypoxia-induced mortality and morbidity in a mouse model of sickle-cell disease. J Clin Invest 2008;118:1924–1933.

43 Mouthon L, Mehrenberger M, Teixeira L, Fakhouri F, Berezne A, Guillevin L, Noel LH: Endothelin-1 expression in scleroderma renal crisis. Hum Pathol 2011;42:95–102.
44 Nakamura T, Ebihara I, Fukui M, Osada S, Tomino Y, Masaki T, Goto K, Furuichi Y, Koide H: Renal expression of mRNAs for endothelin-1, endothelin-3 and endothelin receptors in NZB/W F1 mice. Ren Physiol Biochem 1993;16:233–243.
45 Nakamura T, Ebihara I, Tomino Y, Koide H: Effect of a specific endothelin A receptor antagonist on murine lupus nephritis. Kidney Int 1995;47:481–489.
46 Yoshio T, Masuyama J, Mimori A, Takeda A, Minota S, Kano S: Endothelin-1 release from cultured endothelial cells induced by sera from patients with systemic lupus erythematosus. Ann Rheum Dis 1995;54:361–365.
47 Barton M: Reversal of proteinuric renal disease and the emerging role of endothelin. Nat Clin Pract Nephrol 2008;4:490–501.
48 Luft FC, Mervaala E, Muller DN, Gross V, Schmidt F, Park JK, Schmitz C, Lippoldt A, Breu V, Dechend R, Dragun D, Schneider W, Ganten D, Haller H: Hypertension-induced end-organ damage: a new transgenic approach to an old problem. Hypertension 1999;33:212–218.
49 Neuhofer W, Pittrow D: Endothelin receptor selectivity in chronic kidney disease: rationale and review of recent evidence. Eur J Clin Invest 2009;39(Suppl 2):50–67.
50 Kohan DE: Endothelin, hypertension and chronic kidney disease: new insights. Curr Opin Nephrol Hypertens 2010;19:134–139.
51 Barton M: Therapeutic potential of endothelin receptor antagonists for chronic proteinuric renal disease in humans. Biochim Biophys Acta 2010;1802:1203–1213.
52 Davenport AP, Nunez DJ, Brown MJ: Binding sites for 125I-labelled endothelin-1 in the kidneys: differential distribution in rat, pig and man demonstrated by using quantitative autoradiography. Clin Sci 1989;77:129–131.
53 Rebibou JM, He CJ, Delarue F, Peraldi MN, Adida C, Rondeau E, Sraer JD: Functional endothelin 1 receptors on human glomerular podocytes and mesangial cells. Nephrol Dial Transplant 1992;7:288–292.
54 Lopez-Ongil S, Diez-Marques ML, Griera M, Rodriguez-Puyol M, Rodriguez-Puyol D: Crosstalk between mesangial and endothelial cells: angiotensin II down-regulates endothelin-converting enzyme 1. Cell Physiol Biochem 2005;15:135–144.
55 Fogo AB: Animal models of FSGS: lessons for pathogenesis and treatment. Semin Nephrol 2003;23:161–171.
56 Diamond JR, Karnovsky MJ: Focal and segmental glomerulosclerosis following a single intravenous dose of puromycin aminonucleoside. Am J Pathol 1986;122:481–487.
57 Kasinath BS, Fried TA, Davalath S, Marsden PA: Glomerular epithelial cells synthesize endothelin peptides. Am J Pathol 1992;141:279–283.
58 Cybulsky AV, Stewart DJ, Cybulsky MI: Glomerular epithelial cells produce endothelin-1. J Am Soc Nephrol 1993;3:1398–1404.
59 Gagliardini E, Corna D, Zoja C, Sangalli F, Carrara F, Rossi M, Conti S, Rottoli D, Longaretti L, Remuzzi A, Remuzzi G, Benigni A: Unlike each drug alone, lisinopril if combined with avosentan promotes regression of renal lesions in experimental diabetes. Am J Physiol Renal Physiol 2009;297:F1448–F1456.
60 Megyesi J, Price PM, Tamayo E, Safirstein RL: The lack of a functional p21(WAF1/CIP1) gene ameliorates progression to chronic renal failure. Proc Natl Acad Sci USA 1999;96:10830–10835.
61 Di Cunto F, Topley G, Calautti E, Hsiao J, Ong L, Seth PK, Dotto GP: Inhibitory function of p21Cip1/WAF1 in differentiation of primary mouse keratinocytes independent of cell cycle control. Science 1998;280:1069–1072.
62 Dijkman H, Smeets B, van der Laak J, Steenbergen E, Wetzels J: The parietal epithelial cell is crucially involved in human idiopathic focal segmental glomerulosclerosis. Kidney Int 2005;68:1562–1572.
63 Appel D, Kershaw DB, Smeets B, Yuan G, Fuss A, Frye B, Elger M, Kriz W, Floege J, Moeller MJ: Recruitment of Podocytes from Glomerular Parietal Epithelial Cells. J Am Soc Nephrol 2009;20:333–343.

64 Smeets B, Angelotti ML, Rizzo P, Dijkman H, Lazzeri E, Mooren F, Ballerini L, Parente E, Sagrinati C, Mazzinghi B, Ronconi E, Becherucci F, Benigni A, Steenbergen E, Lasagni L, Remuzzi G, Wetzels J, Romagnani P: Renal progenitor cells contribute to hyperplastic lesions of podocytopathies and crescentic glomerulonephritis. J Am Soc Nephrol 2009;20:2593–2603.
65 Yoshimura A, Iwasaki S, Inui K, Ideura T, Koshikawa S, Yanagisawa M, Masaki T: Endothelin-1 and endothelin B type receptor are induced in mesangial proliferative nephritis in the rat. Kidney Int 1995;48:1290–1297.
66 Gomez-Garre D, Largo R, Liu XH, Gutierrez S, Lopez-Armada MJ, Palacios I, Egido J: An orally active ET_A/ET_B receptor antagonist ameliorates proteinuria and glomerular lesions in rats with proliferative nephritis. Kidney Int 1996;50:962–972.
67 Nicolaidou P, Georgouli H, Matsinos Y, Psychou F, Messaritaki A, Gourgiotis D, Zeis P: Endothelin-1 in children with acute poststreptococcal glomerulonephritis and hypertension. Pediatr Int 2003;45:35–38.
68 Kisanuki YY, Emoto N, Ohuchi T, Widyantoro B, Yagi K, Nakayama K, Kedzierski RM, Hammer RE, Yanagisawa H, Williams SC, Richardson JA, Suzuki T, Yanagisawa M: Low blood pressure in endothelial cell-specific endothelin 1 knockout mice. Hypertension 2010;56:121–128.
69 Herman WH, Emancipator SN, Rhoten RL, Simonson MS: Vascular and glomerular expression of endothelin-1 in normal human kidney. Am J Physiol 1998;275:F8–F17.
70 Frank K, Zeier M, Gross ML, Waldherr R, Ritz E, Amann K: Comprehensive immunohistological analysis of the endothelin system in human kidney grafts. Nephrol Dial Transplant 2006;21:1365–1372.
71 Orth SR, Odoni G, Amann K, Strzelczyk P, Raschack M, Ritz E: The ET(A) receptor blocker LU 135252 prevents chronic transplant nephropathy in the 'Fisher to Lewis' model. J Am Soc Nephrol 1999;10:387–391.
72 Lehrke I, Waldherr R, Ritz E, Wagner J: Renal endothelin-1 and endothelin receptor type B expression in glomerular diseases with proteinuria. J Am Soc Nephrol 2001;12:2321–2329.
73 Rastaldi MP, Tunesi S, Ferrario F, Indaco A, Zou H, Napodano P, D'Amico G: Transforming growth factor-beta, endothelin-1, and c-fos expression in necrotizing/crescentic IgA glomerulonephritis. Nephrol Dial Transplant 1998;13:1668–1674.
74 Nakamura T, Ebihara I, Shirato I, Fukui M, Tomino Y, Koide H: Endothelin-1 mRNA expression by peripheral blood monocytes in IgA nephropathy. Lancet 1993;342:1147–1148.
75 Chen HC, Guh JY, Chang JM, Lai YH: Differential effects of FMLP-activated neutrophils from patients with IgA nephropathy enhanced endothelin 1 production of glomerular mesangial cells. Nephron 2001;89:274–279.
76 Nakamura T, Ebihara I, Fukui M, Tomino Y, Koide H: Effect of a specific endothelin receptor A antagonist on glomerulonephritis of ddY mice with IgA nephropathy. Nephron 1996;72:454–460.
77 Nett PC, Teixeira MM, Candinas D, Barton M: Recent developments on endothelin antagonists as immunomodulatory drugs – from infection to transplantation medicine. Recent Pat Cardiovasc Drug Discov 2006;1:265–276.
78 Saleh MA, Boesen EI, Pollock JS, Savin VJ, Pollock DM: Endothelin receptor A-specific stimulation of glomerular inflammation and injury in a streptozotocin-induced rat model of diabetes. Diabetologia 2011;54:979–988.
79 Sasser JM, Sullivan JC, Hobbs JL, Yamamoto T, Pollock DM, Carmines PK, Pollock JS: Endothelin A receptor blockade reduces diabetic renal injury via an anti-inflammatory mechanism. J Am Soc Nephrol 2007;18:143–154.
80 Boesen EI, Sasser JM, Saleh MA, Potter WA, Woods M, Warner TD, Pollock JS, Pollock DM: Interleukin-1beta, but not interleukin-6, enhances renal and systemic endothelin production in vivo. Am J Physiol Renal Physiol 2008;295:F446–453.
81 Ebihara I, Nakamura T, Tomino Y, Koide H: Effect of a specific endothelin receptor A antagonist on murine lupus nephritis and IgA nephropathy. Contrib Nephrol 1996;118:155–163.

82 Rubin LJ, Badesch DB, Barst RJ, Galie N, Black CM, Keogh A, Pulido T, Frost A, Roux S, Leconte I, Landzberg M, Simonneau G: Bosentan therapy for pulmonary arterial hypertension. N Engl J Med 2002;346:896–903.
83 Cozzi F, Montisci R, Marotta H, Bobbo F, Durigon N, Ruscazio M, Sfriso P, Iliceto S, Todesco S: Bosentan therapy of pulmonary arterial hypertension in connective tissue diseases. Eur J Clin Invest 2006;36(Suppl 3):49–53.
84 Suzuki E, Kyozuka H, Nishida T, Kanno T, Ohira H: Systemic lupus erythematosus, complicated with refractory skin ulcers, treated successfully with bosentan. Mod Rheumatol 2009;19:447–451.
85 Julkunen H, Saijonmaa O, Gronhagen-Riska C, Teppo AM, Fyhrquist F: Raised plasma concentrations of endothelin-1 in systemic lupus erythematosus. Ann Rheum Dis 1991;50:526–527.
86 Karet FE, Davenport AP: Localization of endothelin peptides in human kidney. Kidney Int 1996;49:382–387.
87 Dhaun N, Lilitkarntakul P, Macintyre IM, Muilwijk E, Johnston NR, Kluth DC, Webb DJ, Goddard J: Urinary endothelin-1 in chronic kidney disease and as a marker of disease activity in lupus nephritis. Am J Physiol Renal Physiol 2009;296:F1477–1483.
88 Skolovsky M, Galron R, Kloog Y, Bdolah A, Indig FE, Blumberg S, Fleminger G: Endothelins are more sensitive than sarafotoxins to neutral endopeptidase: possible physiological significance. Proc Natl Acad Sci USA 1990;87:4702–4706.
89 Edwards RM, Pullen M, Nambi P: Distribution of neutral endopeptidase activity along the rat and rabbit nephron. Pharmacology 1999;59:45–50.
90 Janas J, Sitkiewicz D, Pulawska MF, Warnawin K, Janas RM: Purification of endothelin-1-inactivating peptidase from the rat kidney. J Hypertens 1994;12:375–382.
91 Chen CL, Fang HC, Chou KJ, Lee JC, Lee PT, Chung HM, Wang JS: Increased endothelin 1 expression in adult-onset minimal change nephropathy with acute renal failure. Am J Kidney Dis 2005;45:818–825.
92 Orisio S, Benigni A, Bruzzi I, Corna D, Perico N, Zoja C, Benatti L, Remuzzi G: Renal endothelin gene expression is increased in remnant kidney and correlates with disease progression. Kidney Int 1993;43:354–358.
93 Benigni A, Zoja C, Corna D, Orisio S, Longaretti L, Bertani T, Remuzzi G: A specific endothelin subtype A receptor antagonist protects against injury in renal disease progression. Kidney Int 1993;44:440–444.
94 Nakamura T, Fukui M, Ebihara I, Osada S, Takahashi T, Tomino Y, Koide H: Effects of a low-protein diet on glomerular endothelin family gene expression in experimental focal glomerular sclerosis. Clin Sci (Lond) 1995;88:29–37.
95 Terzi F, Henrion D, Colucci-Guyon E, Federici P, Babinet C, Levy BI, Briand P, Friedlander G: Reduction of renal mass is lethal in mice lacking vimentin. Role of endothelin-nitric oxide imbalance. J Clin Invest 1997;100:1520–1528.
96 Hocher B, Lun A, Priem F, Neumayer HH, Raschack M: Renal endothelin system in diabetes: comparison of angiotensin-converting enzyme inhibition and endothelin-A antagonism. J Cardiovasc Pharmacol 1998;31(Suppl 1):S492–S495.
97 Ding SS, Qiu C, Hess P, Xi JF, Zheng N, Clozel M: Chronic endothelin receptor blockade prevents both early hyperfiltration and late overt diabetic nephropathy in the rat. J Cardiovasc Pharmacol 2003;42:48–54.
98 Wolkart G, Stessel H, Saad Z, Kirchengast M, Brunner F: Cardioprotective effects of atrasentan, an endothelin-A receptor antagonist, but not of nitric oxide in diabetic mice with myocyte-specific overexpression of endothelial nitric oxide synthase. Br J Pharmacol 2006;148:671–681.
99 Barton M, Mullins JJ, Bailey MA, Kretzler M: Role of endothelin receptors for renal protection and survival in hypertension: waiting for clinical trials. Hypertension 2006;48:834–837.
100 Ohta K, Hirata Y, Shichiri M, Kanno K, Emori T, Tomita K, Marumo F: Urinary excretion of endothelin-1 in normal subjects and patients with renal disease. Kidney Int 1991;39:307–311.

101 Tharaux PL, Hagege I, Placier S, Vayssairat M, Kanfer A, Girot R, Dussaule JC: Urinary endothelin-1 as a marker of renal damage in sickle cell disease. Nephrol Dial Transplant 2005;20:2408–2413.
102 Genovese G, Friedman DJ, Ross MD, Lecordier L, Uzureau P, Freedman BI, Bowden DW, Langefeld CD, Oleksyk TK, Uscinski Knob AL, Bernhardy AJ, Hicks PJ, Nelson GW, Vanhollebeke B, Winkler CA, Kopp JB, Pays E, Pollak MR: Association of trypanolytic ApoL1 variants with kidney disease in African Americans. Science 2010;329:841–845.
103 Dhaun N, Macintyre IM, Melville V, Lilitkarntakul P, Johnston NR, Goddard J, Webb DJ: Blood pressure-independent reduction in proteinuria and arterial stiffness after acute endothelin-A receptor antagonism in chronic kidney disease. Hypertension 2009;54:113–119.
104 Dhaun N, Macintyre IM, Melville V, Lilitkarntakul P, Johnston NR, Goddard J, Webb DJ: Effects of endothelin receptor antagonism relate to the degree of renin-angiotensin system blockade in chronic proteinuric kidney disease. Hypertension 2009;54:e19–e20.
105 Deng LY, Schiffrin EL: Endothelin-1 gene expression in blood vessels and kidney of spontaneously hypertensive rats (SHR), L-NAME-treated SHR, and renovascular hypertensive rats. J Cardiovasc Pharmacol 1998;31(Suppl 1):S380–383.

Pierre-Louis Tharaux, MD, PhD
Inserm U.970 Paris Centre de recherche Cardiovasculaire – PARCC
Lab 219, 56 rue Leblanc
FR–75015 Paris (France)
Tel. +33 1 53 98 8051
E-Mail pierre-louis.tharaux@inserm.fr

Barton M, Kohan DE (eds): Endothelin in Renal Physiology and Disease.
Contrib Nephrol. Basel, Karger, 2011, vol 172, pp 139–148

Endothelin in Diabetic Renal Disease

Kerstin Benz · Kerstin Amann

Departments for Pediatrics and Nephropathology, University of Erlangen-Nürnberg, Erlangen, Germany

Abstract

Today diabetic nephropathy is the leading course of end-stage renal disease; the incidence and prevalence of diabetic renal disease is still continuing to increase, particularly in the Western world. Despite improvements in diagnosis and treatment of diabetic nephropathy, only partial renal protection is reached with the current standard therapy regiments, including angiotensin-converting enzyme inhibitors and/or angiotensin receptor blockers. Against this background, it is of particular importance to learn more about the pathogenesis of diabetic nephropathy and to find new therapeutic approaches which alone or in combination with standard therapy have the potential to prevent or delay the progression of diabetic nephropathy, thus improving kidney and patient survival. Among others, endothelin (ET) receptor blockers have emerged as a potential therapeutic option that operates on the basis of physiological and pathophysiological effects of endothelin. Of note, the ET system was shown to be involved in hypertension and kidney disease, particularly proteinuric nephropathies, and there is good experimental data indicating a specific role of ET in the pathogenesis and progression of diabetic nephropathy. ET receptor blockers have been shown to be nephroprotective in animal models of type 1 and type 2 diabetes mellitus with the effects partly independent of blood pressure lowering. In patients with hypertension and diabetic nephropathies, the data is controversial and depends on the stage of the disease and the drug used. It was only recently that a large international clinical study (ASCEND) provided evidence for beneficial effects of ET antagonist treatment, i.e. reduction in proteinuria. Due to the premature termination of the study, however, hard endpoints like death could no longer be assessed. Another very recent randomized, double-blind, placebo-controlled trial of subjects with diabetic nephropathy also provided evidence for a specific antiproteinuric effect of the ET receptor antagonist atrasentan on top of an already existing blockade of the renin-angiotensin system. Thus, it appears currently of great scientific and clinical interest to shed some light on the role of the ET system and its blockade in diabetic nephropathy.

Endothelin and the Endothelin System

The Endothelin System in the Kidney

Endothelins (ETs) were first identified in 1988 as a family of powerful vasoconstrictors and growth hormones with ET-1 being the most potent family member [1]. In the kidney, ET-1 exerts a number of physiological functions including control of water and sodium (Na^+) homeostasis. Very similar to the renin-angiotensin system (RAS), the ET system is complex, comprising a converting enzyme and two functionally active receptors, i.e. the ET_A receptor (ET_A-R) and the ET_B (ET_B-R). The nonselective ET_A-R primarily mediates vasoconstriction, thereby playing an important role in the pathogenesis of hypertension, endothelial dysfunction, insulin resistance, inflammation, and fibrosis. The peptide-selective ET_B-R is mainly expressed in vascular endothelial cells, thus mediating vasodilatation via release of nitric oxide and dilatory prostanoids, and finally inhibiting cell proliferation [2].

From a kidney perspective, both receptors are equally important since both are present in the kidney, albeit with different localization and function: the ET_A-R is localized to renal vessels whereas the ET_B-R is present mainly in the medulla, but also in glomerular cells, i.e. mesangial cells (MCs), endothelial cells, and podocytes [3, 4]. Of note, there are marked species differences between man and rat with respect to ET-1 expression and the presence and distribution of its receptors in the kidney [5, 6]. Under basal conditions, more prominent ET-1 expression and a preponderance of the ET_A-R is observed in the rat as compared to the human kidney. In the rat kidney, both receptors were found in interlobular arteries as well as afferent and efferent arterioles, but only ET_A-R was present in interlobar and arcuate arteries. ET_B-R immunoreactivity was sparse on endothelial cells of renal arteries, whereas strong labeling of peritubular and glomerular capillaries as well as vasa recta endothelium was seen. ET_A-R was evident on glomerular MCs and pericytes of descending vasa recta bundles. In the renal tubular system, ET_B-R were located in epithelial cells of proximal tubules and inner medullary collecting ducts, whereas ET(A) receptors were found in distal tubules and cortical collecting ducts [7]. In normal human kidney, ET-1 and its receptors are confined mainly to vascular tissue and only to a lesser extent to glomerular structures [5].

Tubular and Glomerular Effects of Endothelin-1

ET-1 has specific growth promoting, proinflammatory, and profibrotic roles in the kidney apart from its vasoconstrictive effects [8–10]. In the collecting duct, the ET system is involved in the physiologic regulation of renal Na^+, water, and acid excretion. ET_B-R activation on principal tubular cells causes inhibition of Na^+ transport. Activation of the ET_B-R in tubular cells also causes inhibition of water transport. ET_B-R on tubular intercalated cells augments acid secretion, possibly through nitric oxide-dependent mechanisms. The role of ET_A-R in the collecting duct remains unclear; however, recent evidence suggests natriuretic and diuretic effects [11]. ET-1 also inhibits the epithelial Na^+ channel which is responsible for Na^+ reabsorption in the cortical collecting duct by interacting with the 14–3/Nedd4–2 pathway [12]. This of course could represent an important link between renal ET expression, urinary ET excretion, and the pathogenesis of hypertension. Thus, deranged ET-1 production or impaired renal clearance of ET-1 in the setting of kidney disease may cause inappropriate Na^+ and water retention, also contributing to the development and maintenance of hypertension.

In the glomerulus, specific effects of ET-1 on MCs and podocytes by a rapid and transient intracellular [Ca^{2+}] have been shown [3, 13, 14]. MCs are capable of releasing ET-1 in response to a variety of factors, particularly during glomerular injury like hypertension, or an increase in glucose concentration as in diabetes mellitus. MCs also express ET receptors which upon activation can induce MC hypertrophy, MC proliferation, MC contraction, and extracellular matrix production. In podocytes, upregulation of ET-1 gene expression was also described in response to various injuries [13, 15, 16].

In summary, ET-1 and its receptors are involved in the physiological control of Na homeostasis in the kidney and thus in local and systemic blood pressure control. In addition, increased expression of ET-1 with subsequent activation of the two ET receptors in renal cells leads to a complex signaling cascade resulting in MC hypertrophy, proliferation, contraction, extracellular matrix accumulation, podocyte dysfunction, and activation of proinflammatory and profibrotic pathways which may favor the onset and progression of renal disease such as diabetic nephropathy.

Diabetic Nephropathy

Involvement of the kidney, i.e. diabetic nephropathy, is a major complication of diabetes mellitus and certainly one of the most devastating [17]. Renal dysfunction develops in about one third of diabetic patients yielding an estimated 150 million patients worldwide, with the number expected to increase to 300 million in the year 2025 [18]. Renal involvement in the two major types of diabetes mellitus, i.e. type 1 or insulin-dependent diabetes mellitus and type 2 or non-insulin-dependent diabetes mellitus, is morphologically similar, but occurs with different incidence and follows a different time course. About 15–20% of type 1 diabetic patients and 30–40% of type 2 diabetic patients will eventually develop end-stage renal failure.

The natural history of diabetic nephropathy is characterized by 5 stages (see below) [19]. Hypertension, hyperglycemia, dyslipidemia, obesity, and genetic factors are known risk factors for diabetic nephropathy [20]. On the other hand, hypertension (due to loss of functional renal tissue or atherosclerosis of the renal artery) and dyslipidemia (caused by compensatory increases in hepatic protein production due to renal protein loss) are also consequences of diabetic nephropathy. Clinical trials have shown that strict control of hyperglycemia and hypertension can slow the progression of diabetic nephropathy. These trials have also shown that insulin resistance correlates with the onset and severity of albuminuria. Moreover, there is recent evidence that altered insulin signaling in glomerular cells and podocytes in particular may contribute to the development of diabetic nephropathy [21, 22].

Endothelin in Diabetes Mellitus and Diabetic Nephropathy

Shortly after the ET system was discovered, speculation began that interference with the ET system either alone or in combination with blockade of the RAS might be a new therapeutic regiment in hypertension and renal diseases where systemic ET-1 levels and urinary ET-1 excretion were elevated [23, 24]. In patients with diabetes mellitus, this is explained by hyperglycemia-stimulated ET-1 production. Insulin was also found to increase renal ET expression, which was seen as further evidence of a pathophysiological role of the ET system in diabetic nephropathy [25, 26]. This was even more probable since a role of ET in metabolic control had already been shown [27]. In particular, an interplay between insulin, insulin resistance, and ET has been documented and claimed to be pathophysiologically relevant in diabetic complications [28–31].

Pathophysiological Roles of Endothelin-1 in the 5 Stages of Diabetic Nephropathy
In stages 1 and 2, i.e. the prediabetic stages which are characterized by renal hypertrophy and hyperfiltration, growth promoting and vasoactive effects of ET-1 may be operative. In experimental models of early type 1 diabetic nephropathy in rats and rabbits [22, 32], increased ET(A) receptor-binding sites were found in the renal cortex and medulla. The authors speculated that in these early stages, ET-1 may act in a paracrine fashion on the glomeruli where it may regulate glomerular filtration rate through an autocrine-paracrine pathway on MCs and podocytes [3]. This may then lead to stage 3 of diabetic nephropathy, i.e. incipient diabetic nephropathy, which is functionally characterized by microalbuminuria and hypertension. Here activation of ET receptors on MCs and podocytes may promote morphological alterations, i.e. glomerular basement thickening, MC hyperplasia, mesangial matrix expansion (fig. 1a), and podocyte enlargement with loss of foot processes, as well as functional changes such as increased permeability of the glomerular filtration barrier to albumin and elevated proinflammatory markers, possibly via nephrin loss in the glomerular slit membrane

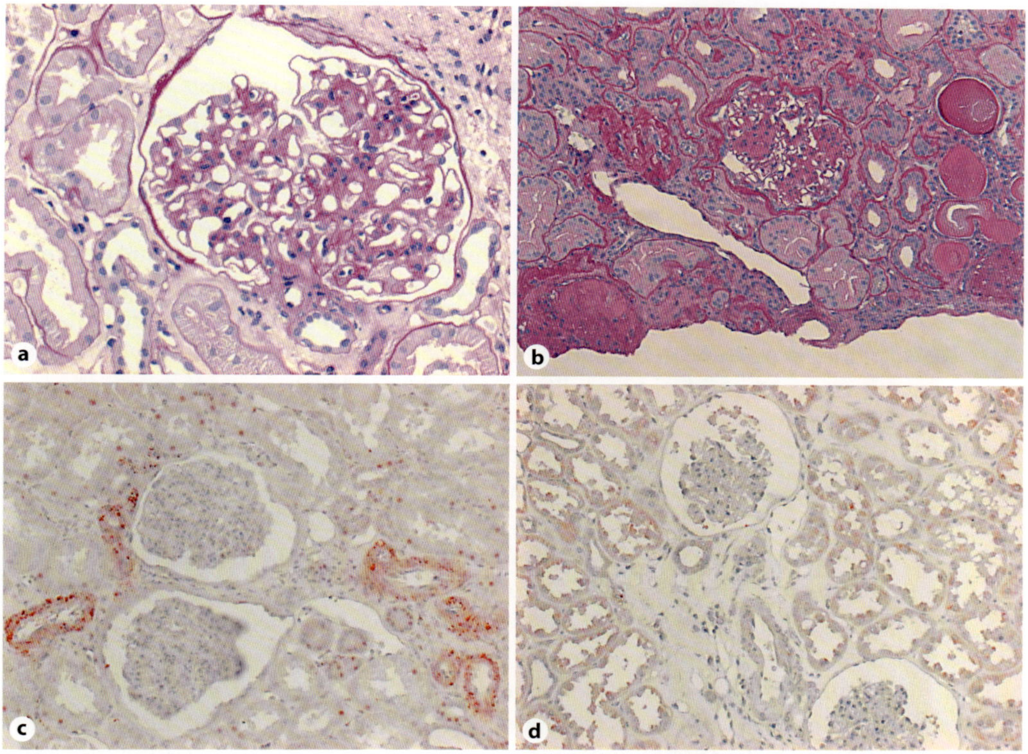

Fig. 1. Typical histological alterations in early (**a**) and late (**b**) diabetic nephropathy with either diffuse (**a**) or nodular type (**b**) glomerulosclerosis and advanced tubulointerstitial damage (**b**). PAS stain, magnifications ×40 (**a**) and ×20 (**b**). Increased expression of ET-1 in advanced diabetic nephropathy (**c**) compared to normal kidney tissue (**d**). Immunohistochemistry, magnification ×20.

[10]. Microalbuminuria or proteinuria per se then induces a specific genetic program in tubular epithelial cells which orchestrate tubular atrophy, interstitial inflammation, and fibrosis [33]. Among these programs, increased expression of ET-1 has been shown [34] which may then favor progression to stage 4, i.e. overt diabetic nephropathy (fig. 1b), with proteinuria and/or the nephrotic syndrome and a marked decrease in glomerular filtration rate. In stage 5, i.e. end-stage diabetic nephropathy, ET-1 may further promote glomerular and tubulointerstitial fibrosis by stimulation of mesangial matrix expansion and remodeling [35] as well as collagen production of interstitial fibroblasts [19, 36, 37].

Thus, the above data support the notion that in diabetic nephropathy, increased renal ET-1 expression via activation of its receptors directly increases glomerular permeability to albumin, renal inflammation, and fibrosis.

Endothelin A Receptor Blockade and Diabetic Nephropathy

Early on, the particular properties of ET receptor antagonists to reduce proteinuria and to delay progression of renal damage, even independent of blood pressure lowering, have been recognized [10, 38, 39].

Experimental Studies

Marked beneficial effects of ET receptor blockade were shown in nephropathy of several experimental models of type 1 and type 2 diabetes mellitus [32, 40–42]. Although all animal models of diabetic nephropathy, in particular the model of streptozotocin-induced diabetes, have their limitations with respect to compatibility to human diabetic nephropathy, these studies have demonstrated increased ET-1 expression and positive effects of ET receptor blockade alone or in combination with angiotensin-converting enzyme inhibition on progression of renal damage as well as beneficial effects on development of diabetic arteriosclerosis. Against this background it is important that also in animal models of true autoimmune diabetes mellitus, i.e. the nonobese diabetic mouse, regulation of renal ET-converting enzymes and preproET-1 expression by diabetes mellitus and ET_A-R treatment [43] has been shown. In addition and potentially of even more clinical relevance, retardation of autoimmune diabetes onset by ET_A-R treatment could be documented [44]. Of further note, in the streptozotocin-induced model of early diabetes mellitus type 1 [32], only ET_A-R blockade, but not angiotensin-converting enzyme inhibition, completely restored the number of glomeruli per kidney to the normal expected number and preserved the number of podocytes per glomerulus. This suggests that in this diabetic animal model, ET blockade has specific glomeruloprotective effects [15].

Human Data

Based on the mostly beneficial results of the above experimental studies, the renal effects of ET receptor blockers were also tested in patients [45]. In hypertensive patients with chronic renal failure, ET_A-R blockade was shown to lower blood pressure and increase renal blood flow [46]. In a randomized, placebo-controlled, double-blind, parallel-design, dosage-range study in diabetic patients with macroalbuminuria, the ET receptor blocker avosentan in addition to standard therapy with angiotensin-converting enzyme inhibition/angiotensin receptor blockade decreased the mean relative urinary albumin excretion rate compared to placebo [47]. In the even more recent long-term study in overt diabetic nephropathy by Mann et al. [48], the effects of the same drug, avosentan, was investigated on progression of overt diabetic nephropathy in a multicenter, multinational, double-blind, placebo-controlled trial in 1,392 patients with type 2 diabetes. As in the previous study [47], oral avosentan or placebo was given in addition to continued angiotensin-converting enzyme inhibition and/or angiotensin receptor blockade. A composite primary outcome was the time to doubling of serum creatinine, end-stage renal disease, or death. Secondary outcomes included changes in urinary albumin-to-creatinine ratio and cardiovascular outcomes. Of note, the trial was terminated prematurely after a follow-up of 4 months because of an excess of cardiovascular events in avosentan-treated patients. A difference in the frequency of the primary outcome between the groups was not detected. However, avosentan significantly reduced the urinary albumin-to-creatinine ratio, but this was paralleled by significant fluid overload in congestive heart failure. This result limited the enthusiasm about the effects of ET receptor blockade in overt diabetic nephropathy [15, 49, 50]. Another very recently published study [51] examined the effects of different doses (0.25, 0.75, or 1.75 mg daily) of the ET_A-R atrasentan on albuminuria in a randomized, double-blind, placebo-controlled trial of 89 subjects with diabetic nephropathy already receiving stable doses of RAS inhibitors. In the 8-week study, the authors found a significant reduction compared to placebo in the urinary albumin-to-creatinine ratio only in the 0.75-mg and 1.75-mg atrasentan groups. Moreover, a significantly higher number of patients in the 0.75-mg atrasentan-treated group compared to the placebo group achieved ≥40% reduction in the urinary albumin-to-creatinine from baseline. Peripheral edema occurred in 9% of subjects receiving placebo and in 14, 18, and 46% of those receiving 0.25, 0.5, and 1.75 mg atrasentan, respectively ($p = 0.007$ for 1.75 mg vs. placebo). Therefore, at the doses tested in this study, atrasentan was found to be generally safe and effective in reducing residual albuminuria in patients with type 2 diabetic nephropathy.

Conclusion

In summary, there is good experimental but limited clinical evidence for a major pathophysiological role of ET in the development and progression of diabetic nephropathy. In contrast to ample experimental evidence in diverse models of either type 1 or type 2 diabetes mellitus where not only protection of the kidney but also reversal of already existing lesions of diabetic nephropathy has been reported [13, 15], the data is not that clear in patients with different stages of diabetic nephropathy. Available clinical data indicate that ET_A-R blockers are safe in low doses and may be used in early stages of proteinuric kidney diseases, i.e. early stages of diabetic nephropathy, in which they may provide even additional benefit to an already existing RAS blockade, but probably not in later stages of chronic kidney disease (stage 3 and 4). Care should also be taken if clear contraindications (i.e. pre-existing heart disease) for the use of ET_A-R blockers are present.

References

1 Yanagisawa M, Kurihara H, Kimura S, Tomobe Y, Kobayashi M, Mitsui Y, Yazaki Y, Goto K, Masaki T: A novel potent vasoconstrictor peptide produced by vascular endothelial cells. Nature 1988;332:411–415.

2 Barton M, Nett P, Amann K, Teixeira M: Anti-Inflammatory Effects of Endothelin Receptor Antagonists and their Importance for Treating Human Disease. Front Cardiovasc Drug Discov 2010;1:236–258.

3 Rebibou JM, He CJ, Delarue F, Peraldi MN, Adida C, Rondeau E, Sraer JD: Functional endothelin 1 receptors on human glomerular podocytes and mesangial cells. Nephrol Dial Transplant 1992;7:288–292.

4 Uchida K, Uchida S, Nitta K, Yumura W, Nihei H: Regulated expression of endothelin converting enzymes in glomerular endothelial cells. J Am Soc Nephrol 1997;8:580–585.

5 Karet FE, Davenport AP: Localization of endothelin peptides in human kidney. Kidney Int 1996;49:382–387.

6 Karet FE: Endothelin peptides and receptors in human kidney. Clin Sci (Lond) 1996;91:267–273.

7 Wendel M, Knels L, Kummer W, Koch T: Distribution of endothelin receptor subtypes ET_A and ET_B in the rat kidney. J Histochem Cytochem 2006;54:1193–1203.

8 Schildroth J, Rettig-Zimmermann J, Kalk P, Steege A, Fahling M, Sendeski M, Paliege A, Lai EY, Bachmann S, Persson PB, Hocher B, Patzak A: Endothelin type A and B receptors in the control of afferent and efferent arterioles in mice. Nephrol Dial Transplant 2011;26:779–789.

9 Kohan DE: Endothelin, hypertension and chronic kidney disease: new insights. Curr Opin Nephrol Hypertens 2010;19:134–139.

10 Saleh MA, Boesen EI, Pollock JS, Savin VJ, Pollock DM: Endothelin-1 increases glomerular permeability and inflammation independent of blood pressure in the rat. Hypertension 2010;56:942–949.

11 Kohan DE: Biology of endothelin receptors in the collecting duct. Kidney Int 2009;76:481–486.

12 Pavlov TS, Chahdi A, Ilatovskaya DV, Levchenko V, Vandewalle A, Pochynyuk O, Sorokin A, Staruschenko A: Endothelin-1 inhibits the epithelial Na+ channel through betaPix/14-3-3/Nedd4-2. J Am Soc Nephrol 2010;21:833–843.

13 Ortmann J, Amann K, Brandes RP, Kretzler M, Munter K, Parekh N, Traupe T, Lange M, Lattmann T, Barton M: Role of podocytes for reversal of glomerulosclerosis and proteinuria in the aging kidney after endothelin inhibition. Hypertension 2004;44:974–981.

14 Sorokin A, Kohan DE: Physiology and pathology of endothelin-1 in renal mesangium. Am J Physiol Renal Physiol 2003;285:F579–F589.

15 Barton M: Reversal of proteinuric renal disease and the emerging role of endothelin. Nat Clin Pract Nephrol 2008;4:490–501.

16 Morigi M, Buelli S, Angioletti S, Zanchi C, Longaretti L, Zoja C, Galbusera M, Gastoldi S, Mundel P, Remuzzi G, Benigni A: In response to protein load podocytes reorganize cytoskeleton and modulate endothelin-1 gene: implication for permselective dysfunction of chronic nephropathies. Am J Pathol 2005;166:1309–1320.

17 Groop PH, Thomas MC, Moran JL, Waden J, Thorn LM, Makinen VP, Rosengard-Barlund M, Saraheimo M, Hietala K, Heikkila O, Forsblom C: The presence and severity of chronic kidney disease predicts all-cause mortality in type 1 diabetes. Diabetes 2009;58:1651–1658.

18 Tang SC: Diabetic nephropathy: a global and growing threat. Hong Kong Med J 2010;16:244–245.

19 Osterby R, Hartmann A, Nyengaard JR, Bangstad HJ: Development of renal structural lesions in type-1 diabetic patients with microalbuminuria. Observations by light microscopy in 8-year follow-up biopsies. Virchows Arch 2002;440:94–101.

20 Dronavalli S, Duka I, Bakris GL: The pathogenesis of diabetic nephropathy. Nat Clin Pract Endocrinol Metab 2008;4:444–452.

21 Jauregui A, Mintz DH, Mundel P, Fornoni A: Role of altered insulin signaling pathways in the pathogenesis of podocyte malfunction and microalbuminuria. Curr Opin Nephrol Hypertens 2009;18:539–545.

22 Khan MA, Dashwood MR, Mumtaz FH, Thompson CS, Mikhailidis DP, Morgan RJ: Upregulation of endothelin A receptor sites in the rabbit diabetic kidney: potential relevance to the early pathogenesis of diabetic nephropathy. Nephron 1999;83:261–267.
23 Bruzzi I, Benigni A: Endothelin is a key modulator of progressive renal injury: experimental data and novel therapeutic strategies. Clin Exp Pharmacol Physiol 1996;23:349–353.
24 Zanatta CM, Gerchman F, Burttet L, Nabinger G, Jacques-Silva MC, Canani LH, Gross JL: Endothelin-1 levels and albuminuria in patients with type 2 diabetes mellitus. Diabetes Res Clin Pract 2008;80:299–304.
25 Hargrove GM, Dufresne J, Whiteside C, Muruve DA, Wong NC: Diabetes mellitus increases endothelin-1 gene transcription in rat kidney. Kidney Int 2000;58:1534–1545.
26 Sudano I, Hermann M, Ruschitzka FT: Endothelin-receptor antagonists in arterial hypertension: further indications? Curr Hypertens Rep 2007;9:59–65.
27 Ak G, Buyukberber S, Sevinc A, Turk HM, Ates M, Sari R, Savli H, Cigli A: The relation between plasma endothelin-1 levels and metabolic control, risk factors, treatment modalities, and diabetic microangiopathy in patients with type 2 diabetes mellitus. J Diabetes Complications 2001;15:150–157.
28 Sarafidis PA, Lasaridis AN: Insulin resistance and endothelin: another pathway for renal injury in patients with the cardiometabolic syndrome? J Cardiometab Syndr 2008;3:183–187.
29 Sarafidis PA, Bakris GL: Review: insulin and endothelin: an interplay contributing to hypertension development? J Clin Endocrinol Metab 2007;92:379–385.
30 Strawbridge AB, Elmendorf JS, Mather KJ: Interactions of endothelin and insulin: expanding parameters of insulin resistance. Curr Diabetes Rev 2006;2:317–327.
31 Lteif A, Vaishnava P, Baron AD, Mather KJ: Endothelin limits insulin action in obese/insulin-resistant humans. Diabetes 2007;56:728–734.
32 Gross ML, El-Shakmak A, Szabo A, Koch A, Kuhlmann A, Munter K, Ritz E, Amann K: ACE-inhibitors but not endothelin receptor blockers prevent podocyte loss in early diabetic nephropathy. Diabetologia 2003;46:856–68.
33 Bruzzi I, Remuzzi G, Benigni A: Endothelin: a mediator of renal disease progression. J Nephrol 1997;10:179–183.
34 Benigni A, Perico N, Remuzzi G: Research on renal endothelin in proteinuric nephropathies dictates novel strategies to prevent progression. Curr Opin Nephrol Hypertens 2001;10:1–6.
35 Kitamura A, Kagami S, Urushihara M, Kondo S, Yoshizumi M, Tamaki T, Kuroda Y: Endothelin-1 is a potent stimulator of alpha1beta1 integrin-mediated collagen matrix remodeling by rat mesangial cells. Biochem Biophys Res Commun 2002;299:555–561.
36 Hocher B, Thone-Reineke C, Rohmeiss P, Schmager F, Slowinski T, Burst V, Siegmund F, Quertermous T, Bauer C, Neumayer HH, Schleuning WD, Theuring F: Endothelin-1 transgenic mice develop glomerulosclerosis, interstitial fibrosis, and renal cysts but not hypertension. J Clin Invest 1997;99:1380–1389.
37 Osterby R, Tapia J, Nyberg G, Tencer J, Willner J, Rippe B, Torffvit O: Renal structures in type 2 diabetic patients with elevated albumin excretion rate. APMIS 2001;109:751–761.
38 Barton M: Therapeutic potential of endothelin receptor antagonists for chronic proteinuric renal disease in humans. Biochim Biophys Acta 2010;1802:1203–1213.
39 Dhaun N, Macintyre IM, Melville V, Lilitkarntakul P, Johnston NR, Goddard J, Webb DJ: Blood pressure-independent reduction in proteinuria and arterial stiffness after acute endothelin-a receptor antagonism in chronic kidney disease. Hypertension 2009;54:113–119.
40 Dhein S, Hochreuther S, Aus Dem Spring C, Bollig K, Hufnagel C, Raschack M: Long-term effects of the endothelin(A) receptor antagonist LU 135252 and the angiotensin-converting enzyme inhibitor trandolapril on diabetic angiopathy and nephropathy in a chronic type I diabetes mellitus rat model. J Pharmacol Exp Ther 2000;293:351–359.

41 Gross ML, Ritz E, Schoof A, Helmke B, Parkman A, Tulp O, Munter K, Amann K: Renal damage in the SHR/N-cp type 2 diabetes model: comparison of an angiotensin-converting enzyme inhibitor and endothelin receptor blocker. Lab Invest 2003;83:1267–1277.
42 Watson AM, Li J, Schumacher C, de Gasparo M, Feng B, Thomas MC, Allen TJ, Cooper ME, Jandeleit-Dahm KA: The endothelin receptor antagonist avosentan ameliorates nephropathy and atherosclerosis in diabetic apolipoprotein E knockout mice. Diabetologia 2010;53:192–203.
43 Ortmann J, Nett PC, Celeiro J, Hofmann-Lehmann R, Tornillo L, Terracciano LM, Barton M: Downregulation of renal endothelin-converting enzyme 2 expression in early autoimmune diabetes. Exp Biol Med (Maywood) 2006;231:1030–1033.
44 Ortmann J, Nett PC, Celeiro J, Traupe T, Tornillo L, Hofmann-Lehmann R, Haas E, Frank B, Terraciano LM, Barton M: Endothelin inhibition delays onset of hyperglycemia and associated vascular injury in type I diabetes: evidence for endothelin release by pancreatic islet beta-cells. Biochem Biophys Res Commun 2005;334:689–695.
45 Honing ML, Hijmering ML, Ballard DE, Yang YP, Padley RJ, Morrison PJ, Rabelink TJ: Selective ET(A) receptor antagonism with ABT-627 attenuates all renal effects of endothelin in humans. J Am Soc Nephrol 2000;11:1498–1504.
46 Goddard J, Johnston NR, Hand MF, Cumming AD, Rabelink TJ, Rankin AJ, Webb DJ: Endothelin-A receptor antagonism reduces blood pressure and increases renal blood flow in hypertensive patients with chronic renal failure: a comparison of selective and combined endothelin receptor blockade. Circulation 2004;109:1186–1193.
47 Wenzel RR, Littke T, Kuranoff S, Jurgens C, Bruck H, Ritz E, Philipp T, Mitchell A: Avosentan reduces albumin excretion in diabetics with macroalbuminuria. J Am Soc Nephrol 2009;20:655–664.
48 Mann JF, Green D, Jamerson K, Ruilope LM, Kuranoff SJ, Littke T, Viberti G: Avosentan for overt diabetic nephropathy. J Am Soc Nephrol 2010;21:527–535.
49 Ritz E, Wenzel R: Endothelin receptor antagonists in proteinuric renal disease: every rose has its thorn. J Am Soc Nephrol 2010;21:392–394.
50 Sarafidis PA, Lasaridis AN: Diabetic nephropathy: endothelin antagonism for diabetic nephropathy. Nat Rev Nephrol 2010;6:447–449.
51 Kohan DE, Pritchett Y, Molitch M, Wen S, Garimella T, Audhya P, Andress DL: Addition of atrasentan to renin-angiotensin system blockade reduces albuminuria in diabetic nephropathy. J Am Soc Nephrol 2011;22:763–772.

Kerstin Amann, MD
Department of Pathology, Nephropathology
Krankenhausstrasse 12
DE–91054 Erlangen (Germany)
Tel. +49 9131 8522291, E-Mail kerstin.amann@uk-erlangen.de

Barton M, Kohan DE (eds): Endothelin in Renal Physiology and Disease.
Contrib Nephrol. Basel, Karger, 2011, vol 172, pp 149–159

Endothelin, Nitric Oxide, and Reactive Oxygen Species in Diabetic Kidney Disease

Jennifer S. Pollock · David M. Pollock

Medical College of Georgia, Georgia Health Sciences University, Augusta, Ga., USA

Abstract

The mechanism(s) of the endothelin (ET) and reactive oxygen species pathways in conjunction with the nitric oxide (NO) pathway that promote and/or blunt the progression of diabetic kidney disease have been the focus of many laboratories' efforts to reveal new therapeutic targets. In both animal models and patients with diabetic nephropathy, pharmacological blockade of ET receptors results in a significant reduction. However, edema has been documented as a persistent side effect. It is unclear whether selective ET_A antagonists or nonselective $ET_{A/B}$ antagonists are preferred in diabetic conditions. We have proposed that ET_B activates the NO pathway to blunt diabetes-induced nephropathy such that ET_A selectivity should be more efficacious. The NO pathway in diabetes facilitates vascular dysfunction while in the renal tubular system, NO serves to blunt disease progression. NO synthase isoform activity is also critically regulated in diabetic kidney disease within the renal vascular and tubular systems through a complex interaction with reactive oxygen species. We will examine the complexities of the ET and NO pathways in diabetic kidney disease to propose novel mechanisms for future investigation.

Several laboratories are beginning to dissect the mechanism(s) of the endothelin (ET) and nitric oxide (NO) pathways that promote and/or blunt the progression of diabetic kidney disease. We have organized the first part of this chapter to describe our understanding of the ET pathway in diabetic kidney disease followed by a discussion focusing on the NO pathway and how it interacts with the ET system. We conclude with a perspective on possible future directions that could further elucidate novel mechanisms and targets for diabetic kidney disease.

Activation of the Endothelin Pathway in Diabetic Kidney Disease

Brownlee and coworkers [1] have proposed a 'unifying hypothesis' that mitochondrial derived superoxide ($O_2^{\bullet-}$) promotes production of reactive oxygen species (ROS) from the other major pathways and is the critical component in diabetes-induced oxidative stress. Glucose is a fuel for the mitochondrial respiratory chain and oxidative phosphorylation and/or cytosolic glycolysis activating cellular oxidases and ROS. Many investigators have postulated that local intracellular sources of oxidative stress are critical in the development of diabetic nephropathy. Asaba et al. [2] have recently reported that apocynin (antioxidant and/or NADPH oxidase inhibitor) reduces advanced oxidation protein product-induced perturbations in inflammatory measures, renal NADPH oxidase activity, and gp91phox expression in the diabetic kidney. Studies using a variety of cell types in culture have established that ROS, in particular H_2O_2 and $O_2^{\bullet-}$, are important stimulators of ET-1 production [3, 4]. ET-1 can act in an autocrine or paracrine fashion, and urinary ET-1 reflects renal ET-1 production [5].

Our group has published the results of an experiment showing urinary ET-1 maximally increased within 2 weeks of induction of diabetes in the rat and remained elevated throughout the course of the experiment to 10 weeks postinduction [6]. Urinary measures of oxidative stress were also increased in a similar time course, suggesting an interaction in these pathways [6]. We initiated a study to determine whether enhanced diabetes-induced oxidative stress stimulated renal ET-1 production. In vivo tempol (free radical scavenger) treatment of diabetic rats significantly blunted urinary ET-1 excretion as well as inner medullary mRNA ET-1 expression [7]. Tempol treatment also blocked ROS production in renal tissues from diabetic rats in comparison to nontreated diabetic rats. Thus, these in vivo data in concert with the in vitro data demonstrate that diabetes-induced ROS activate renal ET-1 production.

Several studies have also shown that ET_A receptors activate NADPH oxidase and other pathways leading to oxidative stress in the vasculature, mesangial cells, and renal tubules in a variety of nondiabetic renal diseases. ET_B receptors have been shown to also activate NADPH oxidase in neuronal cells and stimulate ROS production in models of hypertension. Sasser et al. [6] demonstrated that renal NADPH oxidase expression as well as plasma and urinary measures of oxidative stress were increased in the streptozotocin (STZ)-induced model of diabetes compared to the sham rat, although ET_A receptor antagonism neither ameliorated the oxidative stress nor the increased urinary ET-1 excretion. Moreover, ET_A receptor antagonism significantly reduced the diabetes-induced albuminuria [6], thus demonstrating that diabetes-induced ET_A receptor activation promotes renal injury but not via oxidative stress.

Aberrant inflammatory responses in the kidney, such as infiltration of macrophages and the associated pathways were discovered early in diabetes as well as in established diabetic nephropathy in both human and animal models, and subsequently linked to glomerular and tubular damage as well as renal fibrosis [8–11]. Although these reports support the concept that inflammatory processes promote and/or perpetuate the progression of diabetic nephropathy, the mechanisms involved in diabetes-induced renal inflammation are not well understood. Is renal inflammation during diabetes a response to hyperglycemia or to other metabolic factor(s) that could play a role in driving this process? We sought to address this question by hypothesizing that the diabetes-induced activation of the ET pathway stimulates inflammation. We observed that blockade of the ET_A pathway in the STZ-induced model of diabetes disrupts the infiltration of macrophages and T cells in the renal cortex and medulla, and suppresses accumulation of proinflammatory mediators, especially MCP-1 and TGF-β [6]. MCP-1 expression increases at the site of tissue injury and functions to direct macrophage recruitment to the site [12]. Mechanistically, MCP-1 binds to the inducible CCR2 to promote chemotaxis to a specific area [13]. CCR2 regulates monocyte and macrophage recruitment and is necessary for macrophage-dependent inflammatory responses [14]. While there are a number of studies showing that ET-1 increases the expression of inflammatory molecules in vitro, we have recently obtained in vivo data indicating that ET-1 increases plasma MCP-1 and infiltration of macrophages and T cells in the kidney in the absence of hypertension [15]. Therefore, these data indicate that diabetes-induced ROS activates renal ET production and ET_A receptor-mediated increased inflammation and albuminuria.

Activation of Nitric Oxide Pathway in Diabetic Kidney Disease

All three NO synthase (NOS) isoforms are expressed in the kidney under physiological and diabetic conditions. NOS1 (neuronal NOS or nNOS) was first described in the brain, and is expressed in the macula densa, renal tubules, arteriolar smooth muscle, and vasa recta. NOS2 (inducible NOS or iNOS) was initially found in inflammatory cells under conditions of cytokine induction. However, NOS2 is constitutively expressed in the medullary thick ascending limb (mTAL) and induced in the rat STZ model of diabetes [16]. NOS3 (endothelial NOS or eNOS) was first described in the endothelium of the aorta, and it is expressed in the endothelium of renal arteries and arterioles as well as renal tubules.

Most studies exploring the impact of diabetes on renal NOS activity have centered on the hypothesis that elevated NOS activity contributes to renal hyperfiltration or the eventual development of diabetic nephropathy with a focus only on the vasculature [17, 18]. Indeed, STZ-induced diabetes in NOS3 knockout mice have exacerbated renal injury [19], although it is unknown whether this is a consequence of the elevated blood pressure and/or insulin resistance. Shankar et al. [20] showed that NOS3 knockout mice are insulin resistant at the level of the liver and peripheral tissues, whereas the NOS1 knockout mice are insulin resistant only in peripheral tissues. It has been reported that insulin therapy reversed the renal functional and histological changes due to diabetes in the NOS knockout mice leading to the proposal that endothelial dysfunction is a surrogate marker for nephropathy [21]. As discussed below, this may be an overly simplistic view of NOS in diabetic kidney disease.

The diabetic kidney has high levels of $O_2^{\bullet -}$ leading to an overall reduced NO bioavailability. Strong evidence indicates that the reaction of $O_2^{\bullet -}$ with NO during diabetes has dual consequences: decreased NO bioavailability and generation of peroxynitrite ($ONOO^-$). Stadler et al. [22] recently reported that NOS2 is the major source of radical generation in the kidney of diabetic rats and implicated $ONOO^-$ as the reactive intermediate leading to lipid peroxidation observed in many models of diabetes as well as diabetic patients. Goligorsky's laboratory has shown that chronic ebselen treatment (antioxidant and $ONOO^-$ scavenger) ameliorated proteinuria, inflammation, and prevented 3-nitrotyrosine modification of proteins in renal tissue from Zucker diabetic rats [23]. 'NOS uncoupling' or NOS-dependent $O_2^{\bullet -}$ production represents a significant source of $O_2^{\bullet -}$ in vascular beds during hyperglycemia and is a major determinate in the endothelial dysfunction observed in diabetes [24–26]. It should be noted that in diabetes, endothelial $O_2^{\bullet -}$ production is NOS dependent [24, 25] while mTAL $O_2^{\bullet -}$ production is NOS independent [27]. Therefore, mechanisms regulating NOS function in the endothelium and mTAL during diabetes are vastly different. In particular, endothelial NOS uncoupling during diabetes has been attributed to the loss of the NOS substrate L-arginine via diabetes-induced arginase activity, and/or oxidation of the NOS cofactor tetrahydrobiopterin. Very little is known about the regulation of arginase and tetrahydrobiopterin in the renal tubules, especially in the mTAL during diabetes.

Recently, Palm et al. [28] found that NO regulates renal blood flow and glomerular filtration rate similarly in normal and diabetic kidneys. Yet, specific NOS1 inhibition increased oxygen consumption and transport efficiency only in the diabetic kidney, indicating a switch to NOS1 dependency specifically in the diabetic kidney for maintaining control of tissue oxygenation. Using the STZ model of diabetes, Palm et al. [28] proposed that NOS1 activity is critical as a 'brake' to the increase in transport efficiency in the diabetic kidney. Direct effects of NO on sodium transport are evident in several nephron segments [29, 30]. The nephron segment which has the strongest evidence of NO regulation of NaCl reabsorption is the mTAL [30], where NO decreases net NaCl reabsorption primarily via effects on apical transport mechanisms [31, 32]. While exogenous NO can inhibit Na transport in the normal mTAL, the extent to which chronic increases in NOS activity and endogenous NO production may reduce

Na transport by the mTAL is not known. Furthermore, little is known about the status of the NO system in the renal tubules during diabetes, especially those in the medulla. Our laboratory has recently begun to explore these mechanisms. Indeed, we have documented increases in NOS activity and a 2.5-fold increase in mTAL-derived NO production in the STZ model of type 1 diabetes [16]. Our recent studies revealed increased NOS1 and NOS2 activity in mTALs from diabetic rats with no change in expression of either of these isoforms, indicating possible post-translational enzymatic activity. NOS3 expression and activity were unchanged in the mTAL under diabetic conditions [16]. These data in concert with the findings from Palm's laboratory support the importance of the NOS1 isoform in maintaining renal tubular function during diabetes. However, Anderson's laboratory observed that long-term inhibition of NOS1 in diabetic rats had a modest nephroprotective effect in terms of proteinuria and glomerular injury rather than being deleterious [33]. Interpretation of pharmacological studies such as these is complex given the range of NOS1 function in the body; therefore, more research is necessary to ascertain the differences in the functional role of NOS1 in the renal vasculature and tubular systems during diabetes.

Calcineurin (PP2B) activity and expression are elevated in the renal cortex [34] and medulla of STZ rats, specifically in the mTAL apical membrane [35]. We recently published that PP2B regulates NOS activity, specifically NOS1 and NOS2, and the subsequent increased NO production by the mTAL suggested that post-translational phosphorylation most likely stimulates NOS activity during diabetes [16]. These data indicate that increased PP2B activity in the mTAL during diabetes promotes NO production. Our preliminary data [27] show that total NOS inhibition increases ouabain-sensitive oxygen consumption by mTAL suspensions from STZ diabetic rats, suggesting that endogenously produced NO actively blunts the diabetes-induced increase in Na transport. Yet, chronic inhibition with cyclosporin A abolishes the glomerular hypertrophy and extracellular matrix accumulation in STZ rats [34]. Further research is necessary to elucidate the details of the PP2B pathway in the renal vasculature and tubular systems.

A common feature of immunopathologies, such as diabetes, is the prominent role of type 1 (both CD4+ and CD8+) T cells [36]. Reciprocal regulatory mechanisms between T cell activation and NO production have been documented [37]. Type 1 T cells characteristically produce interferon-γ and other proinflammatory cytokines, which can induce NOS2 in susceptible cells. NO has been shown to regulate T cell induction and activation with low concentrations (5–10 μM) of NO enhancing type 1 T cell proliferation and cytokine production; however, high concentrations (100 μM) of NO are immune suppressive [38]. T cell cGMP levels increase in concert with the enhanced type 1 proliferation independent of antigen-presenting cells [39]. These data indicate that the NO/cGMP pathway directly and selectively activates type 1 T cells. In addition, NO converts CD4+CD25– T cells into CD4+CD25+ T_{reg} cells in vitro [37]. The NO-converted T_{reg} cells are as efficient as the natural T_{reg} cells in suppressing functional immune responses. NO-converted T_{reg} cells do not express Foxp3 distinct from natural T_{reg} cells [37]. We have postulated that the NO/cGMP pathway is an additional signal for the induction of T cell subpopulations by regulating T cell activation to dampen the potentially damaging autoimmune response in the kidney.

Interactions of the Endothelin and Nitric Oxide Pathways in Diabetic Kidney Disease

Several laboratories have documented the ET_A receptor-mediated vasoconstrictor actions as well as endothelial dysfunction in diabetes. Loomis et al. [40] reported that both ET_A and ET_B receptors are responsible for ET-induced $O_2^{\bullet-}$ in isolated endothelium-intact aortic rings ex vivo. Furthermore, this resulted in a significantly enhanced vasoconstriction that was blunted with NADPH oxidase inhibition or supplementation with tetrahydrobiopterin to relieve NOS uncoupling. We have postulated that increased oxidative stress and ET production during diabetes leads to activation of ET receptors in the vasculature, enhancing vasoconstriction and $O_2^{\bullet-}$ production in a positive feedback mechanism.

We have reasoned that the diabetes-induced increase in NO production and activation of the NO/cGMP pathway, at least in renal tubules, represents a potential means of relieving the burden of diabetes-induced renal complications. Thus, we have proposed that ET_B receptor actions in the kidney during diabetes function to counter the actions of the ET_A receptor. Despite the availability of potent and selective receptor antagonists, results are often contradictory and complicated because of possible vasoconstrictor actions of ET_B receptors and the autocrine and paracrine nature of ET-1, making it difficult to assess in vivo activity. Pfab et al. [38] reported that diabetes induced renal ET_B receptor expression without a change in ET_A receptor expression. Furthermore, these authors showed that the lack of ET_B receptor signaling greatly exaggerates diabetes-induced renal injury and hypertension. Future studies are necessary to differentiate the cellular and receptor subtype-dependent activation of ROS or NO in diabetes. Figure 1 depicts the interactions of the ET and NO pathways in diabetic kidney disease as described in this chapter.

The clinical utility of ET_A-selective versus dual $ET_{A/B}$ receptor antagonists remains controversial and will most likely not be resolved by the pharmaceutical companies any time soon as they have committed themselves to one side of the argument or the other [39]. Direct side-by-side comparison studies in patients with diabetic nephropathy are needed to truly resolve the question. In a recent phase 3 trial, the relatively mixed ET antagonist avosentan (ET_A/

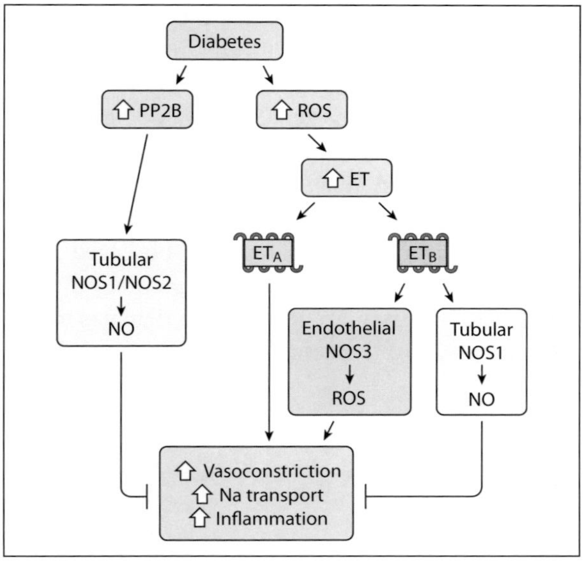

Fig. 1. Scheme depicting interactions of the ET and NO pathways as described in this chapter. On balance, ROS-induced ET-1 production results in elevated ET_A receptor activity that is not sufficiently counterbalanced by ET_B receptor function. The renal tubular ET_B receptor actions are mediated by NO. However, these mechanisms are unable to counteract ET_A-dependent vascular and inflammatory effects.

ET_B affinity ratio of approximately 200 in binding studies) was shown to reduce proteinuria by approximately 50% in patients with diabetic nephropathy after 3 and 6 months of therapy, even though these patients were already being treated with an angiotensin-converting enzyme inhibitor or angiotensin receptor blocker [41]. Unfortunately, the trial was ended early due to serious side effects including heart failure, but new phase 2b trials have been initiated addressing the management of fluid retention. It is unclear at this point whether the development of edema is related to the relatively nonselective characteristics of avosentan or are mechanism-based on ET_A blockade. Another clinical trial is expected to investigate the impact of the nonselective ET antagonist bosentan on the excretion of TNF-α as a marker of inflammation in patients with diabetic nephropathy [42], but the state of this trial is not known. Recently, Kohan et al. [43] reported that atrasentan, a fairly selective ET_A antagonist (ET_A/ET_B affinity ratio of roughly 1800), significantly reduces diabetes-induced albuminuria after 8 weeks of treatment in the presence of angiotensin enzyme inhibitors or angiotensin II blockers in type 2 diabetic patients, and appeared to be well tolerated at low but efficacious doses. These clinical findings argue for the use of ET_A selective antagonists rather than mixed receptor blockers and are consistent with preclinical studies in models of diabetes [44].

We hypothesize that ET_A inhibition will moderate diabetes-mediated vasoconstriction and stimulation of Na transport, as well as blunt the diabetes-induced infiltration and activation of inflammatory cells. Renal cross-transplantation studies have revealed that high dietary Na-induced renal injury and blood pressure responses of ET_B receptor-deficient (*sl/sl*) rats reflect extrarenal effects of ET_B deficiency or exaggerated ET_A receptor activation [45]. These authors did not explore the role of inflammatory cells in their experimental paradigm, nor did they investigate the role of the exacerbated hypertension observed in *sl/sl* rats with diabetes. Future studies utilizing renal cross-transplantation paradigms to determine if the exaggerated hypertensive response to diabetes in *sl/sl* rats is mediated via extrarenal or renal-derived mechanisms are currently being planned.

Perspectives

Severe uncontrolled diabetes can deplete both intracellular and extracellular fluid volumes; however, this situation is relatively rare in patients receiving routine insulin therapy, as it arises primarily during severe hyperglycemia with massive osmotic diuresis. More commonly observed in patients is a situation of sodium retention. Animal models and humans with moderate hyperglycemia in diabetes exhibit increased extracellular fluid volume, increased exchangeable sodium, and an impaired ability to excrete sodium in response to water immersion or isotonic saline infusion [46–48]. The etiology of sodium retention in diabetes is evident in the observations of DiPetrillo et al. [49], who housed rats with STZ-induced diabetes in metabolic cages with free access to food (normal sodium content) and water. While these animals exhibited increases in both sodium excretion and sodium intake, the hyperphagia accompanying diabetes elicited an increase in dietary sodium intake that exceeded the rise in sodium excretion. Thus, a state of positive sodium balance developed even in the face of increased sodium excretion.

Despite considerable investigation, the mechanisms causing dysregulation of renal sodium reabsorption and volume homeostasis in diabetes remain poorly understood [50, 51]. It will be important to include studies investigating whether the ET system is involved in the extrarenal macrophage-dependent regulation of sodium balance and inflammation as suggested by Titze and colleagues [52]. Novel mechanistic targets for diabetic renal disease should include investigations of these extrarenal pathways as well as the renal ET/NO pathways.

References

1 Nishikawa T, Edelstein D, Du XL, Yamagishi S, Matsumura T, Kaneda Y, Yorek MA, Beebe D, Oates PJ, Hammes HP, Giardino I, Brownlee M: Normalizing mitochondrial superoxide production blocks three pathways of hyperglycaemic damage. Nature 2000;404:787–790.

2 Asaba K, Tojo A, Onozato ML, Goto A, Quinn MT, Fujita T, Wilcox CS: Effects of NADPH oxidase inhibitor in diabetic nephropathy. Kidney Int 2005;67:1890–1898.

3 Chen HC, Guh JY, Shin SJ, Tsai JH, Lai YH: Reactive oxygen species enhances endothelin-1 production of diabetic rat glomeruli in vitro and in vivo. J Lab Clin Med 2000;135:309–315.

4 Hughes AK, Stricklett PK, Padilla E, Kohan DE: Effect of reactive oxygen species on endothelin-1 production by human mesangial cells. Kidney Int 1996;49:181–189.

5 Ohuchi T, Yanagisawa M, Gariepy C: Renal tubular effects of endothelin-B receptor signaling: its role in cardiovascular homeostasis and extracellular volume regulation. Curr Opin Nephrol Hypertens 2000;9:435–439.

6 Sasser JM, Sullivan JC, Hobbs JL, Yamamoto T, Pollock DM, Carmines PK, Pollock JS: Endothelin A receptor blockade reduces diabetic renal injury via an anti-inflammatory mechanism. J Am Soc Nephrol 2007;18:143–154.

7 Saleh M, Stevens D, Carmines PK, Pollock DM, Pollock JS: Free radical scavenging decreases endothelin-1 (ET-1) excretion and glomerular permeability during diabetes. FASEB J 2010;24:793.2.

8 Sassy-Prigent C, Heudes D, Mandet C, Belair MF, Michel O, Perdereau B, Bariety J, Bruneval P: Early glomerular macrophage recruitment in streptozotocin-induced diabetic rats. Diabetes 2000;49:466–475.

9 Furuta T, Saito T, Ootaka T, Soma J, Obara K, Abe K, Yoshinaga K: The role of macrophages in diabetic glomerulosclerosis. Am J Kidney Dis 1993;21:480–485.

10 Young BA, Johnson RJ, Alpers CE, Eng E, Gordon K, Floege J, Couser WG, Seidel K: Cellular events in the evolution of experimental diabetic nephropathy. Kidney Int 1995;47:935–944.

11 Yozai K, Shikata K, Sasaki M, Tone A, Ohga S, Usui H, Okada S, Wada J, Nagase R, Ogawa D, Shikata Y, Makino H: Methotrexate prevents renal injury in experimental diabetic rats via anti-inflammatory actions. J Am Soc Nephrol 2005;16:3326–3338.

12 Eardley KS, Zehnder D, Quinkler M, Lepenies J, Bates RL, Savage CO, Howie AJ, Adu D, Cockwell P: The relationship between albuminuria, MCP-1/CCL2, and interstitial macrophages in chronic kidney disease. Kidney Int 2006;69:1189–1197.

13 Proudfoot AE: Chemokine receptors: multifaceted therapeutic targets. Nat Rev Immunol 2002;2:106–115.

14 Weisberg SP, Hunter D, Huber R, Lemieux J, Slaymaker S, Vaddi K, Charo I, Leibel RL, Ferrante AW: CCR2 modulates inflammatory and metabolic effects of high-fat feeding. J Clin Invest 2006;116:115–124.

15 Saleh M, Boesen EI, Pollock JS, Savin V, Pollock DM: Endothelin-1 increases glomerular permeability and inflammation independent of blood pressure in the rat. Hypertension 2010;56:942–949.

16 Foster JM, Carmines PK, Pollock JS: PP2B-dependent NO production in the medullary thick ascending limb during diabetes. Am J Physiol Renal Physiol 2009;297:F471–F480.

17 Schwartz D, Schwartz IF, Blantz RC: An analysis of renal nitric oxide contribution to hyperfiltration in diabetic rats. J Lab Clin Med 2001;137:107–114.

18 Veelken R, Hilgers KF, Hartner A, Haas A, Böhmer KP, Sterzel RB: Nitric oxide synthase isoforms and glomerular hyperfiltration in early diabetic nephropathy. J Am Soc Nephrol 2000;11:71–79.

19 Nakagawa T, Sato W, Glushakova O, Heinig M, Clarke T, Campbell-Thompson M, Yuzawa Y, Atkinson MA, Johnson RJ, Croker B: Diabetic endothelial nitric oxide synthase knockout mice develop advanced diabetic nephropathy. J Am Soc Nephrol 2007;18:539–550.

20 Shankar RR, Wu Y, Shen HQ, Zhu JS, Baron AD: Mice with gene disruption of both endothelial and neuronal nitric oxide synthase exhibit insulin resistance. Diabetes 2000;49:684–687.

21 Shimokawa H, Tsutsui M: Nitric oxide synthases in the pathogenesis of cardiovascular disease: lessons from genetically modified mice. Pflugers Arch 2010;459:959–967.
22 Stadler K, Bonini MG, Dallas S, Jiang J, Radi R, Mason RP, Kadiiska MB: Involvement of inductible nitric oxide synthase in hydroxyl radical-mediated lipid peroxidation in streptozotocin-induced diabetes. Free Radic Biol Med 2008;45:866–874.
23 Chader PN, Gealekman O, Brodsky SV, Elitok S, Tojo A, Crabtree M, Gross SS, Goligorsky MS: Nephropathy in Zucker diabetic fat rat is associated with oxidative and nitrosative stress: prevention by chronic therapy with a peroxynitrite scavenger ebselen. J Am Soc Nephrol 2004;15:2391–2403.
24 Xia Y, Tsai AL, Berka V and Zweier JL: Superoxide generation from endothelial nitric-oxide synthase: a Ca^{2+}/calmodulin-dependent and tetrahydrobiopterin regulatory process. J Biol Chem 1998;273:25804–25808.
25 Xia Y, Dawson VL, Dawson TM, Snyder SH, Zweier JL: Nitric oxide synthase generates superoxide and nitric oxide in arginine-depleted cells leading to peroxynitrite-mediated cellular injury. Proc Natl Acad Sci USA 1996;93:6770–6774.
26 Hink U, Li H, Mollnau H, Oelze M, Matheis E, Hartmann M, Skatchkov M, Thaiss F, Stahl RAK, Warnholtz A, Meinertz T, Griendling K, Harrison DG, Förstermann U, Munzel T: Mechanisms underlying endothelial dysfunction in diabetes mellitus. Circ Res 2001;88:e14–e22.
27 Foster JM, Carmines PK, Pollock JS: Diabetes-induced NOS1 and NOS2 activity blunts oxygen consumption in renal medullary thick ascending limbs. FASEB J 2010;24:812.10.
28 Palm F, Fasching A, Hansell P, Kallskog O: Nitric oxide originating from NOS1 controls oxygen utilization and electrolyte transport efficiency in the diabetic kidney. Am J Physiol Renal Physiol 2010;298:F416–F420.
29 Ortiz PA, Garvin JL: Role of nitric oxide in the regulation of nephron transport. Am J Physiol Renal Physiol 2002;282:F777–F784.
30 Herrera M, Ortiz PA, Garvin JL: Regulation of thick ascending limb transport: role of nitric oxide. Am J Physiol Renal Physiol 2006;290:F1279–F1284.
31 Garvin JL, Hong NJ: Nitric oxide inhibits sodium/hydrogen exchange activity in the thick ascending limb. Am J Physiol Renal Physiol 1999;277:F377–F382.
32 Ortiz PA, Hong NJ, Garvin JL: NO decreases thick ascending limb chloride absorption by reducing Na^+-K^+-$2Cl^-$ cotransporter activity. Am J Physiol Renal Physiol 2001;281:F819–F825.
33 Komers R, Lindsley JN, Oyama TT, Anderson S: Effects of long-term inhibition of neuronal nitric oxide synthase (NOS1) in uninephrectomized diabetic rats. Nitric Oxide 2004;11:147–155.
34 Gooch JL, Barnes JL, Garcia S, Abboud HE: Calcineurin is activated in diabetes and is required for glomerular hypertrophy and ECM accumulation. Am J Physiol Renal Physiol 2003;284:F144–F154.
35 Gooch JL, Pergola PE, Guler RL, Abboud HE, Barnes JL: Differential expression of calcineurin A isoforms in the diabetic kidney. J Am Soc Nephrol 2004;15:1421–1429.
36 Fietta P, Delsante G: The effector T helper cell triade. Riv Biol 2009;102:61–74.
37 Niedbala W, Cai B, Liew FY: Role of nitric oxide in the regulation of T cell functions. Annals of the Rheumatic Diseases 2006;65:iii37–iii40.
38 Pfab T, Thone-Reineke C, Theilig F, Lange I, Witt H, Maser-Gluth C, Bader M, Stasch J, Ruiz P, Bachmann S, Yanagisawa M, Hocher B: Diabetic endothelin B receptor deficient rats develop severe hypertension and progressive renal failure. J Am Soc Nephrol 2006;17:1082–1089.
39 Dhaun N, Pollock DM, Goddard J, Webb DJ: Selective and mixed endothelin receptor antagonism in cardiovascular disease. Trends Pharmacol Sci 2007;28:573–579.
40 Loomis ED, Sullivan JC, Osmond D, Pollock DM, Pollock JS: Endothelin mediates superoxide production and vasoconstriction through activation of NADPH oxidase and uncoupled nitric oxide synthase in the rat aorta. J Pharmacol Exp Ther 2005;315:1058–1064.
41 http://www.speedel.com/assets/2008_SPP301ASCENDStudy.pdf.
42 ClinicalTrials.gov: Bosentan use in patients with diabetic nephropathy. http://clinicaltrials.gov/ct2/show/NCT00638131?intr=%22Bosentan%22&rank=13.

43 Kohan DE, Pritchett Y, Molitch M, Wen S, Garimella T, Audhya P, Andress DL: Addition of atrasentan to renin-angiotensin system blockade reduces albuminuria in diabetic nephropathy. J Am Soc Nephrol 2011;22:763–772.
44 Saleh M, Pollock JS, Pollock DM: Distinct actions of ET_A selective versus combined ET_A/ET_B receptor antagonists in early diabetic kidney disease. J Pharmacol Exp Ther 2011, E-pub ahead of print.
45 Ohkita M, Wang Y, Nguyen NDT, Tsai Y-H, Williams SC, Wiseman RC, Killen PD, Li S, Yanagisawa M, Gariepy CE: Extrarenal ET_B plays a significant role in controlling cardiovascular responses to high dietary sodium in rats. Hypertension 2005;45:940–946.
46 Roland JM, O'Hare JP, Walters G, Corrall RJ: Sodium retention in response to saline infusion in uncomplicated diabetes mellitus. Diabetes Res 1986;3:213–215.
47 O'Hare JA, Ferriss JB, Brady D, Twomey B, O'Sullivan DJ: Exchangeable sodium and renin in hypertensive diabetic patients with and without nephropathy. Hypertension 1985;7:II43–II48.
48 Trevisan R, Fioretto P, Semplicini A, Opocher G, Mantero F, Rocco S, Remuzzi G, Morocutti A, Zanette G, Donadon V: Role of insulin and atrial natriuretic peptide in sodium retention in insulin-treated IDDM patients during isotonic volume expansion. Diabetes 1990;39:289–298.
49 DiPetrillo K, Coutermarsh B, Gesek FA: Urinary tumor necrosis factor contributes to sodium retention and renal hypertrophy during diabetes. Am J Physiol Renal Physiol 2003;284:F113–F121.
50 Wongmekiat O, Johns EJ: Endothelin as a causative factor of blunted volume reflex in diabetic rats. Br J Pharmacol 2003;138:1403–1410.
51 Valentin J-P, Sechi LA, Humphreys MH: Blunted effect of ANP on hematocrit and plasma volume in streptozotocin-induced diabetes mellitus in rats. Am J Physiol Regul Integr Comp Physiol 1994;266:R584–R591.
52 Machnik A, Neuhofer W, Jantsch J, Dahlmann A, Tammela T, Machura K, Park JK, Beck FX, Müller DN, Derer W, Goss J, Ziomber A, Dietsch P, Wagner H, van Rooijen N, Kurtz A, Hilgers KF, Alitalo K, Eckardt KU, Luft FC, Kerjaschki D, Titze J: Macrophages regulate salt-dependent volume and blood pressure by a vascular endothelial growth factor-C-dependent buffering mechanism. Nat Med 2009;15:545–552.

Prof. Jennifer S. Pollock, PhD
Georgia Health Sciences University, Vascular Biology Center
CB-3213B, 1459 Laney Walker Blvd.
Atlanta, GA 30912 (USA)
Tel. +1 706 721 8514, E-Mail jpollock@georgiahealth.edu

Barton M, Kohan DE (eds): Endothelin in Renal Physiology and Disease.
Contrib Nephrol. Basel, Karger, 2011, vol 172, pp 160–170

Endothelin in Renal Inflammation and Hypertension

Mohamed A. Saleh · David M. Pollock

Vascular Biology Center, Department of Surgery, Department of Pharmacology and Toxicology, Georgia Health Sciences University, Augusta, Ga., USA

Abstract

Over the years, a very large amount of evidence has accumulated indicating that endothelin (ET)-1 is an important stimulus for inflammation. This is true for a wide range of organ system diseases, including chronic kidney disease. Nonetheless, our understanding of the role and mechanisms by which ET-1 promotes the activation of both the innate and adaptive immune systems is not understood. ET-1 can directly activate neutrophils as well as endothelial cells to stimulate production of chemoattractant factors, such as monocyte chemoattractant factor-1, and increase synthesis of cell adhesion molecules, such as soluble ICAM-1. The mechanisms that trigger these events, however, are less clear. Elevated blood pressure as well as hyperglycemia could be important factors that facilitate ET-1-dependent inflammation. While renal inflammation has not been used as an endpoint for drug development, the rationale for the use of ET antagonists as anti-inflammatory agents in chronic kidney disease is quite strong, based on animal studies and at least one study in humans with nondiabetic nephritis. While the preponderance of evidence suggests that ET_A selective antagonists are advantageous over combined $ET_{A/B}$ receptor blockers, considerably more work needs to be done in order to understand the complex role of ET in renal inflammation.

The discovery of endothelin (ET) as a potent vasoconstrictor and mitogen over 20 years ago immediately led to speculation that it could be involved in the pathogenesis of organ failure associated with a variety of diseases, including those of the kidney. Perhaps as important, many academics as well as pharmaceutical companies began large programs aimed at developing synthesis inhibitors and receptor antagonists. These pharmacological agents have helped to establish an important role for the ET system as a contributor to the pathophysiology of chronic kidney disease, a complex syndrome that has derangements in

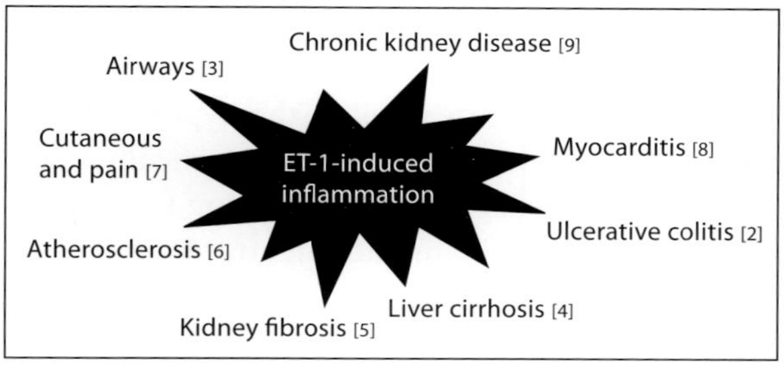

Fig. 1. Evidence of ET-induced inflammation in a variety of organ system diseases.

both hemodynamic and tubular function leading to organ system failure. More recent studies have provided evidence for ET contributing to the inflammation, immune system activation, and fibrosis that is associated with chronic kidney disease [1]. This role is further complicated by the fact that two distinct receptor subtypes, ET_A and ET_B (which frequently have opposing functions), mediate the actions of ET-1.

Endothelin as a General Proinflammatory Agent

The ET system appears to contribute to inflammation and fibrosis in a variety of renal and nonrenal diseases (fig. 1) [2–9]. Combined $ET_{A/B}$ receptor antagonists can inhibit the progression of atherosclerosis in a genetic model of hyperlipidemia, where macrophage foam-cell formation was also attenuated [10]. Additionally in atherosclerosis, nitric oxide (NO)-mediated endothelium-dependent vasodilation is restored by long-term treatment with an ET_A receptor antagonist and is associated with improvement in plasma nitrate levels [6]. In the airways, ET-1 synthesis has been reported to occur just prior to the production of proinflammatory cytokines in a rat model of inflammation while a combined $ET_{A/B}$ receptor antagonist inhibited cytokine production [11]. It cannot be presumed that all of these effects are ET_A-mediated. For example, Griswold et al. [7] observed that mice lacking a functional ET_B receptor were relatively protected against pain, tissue swelling, and inflammatory cell infiltration produced by phenylbenzoquinone-induced algesia or cutaneous treatment with arachidonic acid. Therefore, the location and receptor subtype are important variables that must be considered in determining the mechanism of ET-dependent inflammation. The complexity continues when one considers that ET-1 is produced not only by endothelial cells, but also by renal epithelial cells as well as macrophages and monocytes [12].

One of the potential mechanisms that may explain ET involvement in inflammation is by functioning as a chemoattractant factor to stimulate adhesion molecule expression and synthesis of proinflammatory cytokines. One of the earliest studies on this subject demonstrated that ET-1 enhanced neutrophil accumulation in the isolated rabbit heart, thus leading to the hypothesis that ET-1 may stimulate myocardial inflammation [13]. More specifically, ET-1 stimulated neutrophil adhesion to endothelial cells by increasing the expression of CD18/CD11b on neutrophils, and it augmented neutrophil adhesion in the myocardium under both in vivo and in vitro conditions [13]. Using a modified membrane filter technique, Achmad and Rao [14] also demonstrated that ET-1 is a strong chemoattractant for blood monocytes, a process that is Ca^{2+}-dependent. Further involvement in the inflammatory response comes from observations that ET-1 increases the release of elastase from activated neutrophils [15]. Others reported that ET-1 increased the expression of CD11b/CD18 and CD45 on neutrophil granulocytes, which was inhibited by platelet-activating factor antagonists as well as ET_A, but not ET_B, antagonists [16]. These antagonists also blocked neutrophil adhesion to human endothelial cells.

Another proinflammatory mechanism comes from ET-1-induced increases in endothelial cell production of monocyte chemoattractant protein (MCP-1), a potent chemoattractant for monocytes, through ET_A receptor activation – an effect augmented by TNF-α and IL-1β [17]. MCP-1 is a chemokine that plays a key role in the recruitment of monocytes/macrophages into the kidney and is secreted by mononuclear and various nonleukocytic cells, including renal resident cells [18]. Amiri et al. [19] recently developed a mouse that overexpresses the preproET-1 gene in endothelial cells and observed an increase in a range of inflammatory components within the vasculature including MCP-1, macrophages, adhesion molecules, and other transcription factors completely independent of any change in systemic arterial pressure.

Both ET_A and ET_B receptors may be involved in ET-1-mediated inflammation. Previous studies have shown that ET-1 induces neutrophil adhesion through ICAM-1 expression on cardiac myocytes via the ET_A receptor, and on aortic endothelial cells via the ET_B receptor [20]. Most studies that have examined receptor subtype specific actions seem to indicate a more frequent proinflammatory role for the ET_A receptor. Chen et al. [17] demonstrated that BQ-610, an ET_A selective antagonist, blocked ET-1-induced MCP-1 expression, but the ET_B-selective antagonist BQ-788 had no effect. While these studies suggest a role for the ET_A receptor in ET-1-mediated inflammation, there is no agreement regarding the underlying signaling pathways. As mentioned above, transgenic mice that overexpress ET-1 have elevated markers of inflammation [21]. Blockade of both ET_A and ET_B receptors with a mixed $ET_{A/B}$ antagonist, LU420627, reduced cardiac inflammatory cell infiltration, improved hemodynamic function, and prolonged survival in these mice [21]. However, the ET_A selective antagonist LU135252 was without effect, indicating that the ET_B receptor may be proinflammatory in the mouse heart and could account for the ET-1-

induced cardiomyopathy observed in this model. More direct analysis is needed with specific targeting of the ET_B receptor.

Endothelin-1 in Renal Inflammation

Dhaun et al. [22] have recently provided evidence in humans with nephritis that urinary ET-1 may be a sensitive marker of renal inflammation in the early stages of inflammatory kidney disease, such as lupus nephritis, prior to any decline in renal function. Studies in support of this hypothesis have been accumulating in a range of animal models.

Both ET_A selective and combined $ET_{A/B}$ receptor antagonists can lower arterial pressure in a variety of models of hypertension [23]. In general, these antagonists are more effective in salt-dependent forms of hypertension. Most of these models are associated with an inflammatory response in a variety of organ systems, including the kidney. Chronic administration of mineralocorticoids and high-salt intake in the rat produces hypertension and is also associated with a strong renal inflammatory response including macrophage infiltration and increased DNA binding activity of the nuclear factor AP-1 [24]. These changes can all be reduced by ET_A receptor blockade [24].

A similar benefit of ET_A receptor antagonism has been observed in hypertension produced by chronic angiotensin infusion [25, 26]. Our group has preliminary data indicating that an ET_A selective antagonist attenuates renal T cell infiltration in the chronic angiotensin II infusion model. This effect appears to be blood pressure-dependent within the renal medulla, but pressure-independent in the cortex, suggesting direct activation of proinflammatory pathways by ET-1 in this model. In the transgenic rat that overexpresses human renin and angiotensinogen (known as dTGR), the mixed antagonist bosentan can prevent a variety of inflammatory responses in this model including reducing adhesion molecule expression and inhibition of NF-κB activation in the kidney [27]. ET receptor blockade was more effective than hydralazine despite similar reductions in blood pressure, again supporting the idea that ET exerts proinflammatory effects independent of hypertension.

In a model of hypercholesterolemia and atherosclerosis, ET_A receptor blockade is associated with a variety of beneficial effects in the kidney including reduced oxidative stress, increased eNOS, decreased endothelial LOX-1 expression, decreased tubular iNOS, nitrotyrosine, NF-κB, TGF-β, and decreased perivascular and tubulointerstitial injury [28]. These findings indicate that the endogenous ET system is activated in hypercholesterolemia and contributes to renal inflammation and fibrosis.

Renal inflammation is also a hallmark of ischemia/reperfusion injury in the kidney. Several studies have demonstrated improvement in renal function following ET receptor blockade in the ischemic kidney, but the degree of improvement

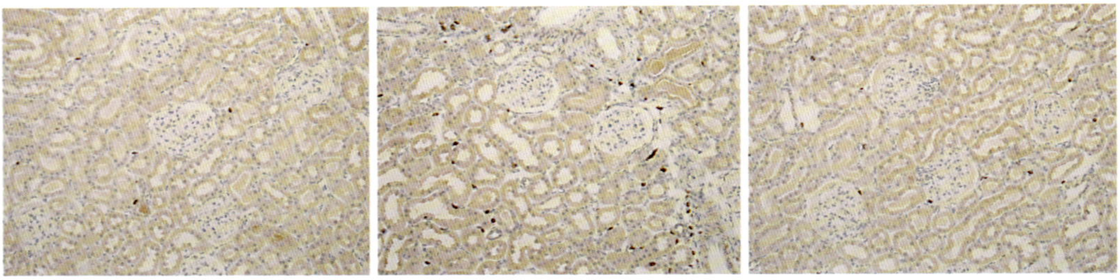

Fig. 2. Immunostaining of CD3-positive cells in the rat kidney. The increase in CD3-positive cells observed with chronic ET-1 infusion is attenuated by treatment with an ET_A receptor antagonist, ABT-627. Adapted from Saleh et al. [9].

is relatively small [29; 30]. Similarly, administration of an anti-ET-1-antibody improved renal function, and furthermore, reduced neutrophil accumulation in the postischemic kidney [31]. In the same study, these investigators observed that ET-1 increased accumulation of labeled neutrophils in the isolated-perfused kidney and increased the expression of CD18 antigen in neutrophils.

Direct Endothelin-1-Induced Inflammation

An important question in many of these studies, especially those in models of hypertension, is whether the reduction in renal inflammation is related to lower arterial pressure or a more direct involvement of the ET system on proinflammatory mechanisms. Our group has utilized a model of chronic ET-1 infusion to test the hypothesis that chronic elevations of ET-1 result in ET_A receptor-dependent inflammation in the kidney [9]. Chronic ET-1 infusion in rats stimulates increases in a variety of inflammatory markers including MCP-1 and soluble intercellular adhesion molecule (sICAM-1) as well as increases in CD3- (fig. 2) and CD68-positive cells, markers for T cells and macrophages, respectively [9]. Administering an ET_A selective antagonist to the animal can block all of these effects of chronic ET-1.

Importantly, when maintained on a normal salt diet, chronic ET-1 does not elevate blood pressure similar to what has been reported in transgenic mice overexpressing the human ET-1 gene. The latter model of lifelong elevated ET-1 is characterized by glomerulosclerosis and interstitial fibrosis without systemic hypertension [5]. These investigators also observed that ET-1 overexpression increased recruitment of inflammatory cells into the kidney, including CD4+ T cells and macrophages [5]. While systemic arterial pressure does not change in these models, one must remain cautious in the interpretation because we do not know what happens locally to glomerular capillary pressure. Similar findings have also been observed in ET-2 transgenic rats, but we would presume the mechanisms would be similar to the ET-1 transgenics since these peptides have a similar affinity for ET receptors [32].

Amiri et al. [19] also reported that transgenic mice overexpressing human preproET-1 specifically in endothelial cells display a similar range of vascular inflammatory responses, including macrophage infiltration, activation of transcription factors AP-1 and NF-κB, and increases in VCAM-1 expression independent of blood pressure elevation. A follow-up study in this overexpression model revealed that these changes were also blocked by ABT-627, consistent with the proinflammatory effects of ET-1 being ET_A receptor-mediated, similar to the chronic infusion model [33]. Therefore, it is very clear in a number of different animal model systems that ET-1 directly triggers the initial phase of renal inflammation by increasing expression of key mediators such as chemoattractant and adhesion molecules. While it is presumed that these proinflammatory actions of ET-1 lead to detrimental tubular and/or hemodynamic function, it remains to be determined whether this pathway can be used as a target for future therapeutic intervention.

Endothelin-1 as a Proinflammatory Agent in Diabetic Nephropathy

ET-1 as a target for treatment of diabetic nephropathy has gained interest in recent years with the somewhat positive findings from a recent phase 3 trial examining the effect of ET receptor blockade on renal function and proteinuria [34, 35]. This work was supported by strong preclinical data and although there were positive effects on reducing proteinuria, there were issues with fluid retention and the study was terminated prematurely. As these studies are discussed in detail elsewhere in this issue, the discussion here will focus on inflammation.

Diabetic nephropathy is associated with considerable renal inflammation in terms of chemical mediators as well as infiltrating cells of the innate and adaptive immune systems [36].

In a model of type 1 diabetes, the streptozotocin-treated rat, we observed a significant elevation in renal macrophages and T cells that were inhibited by chronic treatment with the ET_A receptor antagonist atrasentan [37]. In a follow-up study, we used the same model to profile the relative time course of early markers of inflammation, MCP-1 and sICAM-1 [38]. The initial phase of type 1 diabetes is associated with hyperfiltration and proteinuria. After 6 weeks of hyperglycemia induction, plasma and glomerular sICAM-1 and MCP-1 concentrations were significantly elevated, while no changes were observed at the 3-week time point [38]. Infiltration of macrophages and T lymphocytes are a hallmark of the early phase of diabetic nephropathy, but the mechanisms that trigger this process are poorly understood [39, 40]. Chemoattractants, cell adhesion molecules, and activating cytokines can all precede the infiltration process.

Therefore, the observations that ET antagonists prevent the increase in the early stages of diabetic nephropathy suggest an important role for ET-1 in the disease process.

Infiltrating monocytes and macrophages release a variety of cytokines, lysosomal enzymes, reactive oxygen intermediates, and profibrotic factors such as TGF-β, representing the early stages of renal damage [41]. Many investigators believe that this process of inflammatory cell activation also accelerates mesangial cell proliferation during the development of diabetic nephropathy. Detailed molecular mechanisms that direct macrophage migration are not understood although chemokines, chemokine receptors, and adhesion molecules are involved in the early phases of this process [42–45]. Reducing the infiltration of macrophages is associated with reduced expression of TGF-β and collagen IV in glomeruli, reduced urinary albumin excretion, and reduced glomerular hypertrophy as well as reduced renal injury in diabetic ICAM-1-deficient mice [46].

The extent of ET-1 involvement in renal inflammation seen in type 2 diabetes is not known. Clinical studies investigating the use of ET antagonists in diabetic nephropathy have not explored the effect on renal inflammation [35]. However, a clear rationale for additional clinical trials can be made given the evidence for ET stimulation of factors involved in the early inflammatory response such as ICAM-1 and MCP-1. Patients with diabetic nephropathy have elevated urinary MCP-1 levels that were correlated to the number of CD68-positive infiltrating macrophages in the interstitium [47]. In human mesangial cells, high glucose concentrations enhance expression of MCP-1 via NF-κB activation, an effect that has been confirmed in many in vitro studies [48]. Renal biopsies from patients with type 2 diabetes and overt nephropathy had increased MCP-1 expression primarily in the tubular cells, which also had increased NF-κB activation [49]. MCP-1 can enhance secretion of TGF-β1 in cultured macrophages, thus facilitating a cascade of profibrotic pathways in addition to proinflammatory factors [50]. Therefore, factors that inhibit MCP-1, such as ET_A receptor antagonists, may provide a means of reducing inflammation in diabetic nephropathy.

An ongoing debate that has prevailed for many years in the ET field is the question of whether ET_A selective or mixed $ET_{A/B}$ receptor antagonists would be more effective in treating diabetic nephropathy. One could argue for ET_A receptor selectivity since rats with nonfunctional ET_B receptors develop hypertension and more aggressive renal injury in response to streptozotocin treatment [51]. ET_B receptors are involved in the synthesis of the vasodilator NO by endothelial and epithelial cells [52]. Inhibition of NO has also been associated with an increase in leukocyte adhesion to mesenteric venules; an indication that the increases in monocyte/macrophage infiltration are due to a decrease in NO production [53]. However, mixed $ET_{A/B}$ antagonists have been shown to produce beneficial effects in this model similar to the selective ET_A antagonists [54, 55]. Surprisingly, there have been only a few studies providing a direct head-to-head comparison in many of the disease models, and none that we are aware of for diabetic nephropathy. Recently, we hypothesized that ET_A selective antagonists would provide an advantage over combined $ET_{A/B}$ antagonists with regard to attenuating glomerular injury and inflammatory mediators in an experimental model of diabetes because of the antifibrotic and anti-inflammatory actions reported for the ET_B receptor [56, 23]. We observed that the mixed $ET_{A/B}$ antagonist, A-182086, was less effective in alleviating the increase in inflammatory molecule expression in 7-week streptozotocin-treated rats compared to the ET_A receptor antagonist atrasentan (ABT-627), using both plasma and glomerular markers [57].

Conclusion

The rationale for ET functioning as a proinflammatory peptide is quite clear. What are not clear, however, are the conditions that trigger ET involvement as well as how ET fits into the complex array of mechanisms that characterize the inflammatory response. In the kidney, but perhaps not other organs, the bulk of the evidence would suggest that the inflammatory effects of ET-1 are mediated by the ET_A receptor and that the ET_B receptor may function as a counter-regulatory system. Use of ET_A selective antagonists as a means of reducing renal inflammation require further clinical investigation, but certainly the initial studies are quite promising [22]. Ongoing clinical trials need to include inflammation as secondary endpoints.

References

1 Teder P, Noble PW: A cytokine reborn? Endothelin-1 in pulmonary inflammation and fibrosis. Am J Respir Cell Mol Biol 2000;23:7–10.

2 Murch SH, Braegger CP, Sessa WC, MacDonald TT: High endothelin-1 immunoreactivity in Crohn's disease and ulcerative colitis. Lancet 1992;339:381–385.

3 Kraft M, Beam WR, Wenzel SE, Zamora MR, O'Brien RF, Martin RJ: Blood and bronchoalveolar lavage endothelin-1 levels in nocturnal asthma. Am J Respir Crit Care Med 1994;149:946–952.

4 Rockey DC, Chung JJ: Endothelin antagonism in experimental hepatic fibrosis. Implications for endothelin in the pathogenesis of wound healing. J Clin Invest 1996;98:1381–1388.

5 Hocher B, Thone-Reineke C, Rohmeiss P, Schmager F, Slowinski T, Burst V, Siegmund F, Quertermous T, Bauer C, Neumayer HH, Schleuning WD, Theuring F: Endothelin-1 transgenic mice develop glomerulosclerosis, interstitial fibrosis, and renal cysts but not hypertension. J Clin Invest 1997;99:1380–1389.

6 Barton M, Haudenschild CC, d'Uscio LV, Shaw S, Munter K, Luscher TF: Endothelin ET_A receptor blockade restores NO-mediated endothelial function and inhibits atherosclerosis in apolipoprotein E-deficient mice. Proc Natl Acad Sci USA 1998;95:14367–14372.

7 Griswold DE, Douglas SA, Martin LD, Davis TG, Davis L, Ao Z, Luttmann MA, Pullen M, Nambi P, Hay DW, Ohlstein EH: Endothelin B receptor modulates inflammatory pain and cutaneous inflammation. Mol Pharmacol 1999;56:807–812.
8 Ono K, Matsumori A, Shioi T, Furukawa Y, Sasayama S: Contribution of endothelin-1 to myocardial injury in a murine model of myocarditis: acute effects of bosentan, an endothelin receptor antagonist. Circulation 1999;100:1823–1829.
9 Saleh MA, Boesen EI, Pollock JS, Savin VJ, Pollock DM: Endothelin-1 increases glomerular permeability and inflammation independent of blood pressure in the rat. Hypertension 2010;56:942–949.
10 Babaei S, Picard P, Ravandi A, Monge JC, Lee TC, Cernacek P, Stewart DJ: Blockade of endothelin receptors markedly reduces atherosclerosis in LDL receptor deficient mice: role of endothelin in macrophage foam cell formation. Cardiovasc Res 2000;48:158–167.
11 Finsnes F, Lyberg T, Christensen G, Skjonsberg OH: Effect of endothelin antagonism on the production of cytokines in eosinophilic airway inflammation. Am J Physiol Lung Cell Mol Physiol 2001;280:L659–665.
12 Ehrenreich H, Anderson RW, Fox CH, Rieckmann P, Hoffman GS, Travis WD, Coligan JE, Kehrl JH, Fauci AS: Endothelins, peptides with potent vasoactive properties, are produced by human macrophages. J Exp Med 1990;172:1741–1748.
13 Lopez Farre A, Riesco A, Espinosa G, et al: Effect of endothelin-1 on neutrophil adhesion to endothelial cells and perfused heart. Circulation 1993;88:1166–1171.
14 Achmad TH, Rao GS: Chemotaxis of human blood monocytes toward endothelin-1 and the influence of calcium channel blockers. Biochem Biophys Res Commun 1992;189:994–1000.
15 Halim A, Kanayama N, el Maradny E, Maehara K, Terao T: Activated neutrophil by endothelin-1 caused tissue damage in human umbilical cord. Thromb Res 1995;77:321–327.
16 Zouki C, Baron C, Fournier A, Filep JG: Endothelin-1 enhances neutrophil adhesion to human coronary artery endothelial cells: role of ET_A receptors and platelet-activating factor. Br J Pharmacol 1999;127:969–979.
17 Chen P, Shibata M, Zidovetzki R, Fisher M, Zlokovic BV, Hofman FM: Endothelin-1 and monocyte chemoattractant protein-1 modulation in ischemia and human brain-derived endothelial cell cultures. J Neuroimmunol 2001;116:62–73.
18 Ihm CG: Monocyte chemotactic peptide-1 in diabetic nephropathy. Kidney Int Suppl 1997;60:S20–S22.
19 Amiri F, Paradis P, Reudelhuber TL, Schiffrin EL: Vascular inflammation in absence of blood pressure elevation in transgenic murine model overexpressing endothelin-1 in endothelial cells. J Hypertens 2008;26:1102–1109.
20 Hayasaki Y, Nakajima M, Kitano Y, Iwasaki T, Shimamura T, Iwaki K: ICAM-1 expression on cardiac myocytes and aortic endothelial cells via their specific endothelin receptor subtype. Biochem Biophys Res Commun 1996;229:817–824.
21 Yang LL, Gros R, Kabir MG, Sadi A, Gotlieb AI, Husain M, Stewart DJ: Conditional cardiac overexpression of endothelin-1 induces inflammation and dilated cardiomyopathy in mice. Circulation 2004;109:255–261.
22 Dhaun N, Lilitkarntakul P, Macintyre IM, Muilwijk E, Johnston NR, Kluth DC, Webb DJ, Goddard J: Urinary endothelin-1 in chronic kidney disease and as a marker of disease activity in lupus nephritis. Am J Physiol Renal Physiol 2009;296:F1477–F1483.
23 Kohan DE, Rossi NF, Inscho EW, Pollock DM: Regulation of blood pressure and salt homeostasis by endothelin. Physiol Rev 2011;91:1–77.
24 Tostes RC, Touyz RM, He G, Chen X, Schiffrin EL: Contribution of endothelin-1 to renal activator protein-1 activation and macrophage infiltration in aldosterone-induced hypertension. Clin Sci (Lond) 2002;103(suppl 48):25S–30S.
25 Rajagopalan S, Laursen JB, Borthayre A, Kurz S, Keiser J, Haleen S, Giaid A, Harrison DG: Role for endothelin-1 in angiotensin II-mediated hypertension. Hypertension 1997;30:29–34.

26 Herizi A, Jover B, Bouriquet N, Mimran A: Prevention of the cardiovascular and renal effects of angiotensin II by endothelin blockade. Hypertension 1998;31:10–14.
27 Muller DN, Mervaala EM, Schmidt F, Park JK, Dechend R, Genersch E, Breu V, Loffler BM, Ganten D, Schneider W, Haller H, Luft FC: Effect of bosentan on NF-kappaB, inflammation, and tissue factor in angiotensin II-induced end-organ damage. Hypertension 2000;36:282–290.
28 Chade AR, Best PJ, Rodriguez-Porcel M, Herrmann J, Zhu X, Sawamura T, Napoli C, Lerman A, Lerman LO: Endothelin-1 receptor blockade prevents renal injury in experimental hypercholesterolemia. Kidney Int 2003;64:962–969.
29 Mino N, Kobayashi M, Nakajima A, Amano H, Shimamoto K, Ishikawa K, Watanabe K, Nishikibe M, Yano M, Ikemoto F: Protective effect of a selective endothelin receptor antagonist, BQ-123, in ischemic acute renal failure in rats. Eur J Pharmacol 1992;221:77–83.
30 Stingo AJ, Clavell AL, Aarhus LL, Burnett JCJ: Biological role for the endothelin-A receptor in aortic cross-clamping. Hypertension 1993;22:62–66.
31 Espinosa G, Lopez Farre A, Cernadas MR, Manzarbeitia F, Tan D, Digiuni E, Mosquera JR, Monton M, Millas I, Hernando L, Casado S, Caramelo C: Role of endothelin in the pathophysiology of renal ischemia-reperfusion in normal rabbits. Kidney Int 1996;50:776–782.
32 Hocher B, Liefeldt L, Thone-Reineke C, Orzechowski HD, Distler A, Bauer C, Paul M: Characterization of the renal phenotype of transgenic rats expressing the human endothelin-2 gene. Hypertension 1996;28:196–201.
33 Amiri F, Ko EA, Javeshghani D, Reudelhuber TL, Schiffrin EL: Deleterious combined effects of salt-loading and endothelial cell restricted endothelin-1 overexpression on blood pressure and vascular function in mice. J Hypertens 2010;28:1243–1251.
34 Wenzel RR, Littke T, Kuranoff S, Jurgens C, Bruck H, Ritz E, Philipp T, Mitchell A: Avosentan reduces albumin excretion in diabetics with macroalbuminuria. J Am Soc Nephrol 2009;20:655–664.
35 Mann JF, Green D, Jamerson K, Ruilope LM, Kuranoff SJ, Littke T, Viberti G: Avosentan for overt diabetic nephropathy. J Am Soc Nephrol 2010;21:527–535.
36 Sassy-Prigent C, Heudes D, Mandet C, Belair MF, Michel O, Perdereau B, Bariety J, Bruneval P: Early glomerular macrophage recruitment in streptozotocin-induced diabetic rats. Diabetes 2000;49:466–475.
37 Sasser AK, Mundy BL, Smith KM, Studebaker AW, Axel AE, Haidet AM, Fernandez SA, Hall BM: Human bone marrow stromal cells enhance breast cancer cell growth rates in a cell line-dependent manner when evaluated in 3D tumor environments. Cancer Lett 2007;254:255–264.
38 Saleh MA, Boesen EI, Pollock JS, Savin VJ, Pollock DM: Endothelin receptor A-specific stimulation of glomerular inflammation and injury in a streptozotocin-induced rat model of diabetes. Diabetologia 2011;54:979–988.
39 Furuta T, Saito T, Ootaka T, Soma J, Obara K, Abe K, Yoshinaga K: The role of macrophages in diabetic glomerulosclerosis. Am J Kidney Dis 1993;21:480–485.
40 Shikata K, Makino H: Role of macrophages in the pathogenesis of diabetic nephropathy; in Tomino Y (ed): Type-2 Diabetic Nephropathy in Japan. From Bench to Bedside. Contrib Nephrol. Basel, Karger, 2001, vol 134, pp 46–54.
41 Wolf G: New insights into the pathophysiology of diabetic nephropathy: from haemodynamics to molecular pathology. Eur J Clin Invest 2004;34:785–796.
42 Sugimoto H, Shikata K, Hirata K, Akiyama K, Matsuda M, Kushiro M, Shikata Y, Miyatake N, Miyasaka M, Makino H: Increased expression of intercellular adhesion molecule-1 (ICAM-1) in diabetic rat glomeruli: glomerular hyperfiltration is a potential mechanism of ICAM-1 upregulation. Diabetes 1997;46:2075–2081.
43 Park CW, Kim JH, Lee JH, Kim YS, Ahn HJ, Shin YS, Kim SY, Choi EJ, Chang YS, Bang BK: High glucose-induced intercellular adhesion molecule-1 (ICAM-1) expression through an osmotic effect in rat mesangial cells is PKC-NF-kappa B-dependent. Diabetologia 2000;43:1544–1553.

44 Basta G, Lazzerini G, Massaro M, Simoncini T, Tanganelli P, Fu C, Kislinger T, Stern DM, Schmidt AM, De Caterina R: Advanced glycation end products activate endothelium through signal-transduction receptor RAGE: a mechanism for amplification of inflammatory responses. Circulation 2002;105:816–822.
45 Onozato ML, Tojo A, Goto A, Fujita T: Radical scavenging effect of gliclazide in diabetic rats fed with a high cholesterol diet. Kidney Int 2004;65:951–960.
46 Okada S, Shikata K, Matsuda M, Ogawa D, Usui H, Kido Y, Nagase R, Wada J, Shikata Y, Makino H: Intercellular adhesion molecule-1-deficient mice are resistant against renal injury after induction of diabetes. Diabetes 2003;52:2586–2593.
47 Tashiro K, Koyanagi I, Saitoh A, Shimizu A, Shike T, Ishiguro C, Koizumi M, Funabiki K, Horikoshi S, Shirato I, Tomino Y: Urinary levels of monocyte chemoattractant protein-1 (MCP-1) and interleukin-8 (IL-8), and renal injuries in patients with type 2 diabetic nephropathy. J Clin Lab Anal 2002;16:1–4.
48 Bierhaus A, Schiekofer S, Schwaninger M, Andrassy M, Humpert PM, Chen J, Hong M, Luther T, Henle T, Kloting I, Morcos M, Hofmann M, Tritschler H, Weigle B, Kasper M, Smith M, Perry G, Schmidt AM, Stern DM, Haring HU, Schleicher E, Nawroth PP: Diabetes-associated sustained activation of the transcription factor nuclear factor-kappaB. Diabetes 2001;50:2792–2808.
49 Mezzano S, Aros C, Droguett A, Burgos ME, Ardiles L, Flores C, Schneider H, Ruiz-Ortega M, Egido J: NF-kappaB activation and overexpression of regulated genes in human diabetic nephropathy. Nephrol Dial Transplant 2004;19:2505–2512.
50 Wang SN, LaPage J, Hirschberg R: Role of glomerular ultrafiltration of growth factors in progressive interstitial fibrosis in diabetic nephropathy. Kidney Int 2000;57:1002–1014.
51 Pfab T, Thone-Reineke C, Theilig F, Lange I, Witt H, Maser-Gluth C, Bader M, Stasch JP, Ruiz P, Bachmann S, Yanagisawa M, Hocher B: Diabetic endothelin B receptor-deficient rats develop severe hypertension and progressive renal failure. J Am Soc Nephrol 2006;17:1082–1089.
52 Namiki A, Hirata Y, Ishikawa M, Moroi M, Aikawa J, Machii K: Endothelin-1- and endothelin-3-induced vasorelaxation via common generation of endothelium-derived nitric oxide. Life Sci 1992;50:677–682.
53 Dubey RK, Boegehold MA, Gillespie DG, Rosselli M: Increased nitric oxide activity in early renovascular hypertension. Am J Physiol 1996;270:R118–R124.
54 Arikawa E, Verma S, Dumont AS, McNeill JH: Chronic bosentan treatment improves renal artery vascular function in diabetes. J Hypertens 2001;19:803–812.
55 Ding SS, Qiu C, Hess P, Xi JF, Zheng N, Clozel M: Chronic endothelin receptor blockade prevents both early hyperfiltration and late overt diabetic nephropathy in the rat. J Cardiovasc Pharmacol 2003;42:48–54.
56 Schneider MP, Boesen EI, Pollock DM: Contrasting actions of endothelin ET_A and ET_B receptors in cardiovascular disease. Annu Rev Pharmacol Toxicol 2007;47:731–759.
57 Saleh MA, Pollock JS, Pollock DM: Distinct actions of ET_A selective versus combined $ET_{A/B}$ receptor antagonists in early diabetic kidney disease. J Pharmacol Exptl Therap 2011, E-pub ahead of print.

David M. Pollock, PhD
Vascular Biology Center, Medical College of Georgia
1459 Laney Walker Blvd.
Augusta, GA 30907-2500 (USA)
Tel. +1 706 721 8517, E-Mail dpollock@mcg.edu

Barton M, Kohan DE (eds): Endothelin in Renal Physiology and Disease.
Contrib Nephrol. Basel, Karger, 2011, vol 172, pp 171–184

Endothelin in Chronic Proteinuric Kidney Disease

Elena Gagliardini · Simona Buelli · Ariela Benigni

Mario Negri Institute for Pharmacological Research, Centro Anna Maria Astori, Science and Technology Park Kilometro Rosso, Bergamo, Italy

Abstract

The incidence of chronic kidney diseases is increasing worldwide at an alarming rate. As this is emerging as a global threat to human health, present efforts are concentrated on the identification of new treatments that slow or even reverse the progression of chronic nephropathies. Endothelin (ET)-1 is a potent vasoconstrictor peptide with proinflammatory, mitogenic, and profibrotic effects, and it contributes to both normal renal physiology and pathology. There is robust experimental and clinical evidence for the role of ET-1 and its cognate receptors in many progressive renal disorders. The effectiveness of ET receptor antagonists in improving renal hemodynamics and reducing fibrosis has been largely documented in experimental animals. However, whether selective ET_A or dual ET_A/ET_B receptor antagonists are preferable is still a matter of debate. Combined therapies, including ET receptor antagonists, are promising to maximize partial renoprotection achieved with blockade of the renin-angiotensin system, particularly when treatment is given in the latter phase of the disease. The focus of this review is to explore the role of ET-1 in kidney diseases and to shed light on the novel pharmacological setting in chronic nephropathies.

The number of people suffering from chronic kidney diseases (CKD) has continued to grow, and this growth has been accompanied by clinical, economic, and ethical consequences. Nowadays, around 1.8 million patients with CKD enter programs for renal replacement therapy (RRT) consisting of renal transplantation and hemo/peritoneal dialysis [1]. In wealthy countries, RRT has been undertaken in the last decades on a large scale, consequently placing a great burden on health care budgets; however, these expensive treatments are still limited or completely lacking in developing and undeveloped countries. According to a forecast analysis based on data from the US Renal Data System and Medicare, it was estimated in 2007 that by 2010, 650,000 patients progressing to end-stage renal disease would require RRT at an estimated public expenditure of USD 28 million [2]. Treatment of a relatively limited number of patients with RRT is a major social commitment because of the great expense maintenance of patients in dialysis has. The costs are so prohibitive that in the near future most of the global population may not have access to it. In the absence of definitive therapies for most acquired kidney diseases, it would be best to focus efforts on prevention of renal disease progression with pharmacological and nonpharmacological approaches, and promote further investigations on the mechanisms and mediators contributing to CKD.

Determinants of Renal Disease Progression

In the early 1980s, micropuncture and morphological studies from Dr. Barry Brenner at the Brigham and Women's Hospital in Boston suggested that control of glomerular hypertension limits glomerular damage in rat progressive nephropathy, indicating glomerular capillary hypertension as a major determinant for the perpetuation of scarring events [3]. The adaptive hemodynamic changes that occur after an initial nephron loss enhance the filtration capacity of the remaining nephrons to maintain GFR. However, they ultimately induce a self-perpetuating detrimental circle determining the progressive deterioration of the hyperfiltrating remnant nephrons. High intraglomerular capillary pressure affects the filtration process by impairing the glomerular barrier size-selectivity and allowing extensive protein passage throughout the glomerular capillary wall [4]. Abnormally ultrafiltered plasma proteins resulting in proteinuria do not simply reflect the severity of glomerular injury, but they bear an intrinsic renal toxicity [5]. Protein overload of the proximal tubular cells activates intracellular signals triggering the production of proinflammatory and profibrotic mediators [6]. These substances, released into the renal interstitium, promote the infiltration of inflammatory cells that, in turn, are stimulated to produce cytokines/chemokines, growth factors, and extracellular matrix proteins, further contributing to interstitial fibrosis [6]. As a demonstration of the causal role of proteinuria in progressive kidney disorders, treatments with angiotensin-converting enzyme (ACE) inhibitors that reduce proteinuria by preserving glomerular permselectivity, have been shown to be valuable tools to limit end-stage renal disease progression [7]. Among the biological factors induced by protein overabsorption by proximal tubular cells, ET-1 is intimately involved in the mechanisms leading to renal disorders [8].

The purpose of this chapter is to review evidence supporting the role of ET-1 in renal disorders and the most recent findings on the efficacy of ET receptor antagonists, which were once approved for the treatment of pulmonary arterial hypertension, but now largely indicated as possible renoprotective agents.

Endothelin System in Kidney Physiology

There are roughly 3,000 studies on the role of the ET-1 system in kidney physiology and pathophysiology that have been published to date. From these, it is clear that ET-1 autocrinally/paracrinally regulates renal blood flow, glomerular hemodynamics, sodium/water homeostasis, and acid-base balance [9]. ET-1 is synthesized by almost every cell type of the kidney [10]. Under basal conditions, podocytes constitutively express preproET-1 mRNA and synthesize more ET-1 compared to mesangial cells [11], whereas epithelial cells from the collecting duct (CD) are a major tubular source of ET-1 that, in the kidney inner medulla, reaches the highest concentration in the body. Recent studies show that in cultured CD cells the tubular flow induces an increased ET-1 mRNA expression [12]. ET receptors are widely distributed within the kidney. The type A receptor (ET_A-R) mainly localizes to vascular smooth muscle, vasa recta, and arcuate arteries [10]. In the glomerulus, mesangial cells and podocytes express ET_A-R [10]. The type B receptor (ET_B-R) is more diffusely expressed on vascular and glomerular endothelial cells, mesangial cells, proximal/distal epithelial cells, and on cells of the collecting system at high concentration [10]. The relative abundance of ET_A-R and ET_B-R in the kidney depends on the species [13], which complicates the translation of findings from experimental animals to human beings. Of note, CD-derived ET-1 is an important regulator of systemic blood pressure and renal sodium/water excretion. CD-specific ET-1 knocking out (KO) in mice causes hypertension on a normal sodium diet, which is associated with impaired sodium excretion in response to sodium loading [14]. Studies in animal models in which ET receptors are selectively disrupted in the CD demonstrate that natriuretic and diuretic effects of ET-1 mainly occur via ET_B-R [15]. However, combined KO for ET_A-R and ET_B-R causes increased hypertension compared to ET_B-R KO [16], suggesting that ET_A-R might also play a role in sodium and water handling in the kidney.

Role of ET-1 in Chronic Proteinuric Nephropathies

ET-1 is involved in the progression of CKD due to its chemotactic properties and the stimulatory effect on renal cell proliferation and extracellular matrix production [8]. ET-1 mediates interstitial infiltration of blood monocytes and directly increases proinflammatory cytokine secretion by binding to specific receptors on these cells [8].

In mesangial cells, ET-1 stimulates the production of monocyte chemoattractant protein-1, which may be released during glomerular damage, thus promoting inflammatory cell recruitment. Monocyte chemoattractant protein-1 present in the conditioned media of ET-1-stimulated mesangial cells fosters inflammatory cell migration in vitro [17] and promotes extracellular matrix production with an autocrine signaling loop [18]. Podocytes are an important source of ET-1 during glomerular injury. Excess plasma proteins induce a Rho-kinase-dependent cytoskeletal rearrangement in cultured podocytes that is instrumental for the activation of integrin-associated focal adhesion kinase [19]. This translates into transcriptional events that include ET-1 gene overexpression and protein upregulation [19]. Podocyte-derived ET-1 directly contributes to glomerular barrier dysfunction by reducing podocyte expression of nephrin, a slit-diaphragm protein essential for glomerular barrier integrity [20]. ET-1 also causes nephrin shedding from the cell surface and cytoskeletal redistribution via ET_A-R [19–21]. That podocytes are functional targets for ET-1 via ET_A-R is documented by data showing that podocyte cytoskeletal disruption and upregulation of matrix metalloproteinase-9 induced by puromycin aminonucleoside is prevented by selective ET_A-R antagonist or by RNA interference [22]. Since ET-1 acts as a local hormone whose effects are exerted in a paracrine manner, it is possible that podocyte-derived ET-1 may regulate glomerular capillary tone, enhance vascular permeability, and stimulate mesangial contraction.

In vitro studies have suggested a link between tubular protein reabsorption and renal ET-1 upregulation. Protein overloading of cultured proximal tubular epithelial cells with plasma proteins such as albumin, IgG, and transferrin modulates ET-1 synthesis and release in a dose-dependent manner [23]. Tubular ET-1 secretion occurs mainly towards the basolateral side of the cells. Should this operate in vivo, ET-1 could accelerate mechanisms relevant to the tubulointerstitial inflammation in response to proteinuria, specifically recruitment of monocytes in the interstitium and stimulation of proinflammatory cytokine and growth factor production by infiltrating inflammatory cells.

Increased renal gene and protein expression of ET-1 characterizes proteinuria-associated chronic nephropathies from either immune or nonimmune origins. In rats with reduced renal mass, a model characterized by glomerular hemodynamic alterations, renal ET-1 gene expression and the excretion of the corresponding peptide in the urine increase after surgery and correlate with the progressive renal injury in a time-dependent manner [24, 25]. Passive Heymann nephritis, an immune model resembling human membranous nephropathy characterized by chronic proteinuria and glomerular and tubulointerstitial damage, shares with the remnant kidney the feature of excessive ET-1 urinary excretion, which correlates with structural damage [26]. Specifically, in situ ET-1 hybridization in passive Heymann nephritis rats reveals strong renal staining of ET-1 mRNA, mainly in proximal tubules and renal interstitium in animals with heavy proteinuria and glomerulosclerosis [27]. Increased ET-1 and ET_B-R mRNA levels in the kidneys of NZB/WF1 mice, which develop an immunological disease reminiscent of human lupus, have also been reported [28]. Data are also available that indicate a protective role of ET_B-R in progressive renal damage since ET_B-R-deficient animals develop a more severe disease when a reduction of nephron mass or diabetes is induced [29, 30]. That ET-1 contributes to the development of diabetic nephropathy is derived from in vitro and in vivo findings. High glucose concentrations enhance ET synthesis by macro- and microvascular endothelial cells [31, 32] and increase responsiveness of mesangial cells to vasoactive peptides as ET-1 [33]. In experimental diabetes, messenger RNA for ET-1 is overexpressed in glomeruli and tubular cells [34], and ET-1, via ET_A-R, promotes macrophage infiltration and stimulates the production of TGFβ and PGE_2 [35]. Higher than normal expression of ET-1 gene in diabetes is normalized by insulin, which suggests that this abnormality is caused, at least in part, by hyperglycemia [36].

Therapeutic Potential of Endothelin Receptor Antagonism

The development of selective (ET_A) or dual (ET_A/ET_B) receptor antagonists (RA) has contributed to defining the deleterious role of ET-1 in the progressive nephropathies and has helped clarify the therapeutic potential of targeting the ET system to reduce renal damage in CKD.

Experimental Studies

The model of Ren-2 transgenic rats characterized by signs of fulminant hypertension and renal damage was instrumental a few years ago in establishing the efficacy of selective ET_A-RA on renal injury prevention in the absence of no blood pressure lowering effect [37]. Selective ET_A-R blockade reduces urinary protein excretion and podocyte injury to an extent which is higher than that obtained with a dual ET_A/ET_B-R blocker [37]. Renoprotection achieved by ET_A-RA is partially due to the drug's capability to reduce glomerular hypertension, consequent to systemic hypertension, and control renal hemodynamics. Of interest, ET_A-RA is capable of prolonging animal survival, a hard endpoint that is not achieved by the ET_A/ET_B-RA. The effectiveness of ET antagonists has also been observed in experimental models in which CKD develops as a phenomenon not secondary to systemic hypertension. In rats with renal mass ablation, the selective ET_A-RA is effective in normalizing blood pressure, partially reducing proteinuria and preventing deterioration of renal function [38]. The most relevant effect of ET_A-RA is on structural damage, as demonstrated by the absence of glomerulosclerosis and tubulointerstitial damage in diseased rats after treatment [38]. Similar effects using ET_A-R blockade have been shown in other experimental models of CKD including lupus nephritis [39], puromycin-induced nephrosis [40], and IgA nephropathy [41]. In rats developing focal segmental glomerulosclerosis with age, the selective antagonism of ET_A-R reduces proteinuria, and even reverses established glomerulosclerosis and improves structural injury of podocytes [22]. In this latter study, increased DNA synthesis is observed after ET_A-R inhibition in ET-1-treated cultured podocytes, indicating an inhibitory effect of endogenous ET-1 on the podocyte cell cycle [22]. The significance of these observations in vivo needs confirmation in view of the fact that proliferation of such highly differentiated cells, the podocytes, is still a matter of debate [42, 43]. These findings concur to indicate that the kidney should be considered as a possible target of this class of drugs, whose effects are independent of changes in systemic blood pressure.

The effectiveness of ET-RA has also been proven in diabetic nephropathy associated with diabetes. Few studies have demonstrated the efficacy of dual ET_A/ET_B-RA in preventing proteinuria and diabetes-associated histologic alterations, consisting of renal fibrosis and glomerular matrix deposition [44]. These effects are partly due to the ability of these compounds to normalize NADPH oxidase, the main source of reactive oxygen species [45]. However, no data are available on the efficacy of the ET_A/ET_B-RA given at the late stage of the disease. Regarding selective antagonists, a pioneer study revealed that ET_A-RA administered to diabetic rats at the time of the disease induction prevents the development of renal injury and preserves renal function [46]. The protective effect on the glomerulus exerted by selective ET_A-RA could be attributable in part to the prevention of podocyte loss, which is a typical feature of overt diabetes [47]. The selective ET_A-R blockade with avosentan significantly attenuates renal functional and structural injury, including glomerulosclerosis, macrophage infiltration, nitrotyrosine expression, and diabetes-associated atherosclerosis, in streptozotocin-induced diabetic Apoe KO mice, a mouse model of accelerated diabetic nephropathy [48]. The capability of ET-RA in reducing urinary protein excretion has been largely documented; however, whether they interfere on the permselective properties of the glomerular barrier has not been completely clear for a long time. In vitro and ex vivo studies on cultured podocytes and glomeruli from healthy animals demonstrate that ET-1, by promoting changes in the contractile apparatus of the glomerular visceral epithelial cells, induces cell retraction and formation of intercellular gaps [19–21], events that lead to a significant increase in protein permeability across the glomerular filter [49].

Abnormal glomerular permeability induced by ET-1 in vitro has stimulated experiments aimed at evaluating whether ET-RA could possibly alter the glomerular permselective properties in vivo, as it occurs in animals on ACE inhibitors [50]. In this context, we recently found that blockade of ET-1 activity in experimental diabetes by avosentan does not affect the dimension of the pores perforating the glomerular barrier that are increased in size in diseased animals [51]. Despite glomerular pores larger than normal are still present after treatment, avosentan is capable of partially lowering urinary protein excretion, possibly through an effect on renal hemodynamics as documented by the tendency of renal blood flow to ameliorate in avosentan-treated animals [51]. Moreover, in line with data showing that ET_A-RA improves renal microvascular architecture in hypercholesterolemic pigs [52], we ascertained that ET_A-R blockade is associated with an improvement of peritubular capillaries architecture in diabetic animals [51]. It is possible that improved peritubular vascularization ameliorates renal interstitial blood perfusion, which contributes to the preservation of the tubulointerstitial structure and tubular function in terms of enhanced protein reabsorption [51].

As for drugs interfering with the renin-angiotensin system, renoprotection of ET-RA is more robust when given in the early phase of the disease [34, 53]; however, a combined therapy of ET-RA and ACE inhibitor is needed at the later stages [54]. The therapeutic potential of the multipharmacological approach of ET-RA on the top of renin-angiotensin system blockade is subsequently demonstrated when drugs are administered in the advanced phases of the disease. In overt diabetic nephropathy, the simultaneous blockade of ET-1 activity and angiotensin II synthesis normalizes proteinuria. This could be due to the capability of the combined therapy in maintaining glomerular barrier integrity via the restoration of the podocyte number, which is decreased in diseased animals [51].

The mechanism behind the protective effect on podocyte is unknown. The mature podocyte has limited capacity to divide [55]. Possible mechanisms for podocyte replacement are represented by the contribution of bone marrow stem cells that act as a reservoir for glomerular cells [56, 57] or the migration of progenitor cells recently identified in the Bowman's capsule, which could acquire a podocyte phenotype [58]. Combining ET_A-RA to an ACE inhibitor induces the regression of renal lesions, bringing the number of sclerotic glomeruli to values even lower than that observed in diseased animals before the start of the treatment [51]. The effect of drug combination in fully protecting animals from renal injury might be the result of two diverse mechanisms acting synergistically: the ACE inhibitor preserves glomerular permselective properties while the ET antagonist improves peritubular vascularization and perfusion with the consequent amelioration of tubular function (fig. 1) [51].

Human Studies

ET-1 infusion in healthy volunteers significantly reduced renal perfusion without changes in mean arterial pressure, which is in line with studies of animals [59] and humans [60], indicating that the renal vessels are peculiarly sensitive to local changes in plasma ET-1 level – more than the other vasculature. Administration of a selective ET_A-R antagonist to healthy subjects attenuated the effects of ET-1 infusion, including the reduction of renal vascular resistance [61]. The effectiveness of ET-RA has been demonstrated in patients with CKD. In hypertensive patients with CKD, the selective ET_A blockade with BQ-123 is effective in reducing blood pressure, increasing renal blood flow, and reducing renal vascular resistance [62]. Its effectiveness is blunted when combined with BQ-788, a selective ET_B-RA. Of note, BQ-788 administered alone induces renal and systemic vasoconstriction in patients with CKD as well as in healthy subjects, suggesting that the ET_A-RA is preferable to achieve renoprotection [62].

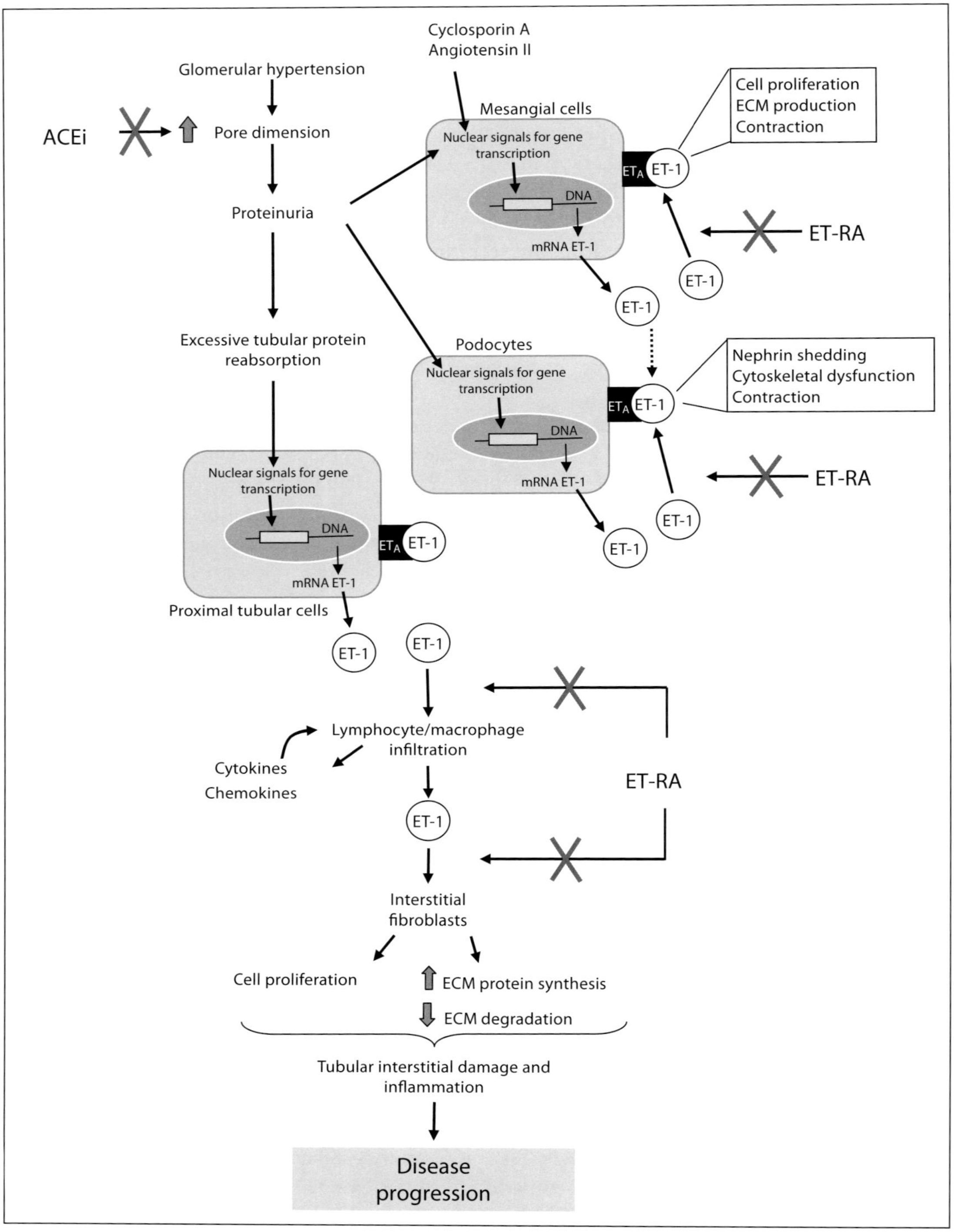
Cyclosporin A
Angiotensin II
Glomerular hypertension
ACEi
Pore dimension
Proteinuria
Mesangial cells
Nuclear signals for gene transcription
DNA
mRNA ET-1
ETA
ET-1
Cell proliferation
ECM production
Contraction
ET-RA
Excessive tubular protein reabsorption
Podocytes
Nephrin shedding
Cytoskeletal dysfunction
Contraction
Proximal tubular cells
Lymphocyte/macrophage infiltration
Cytokines
Chemokines
Interstitial fibroblasts
Cell proliferation
ECM protein synthesis
ECM degradation
Tubular interstitial damage and inflammation
Disease progression

The selective antagonism is also found to be effective in protecting renal function in patients with cardiovascular risk and early atherosclerosis who have been receiving atrasentan for 6 months [63]. Chronic treatment with atrasentan is safe and well tolerated since no increase in hepatic enzymes has been observed in treated patients. Only nasal stuffiness, headache, and edema have been observed after treatment [63]. The selective ET_A-RA also reduces blood pressure, proteinuria, and arterial stiffness in patients with focal segmental glomerulosclerosis and IgA and membranous nephropathy [64]. For patients with chronic proteinuric nephropathy receiving optimal treatment with ACE inhibitors and/or angiotensin receptor blockers, the addition of a selective ET_A-R antagonist, sitaxentan, reduces proteinuria, blood pressure, and arterial stiffness [65].

Few studies are available that describe the effect of ET_A-R antagonists in patients with type 2 diabetic nephropathy. This class of drugs is administered in type 2 diabetes only on top of the standard ACE inhibitor or angiotensin receptor blocker therapy. Avosentan treatment for 12 weeks in 286 patients with diabetes and microalbuminuria decreased urinary protein excretion without affecting creatinine clearance and blood pressure [66]. Patients experienced adverse events including peripheral edema and anemia – possibly as a result of hemodilution – with high drug dosage. In some cases the adverse events led to premature withdrawal from study medication. Subsequently, a larger trial with a hard endpoint had the aim to study the effectiveness of avosentan on the progression of overt diabetic nephropathy. Results from this trial showed that avosentan treatment decreases proteinuria but is associated with the occurrence of serious adverse events. The trial was stopped prematurely after 4 months due to an increased incidence of cardiovascular events, more frequent signs and symptoms of fluid overload, and a trend to a higher mortality rate with avosentan than with placebo [67]. It is also possible that high doses of avosentan, which displays only 50-fold higher affinity for ET_A-R versus ET_B-R, may partially inhibit ET_B-R with consequent significant fluid overload and pulmonary edema [67].

A very recent study rekindled interest in ET_A-R antagonists in diabetic nephropathy. Atrasentan was found to have an acceptable safety profile (although it was monitored for a short period of treatment) and effectively reduced residual albuminuria in diabetic patients who were on stable dosage of renin-angiotensin system inhibitors [68]. In this study, mild peripheral edema was the most common adverse event which showed a dose-response relationship. The low incidence of edema was attributable to the high selectivity of atrasentan for ET_A-R (1,800-fold selective for ET_A-R vs. ET_B-R).

Fig. 1. The combination of ACE inhibitor (ACEi)and ET-RA therapies protects the kidney from progressive renal injury through diverse mechanisms acting synergistically. ECM = Extracellular matrix.

Conclusions

A key role of ET-1 in the pathophysiology of CKD of nondiabetic and diabetic origin has been consistently documented in experimental animals and humans. Blockade of the ET receptor by antagonists has proven to be effective in retarding renal disease progression and even inducing regression of renal lesions in chronic animal models of renal injury. These findings have scarcely been translated into human diseases because of the side effects induced by the chronic administration of ET-RA. Clinical trials have been prematurely stopped due to major adverse events including peripheral edema and congestive heart failure, which calls for the pharmaceutical industry to develop less toxic drugs. Concerns about fluid retention with the use of ET_A-R antagonists show the need for better understanding of how the receptor systems are involved in sodium excretion. ET_B-R has a primary role in regulating sodium/water reabsorption in the kidney [29, 30]; therefore, its inhibition would cause fluid retention. However, ET_A-R blockade may also promote water retention with a mechanism which is poorly understood [16, 68].

The future challenge of renoprotective studies is focused on diabetes, which presently represents the major cause of end-stage renal disease worldwide. Late treatments of the disease involve the use of multiple drugs such as ET-RA on top of ACE inhibitors. A body of evidence demonstrates that ET-1 blockade together with inhibition of angiotensin II synthesis offer renoprotection greater than that obtained with single treatments. These findings highlight the value of a multidrug approach that simultaneously blocks the development of proteinuria with ACE inhibitor/angiotensin receptor blockers by preserving the permselective properties of the glomerular filter as well as the deleterious consequences of excessive ET-1 production generated by protein overabsorption by proximal tubular cells (fig. 1). The combined therapies may represent a new therapeutic approach for the treatment of CKD, especially for patients with late diagnosis or those completely unresponsive to ACE inhibitors.

Since clinical studies with ET antagonists have so far been scarce, it is desirable that in the near future the pharmaceutical industry will give impetus to testing new drugs to be added to the available armamentarium for the treatment of progressive kidney disease.

References

1 Grassmann A, Gioberge S, Moeller S, Brown G: ESRD patients in 2004: global overview of patient numbers, treatment modalities and associated trends. Nephrol Dial Transplant 2005;20:2587–2593.

2 Ruggenenti P, Remuzzi G: Kidney failure stabilizes after a two-decade increase: impact on global (renal and cardiovascular) health. Clin J Am Soc Nephrol 2007;2:146–150.

3 Brenner BM, Meyer TW, Hostetter TH: Dietary protein intake and the progressive nature of kidney disease: the role of hemodynamically mediated glomerular injury in the pathogenesis of progressive glomerular sclerosis in aging, renal ablation, and intrinsic renal disease. N Engl J Med 1982;307:652–659.
4 Remuzzi G, Bertani T: Pathophysiology of progressive nephropathies. N Engl J Med 1998;339:1448–1456.
5 Abbate M, Remuzzi G: Proteinuria as a mediator of tubulointerstitial injury. Kidney Blood Press Res 1999;22:37–46.
6 Abbate M, Zoja C, Remuzzi G: How does proteinuria cause progressive renal damage? J Am Soc Nephrol 2006;17:2974–2984.
7 Perico N, Benigni A, Remuzzi G: Present and future drug treatments for chronic kidney diseases: evolving targets in renoprotection. Nat Rev Drug Discov 2008;7:936–953.
8 Benigni A: Tubulointerstitial disease mediators of injury: the role of endothelin. Nephrol Dial Transplant 2000;15(Suppl 6):50–52.
9 Dhaun N, Goddard J, Webb DJ: The endothelin system and its antagonism in chronic kidney disease. J Am Soc Nephrol 2006;17:943–955.
10 Longaretti L, Benigni A: Endothelin receptor selectivity in chronic renal failure. Eur J Clin Invest 2009;39(Suppl 2):32–37.
11 Cybulsky AV, Stewart DJ, Cybulsky MI: Glomerular epithelial cells produce endothelin-1. J Am Soc Nephrol 1993;3:1398–1404.
12 Lyon-Roberts B, Strait KA, van Peursem E, et al: Flow regulation of collecting duct endothelin-1 production. Am J Physiol Renal Physiol 2011;300:F650–F656.
13 Benigni A, Remuzzi G: Endothelin antagonists. Lancet 1999;353:133–138.
14 Ahn D, Ge Y, Stricklett PK, Gill P, et al: Collecting duct-specific knockout of endothelin-1 causes hypertension and sodium retention. J Clin Invest 2004;114:504–511.
15 Gariepy CE, Ohuchi T, Williams SC, et al: Salt-sensitive hypertension in endothelin-B receptor-deficient rats. J Clin Invest 2000;105:925–933.
16 Ge Y, Bagnall A, Stricklett PK, et al: Combined knockout of collecting duct endothelin A and B receptors causes hypertension and sodium retention. Am J Physiol Renal Physiol 2008;295:F1635–F1640.
17 Ishizawa K, Yoshizumi M, Tsuchiya K, et al: Dual effects of endothelin-1 (1–31): induction of mesangial cell migration and facilitation of monocyte recruitment through monocyte chemoattractant protein-1 production by mesangial cells. Hypertens Res 2004;27:433–440.
18 Simonson MS, Ismail-Beigi F: Endothelin-1 increases collagen accumulation in renal mesangial cells by stimulating a chemokine and cytokine autocrine signaling loop. J Biol Chem 2011;286:11003–11008.
19 Morigi M, Buelli S, Angioletti S, et al: In response to protein load podocytes reorganize cytoskeleton and modulate endothelin-1 gene: implication for permselective dysfunction of chronic nephropathies. Am J Pathol 2005;166:1309–1320.
20 Collino F, Bussolati B, Gerbaudo E, et al: Preeclamptic sera induce nephrin shedding from podocytes through endothelin-1 release by endothelial glomerular cells. Am J Physiol Renal Physiol 2008;294:F1185–F1194.
21 Morigi M, Buelli S, Zanchi C, et al: Shigatoxin-induced endothelin-1 expression in cultured podocytes autocrinally mediates actin remodeling. Am J Pathol 2006;169:1965–1975.
22 Ortmann J, Amann K, Brandes RP, et al: Role of podocytes for reversal of glomerulosclerosis and proteinuria in the aging kidney after endothelin inhibition. Hypertension 2004;44:974–981.
23 Zoja C, Morigi M, Figliuzzi M, et al: Proximal tubular cell synthesis and secretion of endothelin-1 on challenge with albumin and other proteins. Am J Kidney Dis 1995;26:934–941.
24 Benigni A, Perico N, Gaspari F, et al: Increased renal endothelin production in rats with reduced renal mass. Am J Physiol 1991;260:F331–F339.
25 Bruzzi I, Corna D, Zoja C, et al: Time course and localization of endothelin-1 gene expression in a model of renal disease progression. Am J Pathol 1997;151:1241–1247.

26 Zoja C, Benigni A, Camozzi D, et al: Combining lisinopril and L-arginine slows disease progression and reduces endothelin-1 in passive Heymann nephritis. Kidney Int 2003;64:857–863.
27 Zoja C, Liu XH, Abbate M, et al: Angiotensin II blockade limits tubular protein overreabsorption and the consequent upregulation of endothelin 1 gene in experimental membranous nephropathy. Exp Nephrol 1998;6:121–131.
28 Nakamura T, Ebihara I, Fukui M, Osada S, et al: Renal expression of mRNAs for endothelin-1, endothelin-3 and endothelin receptors in NZB/W F1 mice. Ren Physiol Biochem 1993;16:233–243.
29 Tazawa N, Okada Y, Nakata M, et al: Exaggerated vascular and renal pathology in endothelin-B-receptor-deficient rats with subtotal nephrectomy. J Cardiovasc Pharmacol 2004;44(Suppl 1):S467–S470.
30 Pfab T, Thone-Reineke C, Theilig F, et al: Diabetic endothelin B receptor-deficient rats develop severe hypertension and progressive renal failure. J Am Soc Nephrol 2006;17:1082–1089.
31 Yamauchi T, Ohnaka K, Takayanagi R, et al: Enhanced secretion of endothelin-1 by elevated glucose levels from cultured bovine aortic endothelial cells. FEBS Lett 1990;267:16–18.
32 Wu Y, Feng B, Chen S, et al: Glucose-induced endothelin-1 expression is regulated by ERK5 in the endothelial cells and retina of diabetic rats. Can J Physiol Pharmacol 2010;88:607–615.
33 Hua H, Goldberg HJ, Fantus IG, Whiteside CI: High glucose-enhanced mesangial cell extracellular signal-regulated protein kinase activation and alpha1(IV) collagen expression in response to endothelin-1: role of specific protein kinase C isozymes. Diabetes 2001;50:2376–2383.
34 Benigni A, Colosio V, Brena C, et al: Unselective inhibition of endothelin receptors reduces renal dysfunction in experimental diabetes. Diabetes 1998;47:450–456.
35 Sasser JM, Sullivan JC, Hobbs JL, et al: Endothelin A receptor blockade reduces diabetic renal injury via an anti-inflammatory mechanism. J Am Soc Nephrol 2007;18:143–154.
36 Fukui M, Nakamura T, Ebihara I, et al: Gene expression for endothelins and their receptors in glomeruli of diabetic rats. J Lab Clin Med 1993;122:149–156.
37 Opocensky M, Kramer HJ, Backer A, et al: Late-onset endothelin-A receptor blockade reduces podocyte injury in homozygous Ren-2 rats despite severe hypertension. Hypertension 2006;48:965–971.
38 Benigni A, Zoja C, Corna D, et al: A specific endothelin subtype A receptor antagonist protects against injury in renal disease progression. Kidney Int 1993;44:440–444.
39 Nakamura T, Ebihara I, Tomino Y, Koide H: Effect of a specific endothelin A receptor antagonist on murine lupus nephritis. Kidney Int 1995;47:481–489.
40 Ebihara I, Nakamura T, Tomino Y, Koide H: Effect of a specific endothelin receptor A antagonist and an angiotensin-converting enzyme inhibitor on glomerular mRNA levels for extracellular matrix components, metalloproteinases (MMP) and a tissue inhibitor of MMP in aminonucleoside nephrosis. Nephrol Dial Transplant 1997;12:1001–1006.
41 Nakamura T, Ebihara I, Fukui M, et al: Effect of a specific endothelin receptor A antagonist on glomerulonephritis of ddY mice with IgA nephropathy. Nephron 1996;72:454–460.
42 Thorner PS, Ho M, Eremina V, et al: Podocytes contribute to the formation of glomerular crescents. J Am Soc Nephrol 2008;19:495–502.
43 Lasagni L, Ballerini L, Angelotti ML, et al: Notch activation differentially regulates renal progenitors proliferation and differentiation toward the podocyte lineage in glomerular disorders. Stem Cells 2010;28:1674–1685.
44 Cosenzi A, Bernobich E, Trevisan R, et al: Nephroprotective effect of bosentan in diabetic rats. J Cardiovasc Pharmacol 2003;42:752–756.

45 Xu M, Dai DZ, Dai Y: Normalizing NADPH oxidase contributes to attenuating diabetic nephropathy by the dual endothelin receptor antagonist CPU0213 in rats. Am J Nephrol 2009;29:252–256.
46 Nakamura T, Ebihara I, Fukui M, et al: Effect of a specific endothelin receptor A antagonist on mRNA levels for extracellular matrix components and growth factors in diabetic glomeruli. Diabetes 1995;44:895–899.
47 Gross ML, El-Shakmak A, Szabo A, et al: ACE-inhibitors but not endothelin receptor blockers prevent podocyte loss in early diabetic nephropathy. Diabetologia 2003;46:856–868.
48 Watson AM, Li J, Schumacher C, et al: The endothelin receptor antagonist avosentan ameliorates nephropathy and atherosclerosis in diabetic apolipoprotein E knockout mice. Diabetologia 2010;53:192–203.
49 Saleh MA, Boesen EI, Pollock JS, et al: Endothelin-1 increases glomerular permeability and inflammation independent of blood pressure in the rat. Hypertension 2010;56:942–949.
50 Remuzzi A, Perico N, Amuchastegui CS, et al: Short- and long-term effect of angiotensin II receptor blockade in rats with experimental diabetes. J Am Soc Nephrol 1993;4:40–49.
51 Gagliardini E, Corna D, Zoja C, et al: Unlike each drug alone, lisinopril if combined with avosentan promotes regression of renal lesions in experimental diabetes. Am J Physiol Renal Physiol 2009;297:F1448–F1456.
52 Chade AR, Krier JD, Textor SC, et al: Endothelin-A receptor blockade improves renal microvascular architecture and function in experimental hypercholesterolemia. J Am Soc Nephrol 2006;17:3394–3403.
53 Perico N, Amuchastegui SC, Colosio V, et al: Evidence that an angiotensin-converting enzyme inhibitor has a different effect on glomerular injury according to the different phase of the disease at which the treatment is started. J Am Soc Nephrol 1994;5:1139–1146.
54 Benigni A, Corna D, Maffi R, et al: Renoprotective effect of contemporary blocking of angiotensin II and endothelin-1 in rats with membranous nephropathy. Kidney Int 1998;54:353–359.
55 Wiggins RC: The spectrum of podocytopathies: a unifying view of glomerular diseases. Kidney Int 2007;71:1205–1214.
56 Ito T, Suzuki A, Imai E, et al: Bone marrow is a reservoir of repopulating mesangial cells during glomerular remodeling. J Am Soc Nephrol 2001;12:2625–2635.
57 Poulsom R, Forbes SJ, Hodivala-Dilke K, et al: Bone marrow contributes to renal parenchymal turnover and regeneration. J Pathol 2001;195:229–235.
58 Sagrinati C, Netti GS, Mazzinghi B, et al: Isolation and characterization of multipotent progenitor cells from the Bowman's capsule of adult human kidneys. J Am Soc Nephrol 2006;17:2443–2456.
59 Pernow J, Boutier JF, Franco-Cereceda A, et al: Potent selective vasoconstrictor effects of endothelin in the pig kidney in vivo. Acta Physiol Scand 1988;134:573–574.
60 Rabelink TJ, Kaasjager KA, Boer P, et al: Effects of endothelin-1 on renal function in humans: implications for physiology and pathophysiology. Kidney Int 1994;46:376–381.
61 Honing ML, Hijmering ML, Ballard DE, et al: Selective ET(A) receptor antagonism with ABT-627 attenuates all renal effects of endothelin in humans. J Am Soc Nephrol 2000;11:1498–1504.
62 Goddard J, Johnston NR, Hand MF, et al: Endothelin-A receptor antagonism reduces blood pressure and increases renal blood flow in hypertensive patients with chronic renal failure: a comparison of selective and combined endothelin receptor blockade. Circulation 2004;109:1186–1193.
63 Raichlin E, Prasad A, Mathew V, et al: Efficacy and safety of atrasentan in patients with cardiovascular risk and early atherosclerosis. Hypertension 2008;52:522–528.
64 Dhaun N, Macintyre IM, Melville V, et al: Blood pressure-independent reduction in proteinuria and arterial stiffness after acute endothelin-A receptor antagonism in chronic kidney disease. Hypertension 2009;54:113–119.

65 Dhaun N, Macintyre IM, Kerr D, et al: Selective endothelin-A receptor antagonism reduces proteinuria, blood pressure, and arterial stiffness in chronic proteinuric kidney disease. Hypertension 2011;57:772–779.
66 Wenzel RR, Littke T, Kuranoff S, et al: Avosentan reduces albumin excretion in diabetics with macroalbuminuria. J Am Soc Nephrol 2009;20:655–664.
67 Mann JF, Green D, Jamerson K, et al: Avosentan for overt diabetic nephropathy. J Am Soc Nephrol 2010;21:527–535.
68 Kohan DE, Pritchett Y, Molitch M, et al: Addition of atrasentan to renin-angiotensin system blockade reduces albuminuria in diabetic nephropathy. J Am Soc Nephrol 2011;22:763–772.

Ariela Benigni, PhD
Head, Department Molecular Medicine, Mario Negri Institute for Pharmacological Research
Centro Anna Maria Astori, Science and Technology Park Kilometro Rosso
Via Stezzano, 87
IT–24126 Bergamo (Italy)
Tel. +39 035 42131, E-Mail ariela.benigni@marionegri.it

Barton M, Kohan DE (eds): Endothelin in Renal Physiology and Disease.
Contrib Nephrol. Basel, Karger, 2011, vol 172, pp 185–199

Endothelin in Renal Injury due to Sickle Cell Disease

Pierre-Louis Tharaux

Inserm, Université Paris Descartes, Paris, France

Abstract

Sickle cell disease (SCD), the first 'molecular disease' to be identified, has been well characterized as a single amino acid molecular disorder of hemoglobin leading to its pathological polymerization, with resulting red cell rigidity that causes poor microvascular blood flow, with consequent tissue ischemia and infarction. Thus, the manifestations of SCD chronic renal disease have long been considered clinical manifestations of an obstructive vasculopathy of the arterial and capillary microcirculation. Recently, accumulating evidence have indicated that blood vessel functions are affected by SCD, involving abnormal vascular tone and activated endothelium. These abnormalities are particularly prominent in the kidney where specific biochemical conditions in the medulla and papilla favor change in endothelial phenotype and in tubular phenotype that, in turn, may promote dysfunction and destruction of this organ through active endothelin (ET)-1 production and signals. High ET-1 urinary output in SCD subjects at steady state may reflect increased tubular activation of ET-1 production acting on the collecting duct thereby favoring the constant hyposthenuria. Chronically, augmented ET-1 concentrations in the SCD kidney would further aggravate ischemia and sickling through actions on vasa recta and red blood cells. The kidneys suffer multiple ischemic hits during SCD as consequences of vaso-occlusive crisis (VOC). Blockade of ET receptors unraveled the major vasoconstrictive role of ET-1 in the pathophysiology of VOC, stressing the pivotal role of abnormal endothelial phenotype in this hemoglobinopathy and opening potential new therapeutic options. At last, indirect evidence suggest that ET-1 may be involved in the progression of chronic glomerulosclerosis affecting a number of patients. In fact, sickle cell nephropathy is an emerging severe disease that requires pathophysiological studies and development of specific therapies.

Sickle cell disease (SCD) is one the most common inherited blood disorders in the world. SCD is an autosomal recessive disease caused by a point mutation in the hemoglobin (Hb) beta gene found on chromosome 11p15.4. A mutation in the Hb beta gene results in the production of a structurally abnormal Hb, called HbS. Under low oxygen levels or high Hb concentrations, in individuals who are homozygous for HbS, the abnormal HbS molecules cluster together, distorting the red blood cells (RBCs) into sickled shapes. These deformed and rigid RBCs become trapped within small blood vessels and block them, producing pain and eventually damaging organs. Sickle cell nephropathy (SCN) is a prominent complication of SCD, now a well characterized entity with specific manifestations, risk factors, and prognosis [1–9]. In the present short review we will examine evidence for a role of the endothelin (ET) system in the pathophysiology of SCN taken specifically and also as a model for the generalized vasculopathy of SCD.

Defects in Renal Medullary Functions

SCD affects the kidney by acute mechanisms as well as by insidious renal medullary/papillary necrosis, firstly resulting in tubular defects, which increase the risk of dehydration and subsequent sickle crisis. Hyposthenuria is the first clinical evidence of defective medullary tonicity. In preadolescent SCD patients, the maximal urinary concentration is often reduced to 400 mOsmol/kg H_2O [10]. This urine concentrating capacity can be in part restored with multiple transfusions of normal erythrocytes, which shows the reversible nature of the defect [11]. However, in patients older than 15 years, the process is often irreversible [12]. Both vasopressin generation and urinary diluting capacity in SCN patients remain unchanged [8, 13].

The mechanism for this early medullary defect is not fully known. Already in the late 1930s, William Bosworth Castle and his colleague Thomas Hale Ham were studying blood samples obtained from patients with sickle cell anemia and found that the viscosity of the blood increased dramatically as its oxygen content decreased. As Castle later reflected, 'It immediately occurred to us that this was because the elongated, sickled red cells had become tangled up "like haywire"' (cited in [14]). Castle and Ham went on to hypothesize about the pathophysiology of organ damage in SCD anemia in a way that still drives much of our thinking today: 'a vicious cycle of erythrostasis may be set up because the critical oxygen tension for a marked increase of sickling is relatively close to that of the normal venous blood' [15]. This hypothesis predicts that as red cells become deoxygenated in the capillaries, the sickle Hb inside them polymerizes, decreasing the cells' ability to squeeze through the capillaries in single file. Blood flow stops, and the surrounding tissues become ischemic. The hypothesis makes sense for tissues such as the kidney medulla with physiological and pathophysiological low oxygen tension [16–18], where the blood flow is sluggish. Furthermore, due to its important epithelial metabolic activity [19], the renal medulla is very susceptible to hypoxia, and in patients with SCD, it is constantly damaged [7, 20]. In summary, low oxygen tension with acidosis and hyper osmolar conditions encountered in the medulla and the renal papillae are extremely favorable to sickling of RBCs in vasa recta [21], thereby causing microthrombi, infarction, and collateral formation.

When such events occur acutely, the result is renal papillae necrosis and hematuria. These anatomical changes in the medullary circulation induce defects in zonation and countercurrent exchange with concentrating capacity. As a consequence, the capacity to generate negative free water is defective [22, 23] and concomitant medullary sodium loss occurs [23]. The tendency for hypovolemia and unexplained high renal blood flow and glomerular filtration rate (GFR) tend to favor increased proximal tubular activity, finally resulting in normal sodium conservation [24].

Other minor disturbances such as defects in both urinary acidification and potassium excretion are frequent, although overt metabolic acidosis and hyperkalemia occur infrequently. Proximal tubular function is supranormal, as manifested by increased reabsorption of phosphate and increased secretion of creatinine. The former may result in mild hyperphosphatemia, while the latter causes substantial overestimation of the GFR by creatinine clearance. The most prominent feature of medullary damage the patients encounter is impaired urinary concentrating ability, but with an intact diluting capacity.

The Endothelin System and Renal Medullary Functions in Sickle Cell Disease Subjects

This same medullary area is prone to high local ET concentrations. The preproET-1 gene is induced by hypoxia sensitive pathway(s) in the PCT [25] and the DCT [26], the inner medullary collecting duct (CD) [27, 28], and in endothelial cells [29–31] in a VHL/HIF pathway-dependent fashion [32, 33], at least in this latter case. In addition, increased vasoconstrictive responses to ET-1 have been observed in acute and chronically hypoxic rats as well as in isolated human arteries [34–36]. Thus, the naturally hypoxic milieu of the medulla, further aggravated by sickling of erythrocytes and marked rarefaction of peritubular capillaries, would promote increased ET-1 secretion and may also potentiate its potent vasoconstrictive action. Interestingly, luminal shear stress and flow was recently demonstrated to be an additional mechanism stimulating ET-1 production by the mouse and rat CD [37]. This mechanism may sustain ET-1 synthesis in the context of hypothenuria.

In line with these experimental facts, we found elevated ET-1 plasma concentrations and, more prominently, a fourfold increase in ET-1 urinary output in young adult patients with SCD, compared with African-French and African matched controls at steady state [13]. Because most of the plasma ET-1 filtered through glomeruli is subject to degradation by neutral endopeptidase (EC 3.4.24.11) in the proximal tubule and by an acidic, highly specific ET-1-inactivating metalloendopeptidase, ET-1 in urine is of renal origin [38, 39]. Thus, urinary ET-1 reflects the production of this peptide by tubules and is a hallmark of specific pathophysiological processes. These data suggest increased ET-1 concentrations in tubules from SCD patients at steady state. Furthermore, urinary ET-1 output and urine osmolality were significantly negatively correlated. Since ET-1 promotes increased free-water clearance and counters the effects of ADH on tubular functions through ET_A [40–42], the marked increase in urinary ET-1 excretion during SCD at steady state is likely to favor hyposthenuria and subsequent dehydration, a potential trigger of an acute crisis. In addition, increased renal ET-1 production may promote renal hypoperfusion and progressive tissue scarring.

Interestingly, the prevalence of arterial hypertension in SCD patients is not as elevated as one would expect in populations of African descent. Increased medullary production of ET-1 in sickle kidney may in fact protect development of hypertension through slight urinary chronic sodium loss. This has not been investigated in depth, but solid evidence has indicated that CD-derived ET-1 is an important physiologic regulator of renal Na excretion and systemic blood pressure through a predominant ET_B-mediated effect on CD [43, 44].

Another target of the high ET-1 concentration in the kidney could be RBCs. The stimulation of ET receptor type B subtypes favors RBC dehydration through the activation of the Gardos channel, causing cellular volume to decrease and cellular Hb concentration to increase [45, 46]. This increase in erythrocytic Hb concentration leads to an increased rate of HbS polymerization and sickling under hypoxic and hypertonic conditions, for example, in the renal medulla. Indeed, in a model of experimental vaso-occlusive crisis (VOC) in sickle transgenic mice we could observe that dual ET_A/ET_B receptor blockade with bosentan alleviated the level of circulating dehydrated dense RBCs [47]. This may explain part of the observed alleviation of renal ischemia measured in the treated sickle mice.

The Role of the Endothelin System in Acute Renal Insults during Vaso-Occlusive Events

Painful VOC is one of the major and specific manifestations of SCD. VOC is a complex process involving Hb S polymer formation and poorly deformable dense RBCs, leading to vaso-occlusions of capillaries and postcapillary venules due to activated endothelium entrapping hyper adhesive sickle RBCs, reticulocytes, and leukocytes [48–51]. These multiple local ischemic events cause tissue damage and life-threatening complications. Although kidney ischemia is by far more clinically silent and less painful than other VOC complications, there is a lot of experimental evidence indicating that the kidneys are also targeted by VOCs. ET-1 was found produced at a higher level in the renal and pulmonary microvasculature of transgenic sickle SAD mice than of wild-type mice at steady state, consistent with the abnormally high systemic ET-1 levels in sera and urines from SCD patients [13, 52–55]. Notably, most of preproET-1 mRNA expression assessed by in situ hybridization in the lung and kidneys from these animals proved to originate from capillary and arteriolar endothelium [47]. Interestingly, sickle mice chronically treated with bosentan did not display significant systemic or regional hemodynamic changes at steady state, suggesting that endogenous vascular ET-1, even at a high level, does not markedly alter the general hemodynamics, as also seen in other models such as experimental hypertension [56, 57] and transgenic mice overexpressing ET-1 [58].

The cause of increased endothelial production of ET-1 in SCD has not been fully elucidated. Increased endothelial shear stress due to hyperviscosity and flow in the microcirculation may play a role. Pathophysiological conditions associated with SCD, including diminished bioavailability of nitric oxide (NO) [59–61] due to cell-free Hb [62], increased release of erythrocyte-derived arginase [63], increased oxidative stress [59], local hypoxia, and cytokine release, may promote activation of endothelial ET-1 production with unbalanced and excessive ET-1-mediated vasoconstriction as shown in conductance arteries [64] and in renal arterioles [65]. SAD transgenic mice do not display a hemolytic anemia as severe as that observed in most human sickle subjects, which suggests that the shortfall in NO due to cell-free Hb or release of erythrocyte arginase may not be as severe in SAD mice as in patients. This may explain in part why SAD mice experience little spontaneous VOC without a hypoxic challenge. Nevertheless, NO administration has been shown to inhibit VOC-induced lung damage in the same model of H/R-induced VOC [66, 67] although this effect was unfortunately not confirmed among patients with SCD hospitalized with VOC [68]. Altogether, these various observations suggest that an imbalance between ET-1 and NO plays a critical role in SCD with some compensation at steady state.

In acute VOC, sudden falls in blood flow and shear stress may substantially alter endothelial production of NO [69], a potent inhibitor of ET-1 synthesis [70] and its vasoconstrictive effect [71]. Thus, ET receptor inactivation with bosentan in a mouse model leads to reversal or prevention of vaso-occlusive events, including tissue damage, RBC dehydration, leukocyte recruitment, infiltration of organs with activated neutrophils, nitrative stress, and death. Furthermore, this study demonstrated a rapidly reversible vasospastic component in SCD [47], suggesting a novel paradigm for SCD. Using a noninvasive method to measure renal blood flow velocity in the whole kidney [72], it was found that up to 50% of the fall in renal blood flow induced by experimental VOC is reversed within minutes by acute ET receptor antagonism. The fast kinetics of this reversal suggests that only sickle but not normal mice suffered substantial renal vasoconstriction, aggravating blood cell entrapment [47].

Other pathophysiological ET-1-mediated actions may have been antagonized in this model since ET-1 has been suggested to promote entrapment of platelets and neutrophils in the isolated-perfused lung model [73]. These early findings may be relevant in the context of ischemia-reperfusion injury caused by VOC. Also evoking a rapid proinflammatory action of ET-1 in vivo, bolus i.v. injection of ET-1 to anaesthetized guinea-pigs evoked dose-dependent neutropenia with relative lymphocytosis and monocytosis, but did not induce influx of neutrophil granulocytes into pulmonary tissues [74]. Neutropenia is expected to be due to enhanced neutrophil adhesion to the endothelium [75], possibly through modulation of expression of VCAM-1 and ICAM-1 [76] or expression of CD18 and CD11b on the neutrophil surface [77]. Nevertheless, SCD patients and SAD mice display neutrophilia instead of neutropenia [47, 78, 79] that is positively correlated to vascular acute events [79, 80]. Furthermore, the neutrophil count is further increased during VOC, and bosentan administration prevented this increase in SAD mice [47]. Thus, increased ET-1 activity in SCD may affect other leukocyte fates beside adhesion, such as mobilization of neutrophils from the bone marrow, neutrophil apoptosis [81, 82], or impaired clearance of apoptotic neutrophils by macrophages. Sequestered neutrophils coordinate tissue injury by interacting with the endothelium, circulating RBCs, and platelets. These interactions are mediated through the actions of endothelial E-selectin, neutrophil E-selectin ligand-1, and the promiscuous leukocyte integrin $\alpha_M\beta_2$ (also known as CD11b/CD18, Mac-1, and CR3). Using elegant in vivo multichannel fluorescent imaging of the systemic circulation, Hidalgo et al. [83] reported that the E-selectin ligand-1 on neutrophils, by binding to E-selectin on endothelium, signals the polarized activation of surface $\alpha_M\beta_2$ integrin on the neutrophils. In a mouse model of SCD, they showed that $\alpha_M\beta_2$ integrin in turn mediates the heterotypic associations of neutrophils with normal and sickle RBCs [83]. Furthermore, platelet-endothelium interaction could be affected by ET-1. Recent clinical investigation demonstrated that, further to inhibiting endothelium-dependent vasodilatation, ET-1 may also attenuate endothelium-dependent inhibition of platelet activation induced by acetylcholine [84]. Altogether, studies are still needed to evaluate the involvement of the ET system in the complex interplay between neutrophils, platelets, and sickle cells [48, 83, 85].

If bosentan or alternative ET receptor antagonists prove safe and effective for preventing or treating acute vaso-occlusive events in the clinical setting, they may provide major benefits for lung and renal integrity, quality of life, and survival of sickle cell patients.

Potential Involvement of the Endothelin-1 System in Chronic Sickle Cell Nephropathy and Some Remaining Questions

Although acute ischemic insults would impact the kidney architecture and function, SCN is still a progressive and chronic severe complication of SCD. Also, in addition to mild medullary dysfunction, proteinuria and global or focal segmental glomerulosclerosis with renal insufficiency are of concern. In 2005, Powars et al. [86] reported the prevalence of irreversible organ damage in 1,056 Hb SS patients and found a 12% prevalence rate of chronic renal insufficiency with a median age at diagnosis of 37 years [34–39]. Furthermore, sickle chronic kidney disease had the most patients with prior chronic conditions (75%) such as chronic lung disease, priapism, gall bladder disease, avascular necrosis, cerebral vascular accident, leg ulcer, and retinopathy [86]. This may be the cumulative consequence of the lifetime hemolytic anemia [87].

The clustering of organ damage in multiple sites has a profound negative influence on the quality of life and survival [86, 88–92]. The manifestations of chronic renal disease [7, 93, 94], chronic pulmonary disease with pulmonary hypertension [95, 96], retinopathy [97–101], and cerebral microinfarctions [86, 102] have long been considered clinical manifestations of an obstructive vasculopathy of the arterial and capillary microcirculation. In a prospective survival analysis of 964 patients with sickle cell anemia in adults, Platt et al. [89] observed an 18% overall mortality in adult SCD patients with 40% of these (7.6% of the total) manifesting overt renal failure. By multivariate regression analysis, renal failure was identified as the major risk factor for early mortality in adult patients with SCD.

Before these seminal studies, the severity of SCN was underestimated for a long time, in part because the average life expectancy of SCD subjects was short, with many patients not living long enough to develop overt renal failure. Common chronic glomerular hyperfiltration (leading to low creatinine plasma levels) may also lead to underestimation of the glomerular disease. Once progression of the glomerular damage is evident, GFR begins to decrease and return to normal values. Moreover, the prevalence of abnormal Hb, HbAS, and HbAC in an African-American end-stage renal disease cohort was found to be more than twice that of the baseline African-American population in the same geographic region (North Carolina, USA). The high prevalence of these hemoglobinopathies suggests that they may contribute to progression to end-stage renal disease by providing a background of global renal injury [103]. Thus, SCN is no longer confined to the medullary changes per se; it extends to a more complex entity with changes in cortical functions which compensate for a defective water and sodium conservation in the medulla.

de Jong and Statius van Eps [24] proposed that specific renal physiologic derangements in SCD might lead to a chronic glomerulosclerosis [24, 104]. Among key predictors and risk factors for glomerulosclerosis and renal failure, long-term hyperfiltration with glomerular hypertrophy (glomerulomegaly [105]) is suspected from experimental and clinical data [2, 6, 106–112]. Longitudinal studies are needed to reconstitute the natural history of SCN. Nevertheless, high renal blood flow and glomerular hyperfiltration is frequent in children with a high incidence of microalbuminuria [113–118]. Both GFR and renal plasma flow are increased in young patients with SCD [119, 120]. The prevalence of hyperfiltration is as high as 49% in young adults [121] with a SCN prevalence of 84% at the age of 27 years when we consider hyperfiltration together with albuminuria or renal failure as relevant criteria [121]. Early loss of the ultrafiltration coefficient and glomerular permselectivity then occur [122]. Thus, there is a need for novel therapeutic options.

So far, we do not know if increased ET-1 in the sickle kidney participates in potentially deleterious glomerular hyperfiltration. ET receptor blockade is expected to alleviate postglomerular resistances as shown in various experimental and clinical conditions. Thus, ET-1 antagonism may be beneficial by blunting the high glomerular capillary pressure. Nevertheless, high ET-1-mediated vasoconstrictive tone in the kidney does not usually lead to elevated renal blood flow (with subnormal blood pressure and low renal vascular resistances). Intriguingly, prostaglandin inhibitors decrease the GFR [120, 123, 124]. In fact, indomethacin did not change GFR and effective renal blood flow in control subjects; a significant fall in these functions was found after indomethacin administration to SCD subjects [123]. This suggests that prostaglandins are of importance in maintaining the supernormal or normal GFR and effective renal blood flow in SCD patients. Prostaglandins may therefore be relevant with respect to the hyperfiltration in young SCD patients. Because prostaglandin production in the kidney is promoted by various, often vasoconstrictor, stimuli, it is quite conceivable that ischemic damage to the inner medulla in SCD induces synthesis of vasodilator prostaglandins. We hypothesize that ET-1 would be a potent stimulus for prostaglandins and prostacyclin during SCN progression [125, 126]. Further studies will be needed to evaluate the potential for chronic manipulation of the ET system in limiting progression of SCD-associated glomerulopathy. In addition, as discussed about the pathophysiology of VOC, a chronic proinflammatory effect of ET-1 is likely as recently shown in other models of chronic glomerulopathy [127, 128].

In summary, SCD is a severe disease with remarkable activation of ET-1 synthesis in the kidney. In the acute setting of experimental VOC, ET-1-mediated vasoconstriction participates in renal ischemia and overall severity of the vasculopathy. Similarly, we cannot rule out that local ET-1 is involved with infraclinical VOC in the kidney.

At steady state, two prominent chronic functional features of SCN, namely hyposthenuria and glomerular hyperfiltration may be linked to increased local ET signals leading to counteraction of ADH action in the CD and to stimulation of potent vasodilatory molecules acting on the preglomerular vasculature, respectively. Let us hope the clinical challenge will stimulate further preclinical and clinical studies.

References

1 Schlitt LE, Keitel HG: Renal manifestations of sickle cell disease: a review. Am J Med Sci 1960;239:773–778.

2 Pitcock JA, Muirhead EE, Hatch FE, Johnson JG, Kelly BJ: Early renal changes in sickle cell anemia. Arch Pathol 1970;90:403–410.

3 Berman LB: Sickle cell nephropathy. JAMA 1974;228:1279.

4 Buckalew VM Jr, Someren A: Renal manifestations of sickle cell disease. Arch Intern Med 1974;133:660–669.

5 Chaubey BS, Waiker SM, Shivde AV, Grover S: Renal manifestations of sickle cell disease. J Assoc Physicians India 1975;23:171–177.

6 Flanagan G, Packham DK, Kincaid-Smith P: Sickle cell disease and the kidney. Am J Kidney Dis 1993;21:325–327.

7 Falk RJ, Jennette JC: Sickle cell nephropathy. Adv Nephrol Necker Hosp 1994;23:133–147.

8 Saborio P, Scheinman JI: Sickle cell nephropathy. J Am Soc Nephrol 1999;10:187–192.

9 Scheinman JI: Sickle cell disease and the kidney. Semin Nephrol 2003;23:66–76.

10 Statius van Eps LW, Schouten H, Haar Romeny-Wachter CC, La Porte-Wijsman LW: The relation between age and renal concentrating capacity in sickle cell disease and hemoglobin C disease. Clin Chim Acta 1970;27:501–511.

11 Itano HA, Keitel HG, Thompson D: Hyposthenuria in sickle cell anemia: a reversible renal defect. J Clin Invest 1956;35:998–1007.

12 Levitt MF, Hauser AD, Levy MS, Polimeros D: The renal concentrating defect in sickle cell disease. Am J Med 1960;29:611–622.

13 Tharaux PL, Hagege I, Placier S, Vayssairat M, Kanfer A, Girot R, Dussaule JC: Urinary endothelin-1 as a marker of renal damage in sickle cell disease. Nephrol Dial Transplant 2005;20:2408–2413.

14 Platt OS: Preventing stroke in sickle cell anemia. N Engl J Med 2005;353:2743–2745.

15 Ham TH, Castle WB: Relation of increased hypotonic fragility and of erythrostasis to the mechanism of hemolysis in certain anemias. Trans Assoc Am Physicians 1940;55:127–132.

16 Brezis M, Rosen SN, Epstein FH: The pathophysiological implications of medullary hypoxia. Am J Kidney Dis 1989;13:253–258.

17 Epstein FH: Oxygen and renal metabolism. Kidney Int 1997;51:381–385.

18 Prasad PV, Edelman RR, Epstein FH: Noninvasive evaluation of intrarenal oxygenation with BOLD MRI. Circulation 1996;94:3271–3275.

19 Brezis M, Rosen S: Hypoxia of the renal medulla – its implications for disease. N Engl J Med 1995;332:647–655.

20 Lesobre B, Pieron R, Mafart Y, Paillard F, Coudray J, Boulu P: Insufficient urine concentration in sickle cell anemia and S-C hemoglobinopathy. Study of 56 patients (in French). Ann Med Interne (Paris) 1978;129:289–292.

21 Barbedo MM, McCurdy PR: Red cell life span in sickle cell trait. Acta Haematol 1974;51:339–343.

22 Hatch FE, Culbertson JW, Diggs LW: Nature of the renal concentrating defect in sickle cell disease. J Clin Invest 1967;46:336–345.

23 Forrester TE, Alleyne GA: Excretion of salt and water by patients with sickle-cell anaemia: effect of a diuretic and solute diuresis. Clin Sci Mol Med 1977;53:523–527.

24 de Jong PE, Statius van Eps LW: Sickle cell nephropathy: new insights into its pathophysiology. Kidney Int 1985;27:711–717.

25 Ong AC, Jowett TP, Firth JD, Burton S, Karet FE, Fine LG: An endothelin-1 mediated autocrine growth loop involved in human renal tubular regeneration. Kidney Int 1995;48:390–401.

26 Nir A, Clavell AL, Heublein D, Aarhus LL, Burnett JC Jr: Acute hypoxia and endogenous renal endothelin. J Am Soc Nephrol 1994;4:1920–1924.

27 Kohan DE, Padilla E: Endothelin-1 is an autocrine factor in rat inner medullary collecting ducts. Am J Physiol 1992;263:F607–F612.
28 Miller RL, Kohan DE: Hypoxia regulates endothelin-1 production by the inner medullary collecting duct. J Lab Clin Med 1998;131:45–48.
29 Phelan MW, Faller DV: Hypoxia decreases constitutive nitric oxide synthase transcript and protein in cultured endothelial cells. J Cell Physiol 1996;167:469–476.
30 Kourembanas S, Marsden PA, McQuillan LP, Faller DV: Hypoxia induces endothelin gene expression and secretion in cultured human endothelium. J Clin Invest 1991;88:1054–1057.
31 Aversa CR, Oparil S, Caro J, Li H, Sun SD, Chen YF, Swerdel MR, Monticello TM, Durham SK, Minchenko A, Lira SA, Webb ML: Hypoxia stimulates human preproendothelin-1 promoter activity in transgenic mice. Am J Physiol 1997;273:L848–L855.
32 Minchenko A, Caro J: Regulation of endothelin-1 gene expression in human microvascular endothelial cells by hypoxia and cobalt: role of hypoxia responsive element. Mol Cell Biochem 2000;208:53–62.
33 Wykoff CC, Pugh CW, Maxwell PH, Harris AL, Ratcliffe PJ: Identification of novel hypoxia dependent and independent target genes of the von Hippel-Lindau (VHL) tumour suppressor by mRNA differential expression profiling. Oncogene 2000;19:6297–6305.
34 MacLean MR, McCulloch KM: Influence of applied tension and nitric oxide on responses to endothelins in rat pulmonary resistance arteries: effect of chronic hypoxia. Br J Pharmacol 1998;123:991–999.
35 MacLean MR, Randall MD, Hiley CR: Effects of moderate hypoxia, hypercapnia and acidosis on haemodynamic changes induced by endothelin-1 in the pithed rat. Br J Pharmacol 1989;98:1055–1065.
36 McCulloch KM, Docherty C, MacLean MR: Endothelin receptors mediating contraction of rat and human pulmonary resistance arteries: effect of chronic hypoxia in the rat. Br J Pharmacol 1998;123:1621–1630.
37 Lyon-Roberts B, Strait KA, van Peursem E, Kittikulsuth W, Pollock JS, Pollock DM, Kohan DE: Flow regulation of collecting duct endothelin-1 production. Am J Physiol Renal Physiol 2011;300:F650–F656.
38 Janas J, Sitkiewicz D, Januszewicz A, Szczesniak C, Grenda R, Janas RM: Endothelin-1 inactivating peptidase in the human kidney and urine. J Hypertens 2000;18:475–483.
39 Janas J, Sitkiewicz D, Pulawska MF, Warnawin K, Janas RM: Purification of endothelin-1-inactivating peptidase from the rat kidney. J Hypertens 1994;12:375–382.
40 Oishi R, Nonoguchi H, Tomita K, Marumo F: Endothelin-1 inhibits AVP-stimulated osmotic water permeability in rat inner medullary collecting duct. Am J Physiol 1991;261:F951–F956.
41 Nadler SP, Zimpelmann JA, Hebert RL: Endothelin inhibits vasopressin-stimulated water permeability in rat terminal inner medullary collecting duct. J Clin Invest 1992;90:1458–1466.
42 Ge Y, Ahn D, Stricklett PK, Hughes AK, Yanagisawa M, Verbalis JG, Kohan DE: Collecting duct-specific knockout of endothelin-1 alters vasopressin regulation of urine osmolality. Am J Physiol Renal Physiol 2005;288:F912–F920.
43 Ahn D, Ge Y, Stricklett PK, Gill P, Taylor D, Hughes AK, Yanagisawa M, Miller L, Nelson RD, Kohan DE: Collecting duct-specific knockout of endothelin-1 causes hypertension and sodium retention. J Clin Invest 2004;114:504–511.
44 Ge Y, Bagnall A, Stricklett PK, Strait K, Webb DJ, Kotelevtsev Y, Kohan DE: Collecting duct-specific knockout of the endothelin B receptor causes hypertension and sodium retention. Am J Physiol Renal Physiol 2006;291:F1274–F1280.
45 Rivera A, Jarolim P, Brugnara C: Modulation of Gardos channel activity by cytokines in sickle erythrocytes. Blood 2002;99:357–603.
46 Rivera A, Rotter MA, Brugnara C: Endothelins activate Ca(2+)-gated K(+) channels via endothelin B receptors in CD-1 mouse erythrocytes. Am J Physiol 1999;277:C746–C754.

47 Sabaa N, de Franceschi L, Bonnin P, Castier Y, Malpeli G, Debbabi H, Galaup A, Maier-Redelsperger M, Vandermeersch S, Scarpa A, Janin A, Levy B, Girot R, Beuzard Y, Leboeuf C, Henri A, Germain S, Dussaule JC, Tharaux PL: Endothelin receptor antagonism prevents hypoxia-induced mortality and morbidity in a mouse model of sickle-cell disease. J Clin Invest 2008;118:1924–1933.

48 Turhan A, Weiss LA, Mohandas N, Coller BS, Frenette PS: Primary role for adherent leukocytes in sickle cell vascular occlusion: a new paradigm. Proc Natl Acad Sci USA 2002;99:3047–3051.

49 Hebbel RP: Special issue of microcirculation: examination of the vascular pathobiology of sickle cell anemia. Microcirculation 2004;11:99–100.

50 Hebbel RP, Osarogiagbon R, Kaul D: The endothelial biology of sickle cell disease: inflammation and a chronic vasculopathy. Microcirculation 2004;11:129–151.

51 Kato GJ, Hebbel RP, Steinberg MH, Gladwin MT: Vasculopathy in sickle cell disease: Biology, pathophysiology, genetics, translational medicine, and new research directions. Am J Hematol 2009;84:618–625.

52 Hammerman SI, Kourembanas S, Conca TJ, Tucci M, Brauer M, Farber HW: Endothelin-1 production during the acute chest syndrome in sickle cell disease. Am J Respir Crit Care Med 1997;156:280–285.

53 Graido-Gonzalez E, Doherty JC, Bergreen EW, Organ G, Telfer M, McMillen MA: Plasma endothelin-1, cytokine, and prostaglandin E_2 levels in sickle cell disease and acute vaso-occlusive sickle crisis. Blood 1998;92:2551–2555.

54 Rybicki AC, Benjamin LJ: Increased levels of endothelin-1 in plasma of sickle cell anemia patients. Blood 1998;92:2594–2596.

55 Werdehoff SG, Moore RB, Hoff CJ, Fillingim E, Hackman AM: Elevated plasma endothelin-1 levels in sickle cell anemia: relationships to oxygen saturation and left ventricular hypertrophy. Am J Hematol 1998;58:195–199.

56 Tharaux PL, Chatziantoniou C, Casellas D, Fouassier L, Ardaillou R, Dussaule JC: Vascular endothelin-1 gene expression and synthesis and effect on renal type I collagen synthesis and nephroangiosclerosis during nitric oxide synthase inhibition in rats. Circulation 1999;99:2185–2191.

57 Moreau P, Takase H, Kung CF, Shaw S, Luscher TF: Blood pressure and vascular effects of endothelin blockade in chronic nitric oxide-deficient hypertension. Hypertension 1997;29:763–769.

58 Hocher B, Thöne-Reineke C, Rohmeiss P, Schmager F, Slowinski T, Burst V, Siegmund F, Quertermous T, Neumayer HH, Schleuning WD, Theuring F: Endothelin-1 transgenic mice develop glomerulosclerosis, interstitial fibrosis, and renal cysts but not hypertension. J Clin Invest 1997;99:1380–1389.

59 Aslan M, Ryan TM, Adler B, Townes TM, Parks DA, Thompson JA, Tousson A, Gladwin MT, Patel RP, Tarpey MM, Batinic-Haberle I, White CR, Freeman BA: Oxygen radical inhibition of nitric oxide-dependent vascular function in sickle cell disease. Proc Natl Acad Sci USA 2001;98:15215–15220.

60 Aslan M, Ryan TM, Townes TM, Coward L, Kirk MC, Barnes S, Alexander CB, Rosenfeld SS, Freeman BA: Nitric oxide-dependent generation of reactive species in sickle cell disease. Actin tyrosine induces defective cytoskeletal polymerization. J Biol Chem 2003;278:4194–4204.

61 Gladwin MT, Schechter AN, Ognibene FP, Coles WA, Reiter CD, Schenke WH, Csako G, Waclawiw MA, Panza JA, Cannon RO 3rd: Divergent nitric oxide bioavailability in men and women with sickle cell disease. Circulation 2003;107:271–278.

62 Reiter CD, Wang X, Tanus-Santos JE, Hogg N, Cannon RO 3rd, Schechter AN, Gladwin MT: Cell-free hemoglobin limits nitric oxide bioavailability in sickle-cell disease. Nat Med 2002;8:1383–1389.

63 Morris CR, Kato GJ, Poljakovic M, Wang X, Blackwelder WC, Sachdev V, Hazen SL, Vichinsky EP, Morris SM Jr, Gladwin MT: Dysregulated arginine metabolism, hemolysis-associated pulmonary hypertension, and mortality in sickle cell disease. JAMA 2005;294:81–90.

64 Boulanger C, Luscher TF: Release of endothelin from the porcine aorta. Inhibition by endothelium-derived nitric oxide. J Clin Invest 1990;85:587–590.
65 Ito S, Juncos LA, Nushiro N, Johnson CS, Carretero OA: Endothelium-derived relaxing factor modulates endothelin action in afferent arterioles. Hypertension 1991;17:1052–1056.
66 de Franceschi L, Malpeli G, Scarpa A, Janin A, Muchitsch EM, Roncada P, Leboeuf C, Corrocher R, Beuzard Y, Brugnara C: Protective effects of S-nitrosoalbumin on lung injury induced by hypoxia-reoxygenation in mouse model of sickle cell disease. Am J Physiol Lung Cell Mol Physiol 2006;291:L457–465.
67 de Franceschi L, Baron A, Scarpa A, Adrie C, Janin A, Barbi S, Kister J, Rouyer-Fessard P, Corrocher R, Leboulch P, Beuzard Y: Inhaled nitric oxide protects transgenic SAD mice from sickle cell disease-specific lung injury induced by hypoxia/reoxygenation. Blood 2003;102:1087–1096.
68 Gladwin MT, Kato GJ, Weiner D, Onyekwere OC, Dampier C, Hsu L, Hagar RW, Howard T, Nuss R, Okam MM, Tremonti CK, Berman B, Villella A, Krishnamurti L, Lanzkron S, Castro O, Gordeuk VR, Coles WA, Peters-Lawrence M, Nichols J, Hall MK, Hildesheim M, Blackwelder WC, Baldassarre J, Casella JF: Nitric oxide for inhalation in the acute treatment of sickle cell pain crisis: a randomized controlled trial. JAMA 2011;305:893–902.
69 Noris M, Morigi M, Donadelli R, Aiello S, Foppolo M, Todeschini M, Orisio S, Remuzzi G, Remuzzi A: Nitric oxide synthesis by cultured endothelial cells is modulated by flow conditions. Circ Res 1995;76:536–543.
70 Kourembanas S, McQuillan LP, Leung GK, Faller DV: Nitric oxide regulates the expression of vasoconstrictors and growth factors by vascular endothelium under both normoxia and hypoxia. J Clin Invest 1993;92:99–104.
71 Luscher TF, Yang Z, Tschudi M, von Segesser L, Stulz P, Boulanger C, Siebenmann R, Turina M, Buhler FR: Interaction between endothelin-1 and endothelium-derived relaxing factor in human arteries and veins. Circ Res 1990;66:1088–1094.
72 Bonnin P, Sabaa N, Flamant M, Debbabi H, Tharaux PL: Ultrasound imaging of renal vaso-occlusive events in transgenic sickle mice exposed to hypoxic stress. Ultrasound Med Biol 2008;34:1076–1084.
73 Helset E, Ytrehus K, Tveita T, Kjaeve J, Jorgensen L: Endothelin-1 causes accumulation of leukocytes in the pulmonary circulation. Circ Shock 1994;44:201–209.
74 Filep JG, Fournier A, Foldes-Filep E: Acute pro-inflammatory actions of endothelin-1 in the guinea-pig lung: involvement of ET_A and ET_B receptors. Br J Pharmacol 1995;115:227–236.
75 Zouki C, Baron C, Fournier A, Filep JG: Endothelin-1 enhances neutrophil adhesion to human coronary artery endothelial cells: role of ET(A) receptors and platelet-activating factor. Br J Pharmacol 1999;127:969–979.
76 Callera GE, Montezano AC, Touyz RM, Zorn TM, Carvalho MH, Fortes ZB, Nigro D, Schiffrin EL, Tostes RC: ET_A receptor mediates altered leukocyte-endothelial cell interaction and adhesion molecules expression in DOCA-salt rats. Hypertension 2004;43:872–879.
77 Lopez Farre A, Riesco A, Espinosa G, et al.: Effect of endothelin-1 on neutrophil adhesion to endothelial cells and perfused heart. Circulation 1993;88:1166–1171.
78 West MS, Wethers D, Smith J, Steinberg M: Laboratory profile of sickle cell disease: a cross-sectional analysis. The Cooperative Study of Sickle Cell Disease. J Clin Epidemiol 1992;45:893–909.
79 Kinney TR, Sleeper LA, Wang WC, Zimmerman RA, Pegelow CH, Ohene-Frempong K, Wethers DL, Bello JA, Vichinsky EP, Moser FG, Gallagher DM, DeBaun MR, Platt OS, Miller ST: Silent cerebral infarcts in sickle cell anemia: a risk factor analysis. The Cooperative Study of Sickle Cell Disease. Pediatrics 1999;103:640–645.
80 Castro O, Brambilla DJ, Thorington B, Reindorf CA, Scott RB, Gillette P, Vera JC, Levy PS: The acute chest syndrome in sickle cell disease: incidence and risk factors. The Cooperative Study of Sickle Cell Disease. Blood 1994;84:643–649.

81 Witko-Sarsat V, Mocek J, Bouayad D, Tamassia N, Ribeil JA, Candalh C, Davezac N, Reuter N, Mouthon L, Hermine O, Pederzoli-Ribeil M, Cassatella MA: Proliferating cell nuclear antigen acts as a cytoplasmic platform controlling human neutrophil survival. J Exp Med 2010;207:2631–2645.
82 Witko-Sarsat V, Pederzoli-Ribeil M, Hirsh E, Sozzani S, Cassatella MA: Regulating neutrophil apoptosis: new players enter the game. Trends Immunol 2011;32:117–124.
83 Hidalgo A, Chang J, Jang JE, Peired AJ, Chiang EY, Frenette PS: Heterotypic interactions enabled by polarized neutrophil microdomains mediate thromboinflammatory injury. Nat Med 2009;15:384–391.
84 Malmstrom RE, Settergren M, Pernow J: Endothelin attenuates endothelium-dependent platelet inhibition in man. Acta Physiol (Oxf) 2010;198:441–448.
85 Frenette PS: Sickle cell vasoocclusion: heterotypic, multicellular aggregations driven by leukocyte adhesion. Microcirculation 2004;11:167–177.
86 Powars DR, Chan LS, Hiti A, Ramicone E, Johnson C: Outcome of sickle cell anemia: a 4-decade observational study of 1056 patients. Medicine (Baltimore) 2005;84:363–376.
87 Jison ML, Gladwin MT: Hemolytic anemia-associated pulmonary hypertension of sickle cell disease and the nitric oxide/arginine pathway. Am J Respir Crit Care Med 2003;168:3–4.
88 Serjeant GR: The clinical features of sickle cell disease. Baillieres Clin Haematol 1993;6:93–115.
89 Platt OS, Brambilla DJ, Rosse WF, Milner PF, Castro O, Steinberg MH, Klug PP: Mortality in sickle cell disease. Life expectancy and risk factors for early death. N Engl J Med 1994;330:1639–1644.
90 Olney RS: Preventing morbidity and mortality from sickle cell disease. A public health perspective. Am J Prev Med 1999;16:116–121.
91 Abbott KC, Hypolite IO, Agodoa LY: Sickle cell nephropathy at end-stage renal disease in the United States: patient characteristics and survival. Clin Nephrol 2002;58:9–15.
92 Gladwin MT, Sachdev V, Jison ML, Shizukuda Y, Plehn JF, Minter K, Brown B, Coles WA, Nichols JS, Ernst I, Hunter LA, Blackwelder WC, Schechter AN, Rodgers GP, Castro O, Ognibene FP: Pulmonary hypertension as a risk factor for death in patients with sickle cell disease. N Engl J Med 2004;350:886–895.
93 McBurney PG, Hanevold CD, Hernandez CM, Waller JL, McKie KM: Risk factors for microalbuminuria in children with sickle cell anemia. J Pediatr Hematol Oncol 2002;24:473–477.
94 Powars DR, Elliott-Mills DD, Chan L, Niland J, Hiti AL, Opas LM, Johnson C: Chronic renal failure in sickle cell disease: risk factors, clinical course, and mortality. Ann Intern Med 1991;115:614–620.
95 Bhalla M, Abboud MR, McLoud TC, Shepard JA, Munden MM, Jackson SM, Beaty JR, Laver JH: Acute chest syndrome in sickle cell disease: CT evidence of microvascular occlusion. Radiology 1993;187:45–49.
96 Castro O, Hoque M, Brown BD: Pulmonary hypertension in sickle cell disease: cardiac catheterization results and survival. Blood 2003;101:1257–1261.
97 Kimmel AS, Magargal LE, Maizel R, Robb-Doyle E: Proliferative sickle cell retinopathy under age 20: a review. Ophthalmic Surg 1987;18:126–128.
98 Clarkson JG: The ocular manifestations of sickle-cell disease: a prevalence and natural history study. Trans Am Ophthalmol Soc 1992;90:481–504.
99 Penman AD, Talbot JF, Chuang EL, Thomas P, Serjeant GR, Bird AC: New classification of peripheral retinal vascular changes in sickle cell disease. Br J Ophthalmol 1994;78:681–689.
100 Charache S: Eye disease in sickling disorders. Hematol Oncol Clin North Am 1996;10:1357–1362.
101 McLeod DS, Merges C, Fukushima A, Goldberg MF, Lutty GA: Histopathologic features of neovascularization in sickle cell retinopathy. Am J Ophthalmol 1997;124:455–472.

102 Prengler M, Pavlakis SG, Prohovnik I, Adams RJ: Sickle cell disease: the neurological complications. Ann Neurol 2002;51:543–552.

103 Derebail VK, Nachman PH, Key NS, Ansede H, Falk RJ, Kshirsagar AV: High prevalence of sickle cell trait in African Americans with ESRD. J Am Soc Nephrol 2010;21:413–417.

104 Falk RJ, Scheinman J, Phillips G, Orringer E, Johnson A, Jennette JC: Prevalence and pathologic features of sickle cell nephropathy and response to inhibition of angiotensin-converting enzyme. N Engl J Med 1992;326:910–915.

105 Bernstein J, Whitten CF: A histologic appraisal of the kidney in sickle cell anemia. Arch Pathol 1960;70:407–418.

106 Brenner BM, Meyer TW, Hostetter TH: Dietary protein intake and the progressive nature of kidney disease: the role of hemodynamically mediated glomerular injury in the pathogenesis of progressive glomerular sclerosis in aging, renal ablation, and intrinsic renal disease. N Engl J Med 1982;307:652–659.

107 Anderson S, Meyer TW, Rennke HG, Brenner BM: Control of glomerular hypertension limits glomerular injury in rats with reduced renal mass. J Clin Invest 1985;76:612–619.

108 Schmitt F, Martinez F, Brillet G, Giatras I, Choukroun G, Girot R, Bachir D, Galacteros F, Lacour B, Grunfeld JP: Early glomerular dysfunction in patients with sickle cell anemia. Am J Kidney Dis 1998;32:208–214.

109 McCoy RC: Ultrastructural alterations in the kidney of patients with sickle cell disease and the nephrotic syndrome. Lab Invest 1969;21:85–95.

110 Bakir AA, Hathiwala SC, Ainis H, Hryhorczuk DO, Rhee HL, Levy PS, Dunea G: Prognosis of the nephrotic syndrome in sickle glomerulopathy. A retrospective study. Am J Nephrol 1987;7:110–115.

111 Bhathena DB, Sondheimer JH: The glomerulopathy of homozygous sickle hemoglobin (SS) disease: morphology and pathogenesis. J Am Soc Nephrol 1991;1:1241–1252.

112 Verani RR, Conley SB: Sickle cell glomerulopathy with focal segmental glomerulosclerosis. Child Nephrol Urol 1991;11:206–208.

113 Etteldorf JN, Smith JD, Tuttle AH, Diggs LW: Renal hemodynamic studies in adults with sickle cell anemia. Am J Med 1955;18:243–248.

114 Hatch FE Jr, Azar SH, Ainsworth TE, Nardo JM, Culbertson JW: Renal circulatory studies in young adults with sickle cell anemia. J Lab Clin Med 1970;76:632–640.

115 Aoki RY, Saad ST: Microalbuminuria in sickle cell disease. Braz J Med Biol Res 1990;23:1103–1106.

116 Dharnidharka VR, Dabbagh S, Atiyeh B, Simpson P, Sarnaik S: Prevalence of microalbuminuria in children with sickle cell disease. Pediatr Nephrol 1998;12:475–478.

117 Bayazit AK, Noyan A, Aldudak B, Ozel A, Anarat A, Kilinc Y, Sasmaz, Gali E, Anarat R, Dikmen N: Renal function in children with sickle cell anemia. Clin Nephrol 2002;57(2):127–130.

118 Datta V, Ayengar JR, Karpate S, Chaturvedi P: Microalbuminuria as a predictor of early glomerular injury in children with sickle cell disease. Indian J Pediatr 2003;70:307–309.

119 Barros FB, Lima CS, Santos AO, Mazo-Ruiz MF, Lima MC, Etchebehere EC, Costa FF, Saad ST, Camargo EE, Ramos CD: 51Cr-EDTA measurements of the glomerular filtration rate in patients with sickle cell anaemia and minor renal damage. Nucl Med Commun 2006;27:959–962.

120 Thompson J, Reid M, Hambleton I, Serjeant GR: Albuminuria and renal function in homozygous sickle cell disease: observations from a cohort study. Arch Intern Med 2007;167:701–708.

121 Haymann JP, Stankovic K, Levy P, Avellino V, Tharaux PL, Letavernier E, Grateau G, Baud L, Girot R, Lionnet F: Glomerular hyperfiltration in adult sickle cell anemia: a frequent hemolysis associated feature. Clin J Am Soc Nephrol 2010;5:756–761.

122 Guasch A, Cua M, Mitch WE: Early detection and the course of glomerular injury in patients with sickle cell anemia. Kidney Int 1996;49:786–791.

123 de Jong PE, de Jong-Van Den Berg TW, Sewrajsingh GS, Schouten H, Donker AJ, Statius van Eps LW: The influence of indomethacin on renal haemodynamics in sickle cell anaemia. Clin Sci (Lond) 1980;59:245–250.
124 Allon M: Renal abnormalities in sickle cell disease. Arch Intern Med 1990;150:501–504.
125 Silldorff EP, Yang S, Pallone TL: Prostaglandin E_2 abrogates endothelin-induced vasoconstriction in renal outer medullary descending vasa recta of the rat. J Clin Invest 1995;95:2734–2740.
126 Schramek H, Coroneos E, Dunn MJ: Interactions of the vasoconstrictor peptides, angiotensin II and endothelin-1, with vasodilatory prostaglandins. Semin Nephrol 1995;15:195–204.
127 Sasser JM, Sullivan JC, Hobbs JL, Yamamoto T, Pollock DM, Carmines PK, Pollock JS: Endothelin A receptor blockade reduce diabetic renal injury via an anti-inflammatory mechanism. J Am Soc Nephrol 2007;18: 143–154.
128 Saleh MA, Boesen EI, Pollock JS, Savin VJ, Pollock DM: Endothelin-1 increases glomerular permeability and inflammation independent of blood pressure in the rat. Hypertension 2010;56:942–949.

Pierre-Louis Tharaux, MD, PhD
Inserm U.970 Paris Centre de recherche Cardiovasculaire – PARCC
Lab 219, 56 rue Leblanc
FR–75015 Paris (France)
Tel. +33 1 53 98 8051
E-Mail pierre-louis.tharaux@inserm.fr

Barton M, Kohan DE (eds): Endothelin in Renal Physiology and Disease.
Contrib Nephrol. Basel, Karger, 2011, vol 172, pp 200–209

Endothelin in Polycystic Kidney Disease

Ming-Yang Chang[a] · Albert C.M. Ong[b]

[a]Kidney Research Center, Chang Gung Memorial Hospital, Chang Gung University College of Medicine, Taoyuan, Taiwan; [b]Kidney Genetics Group, Academic Nephrology Unit, University of Sheffield Medical School, Sheffield, UK

Abstract

Autosomal dominant polycystic kidney disease (ADPKD) is the most common genetic kidney disease in man and is caused by germline mutations in *PKD1* or *PKD2*. Affected patients develop progressively enlarged kidneys due to the growth of multiple renal epithelial cysts. Several studies have demonstrated marked intrafamilial phenotypic variability in PKD1 or PKD2 pedigrees, indicating the importance of nonallelic factors such as genetic modifying loci in determining individual phenotype. Endothelin (ET)-1 exerts multiple and often opposing effects on different aspects of renal physiology through its major ET receptor subtypes, ET_A and ET_B. Recent studies have reported that EDN1 and EDNRA polymorphisms can influence the age of onset of end-stage renal disease in ADPKD. Both circulating and local ET-1 systems are abnormally activated in human disease and experimental models, and ET_A receptor expression is specifically upregulated in human ADPKD kidneys. Overexpression of ET-1 in transgenic mice is sufficient to trigger cyst initiation. However, studies utilizing selective ET_A and ET_B receptor antagonists to delay cystic disease progression in rodent PKD models have proved disappointing and do not support further extension into clinical trials. A critical balance between ET_A and ET_B action in the cystic kidney appears to be necessary to maintain kidney structure and function. Current evidence suggests that ET-1 and its receptors act as major modifying genes for renal disease progression in ADPKD. The future challenge will be to translate these findings to modify disease severity or for predicting prognosis in man.

Autosomal dominant polycystic kidney disease (ADPKD) is the most common genetic kidney disease in man [1]. Its estimated prevalence is between 1:400 and 1:1,000 and it may lead to end-stage renal disease (ESRD) in 50% of affected individuals by 60 years of age [2, 3]. Almost all cases of ADPKD are due to germline mutations in *PKD1* or *PKD2*. The ADPKD proteins polycystin-1 (460 kDa) and polycystin-2 (110 kDa) constitute a subfamily (TRPP) of the transient receptor potential channel superfamily [2] and play critical roles in maintaining normal tubular structure and function through the regulation of multiple signaling pathways involving Ca^{2+}, mTOR, and cAMP homeostasis [4].

Among chronic kidney diseases (CKD), ADPKD is unique because affected patients develop progressively enlarged kidneys due to the growth of multiple renal epithelial cysts. Renal tubular epithelial cells (in all tubular segments) proliferate at a higher basal rate even at the precystic stage and some authors have even referred to ADPKD as a 'neoplasm in disguise' [5]. The mechanism of cyst initiation and expansion includes abnormalities in cell proliferation, fluid secretion, extracellular matrix formation, apoptosis, angiogenesis, defective planar cell polarity, and cilia-mediated mechanosensation [2]. Molecules involved in these processes could play rate-limiting steps in renal disease progression [6].

Major systemic features of ADPKD include liver and pancreatic cysts as well as intracranial aneurysms; these can lead to early mortality and morbidity in some patients [7]. There is a high prevalence of hypertension (70%) and this is considered a risk factor for disease progression [8]. Other cardiovascular abnormalities in ADPKD include left ventricular hypertrophy, aortic aneurysms, and cardiac valvular abnormalities [8].

Endothelin (ET)-1 has been reported to exert multiple effects on different aspects of renal physiology. Apart from circulating ET-1, there is good evidence that many renal cell types, including tubular cells, synthesize and bind ET-1 [9], indicating that it has important paracrine or autocrine actions. Two major ET receptor subtypes, ET_A and ET_B, have been shown to mediate ET-1 action, but often with opposing effects. The disease-specific role of each receptor subtype in different types of CKD is less clear.

Several studies have demonstrated marked intrafamilial phenotypic variability in PKD1 or PKD2 pedigrees, indicating the importance of nonallelic factors such as genetic modifying loci or environmental triggers in determining individual phenotype [6, 10]. The purpose of this review is to address the question as to whether ET-1 is a major modifying factor in PKD progression.

EDN1 and EDNRA Polymorphisms in Human Autosomal Dominant Polycystic Kidney Disease Progression

The mean age at ESRD is 53.4 years for PKD1 and 72.7 years for PKD2 [11]. However, there is a marked phenotypic variability in individual disease severity which cannot be accounted for by allelic effects. Single-nucleotide polymorphisms (SNPs) of ET-1 (EDN1) and ET_A receptor (EDNRA) genes have been shown to have a modest effect on the age of onset of ESRD in ADPKD. In one study investigating the influence of an EDNRA SNP (C/T polymorphism in exon 8) on disease progression in 193 ADPKD patients (87 males, 106 females), females with the CC genotype reached ESRD at a significantly later age than CT heterozygotes: CC (57.4 ± 8.1 years), CT (53.0 ± 9.1), and TT (54.5 ± 6.4) [12]. This SNP was also associated with a lower pulse pressure and could thus be protective in ADPKD patients [12].

The same group also examined the influence of three SNPs in EDN1 (K198N, 3A/4A, T-1370G) on progression towards ESRD in 205 ADPKD patients [13]. Of interest, carriers of the 4A allele in combination with the 198N allele (4A/4A, 3A/4A + 198KN,NN) had a significantly lower age of onset of ESRD (47.1 ± 8.7 years), 6 years younger than the carriers of other genotypes [13]. The deleterious effect of the combination of 4A and 198N alleles of EDN1 could be due to a higher processing rate of preproET into ET-1 in carriers of the 198N allele, leading to the higher ET-1 plasma levels [13]. These studies suggest that activation of the systemic ET-1 system in human ADPKD could participate in disease progression possibly by increasing systemic blood pressure and/or reducing renal blood flow.

Plasma or Cyst Endothelin-1 Levels in Human Autosomal Dominant Polycystic Kidney Disease

Many renal cell types can synthesize and bind ET-1, indicating its role as an autocrine or paracrine factor [9]. In one study, concentrations of immunoreactive ET-1, cyclic AMP and EGF were found to be elevated in human ADPKD cyst fluid [14]. Another study reported that patients with ADPKD (hypertensive and normotensive) had increased plasma levels of ET-1 in comparison with patients with essential hypertension or healthy subjects [15]. These studies suggest that both circulating and local ET-1 systems are abnormally activated in human ADPKD.

Endothelin-1 and Endothelin Receptor Expression in Polycystic Kidney Disease

Increased levels of ET-1 have been reported in cystic kidneys from experimental models and human disease though there are notable differences in the expression of ET receptor subtypes. In the cpk mouse model of PKD, renal ET-1 mRNA levels were 3.2-fold greater than controls at 1 week and 10-fold greater at 3 weeks when cystic disease was at its height [16]. Similarly, renal ET_A and ET_B mRNA levels increased to 4.2- and 6.3-fold over control kidneys, respectively [16]. In the Han:SPRD rat model of PKD, significantly elevated tissue levels of ET-1 were found in cystic kidneys, but ET_A and ET_B receptor density were reported to be markedly decreased [17]. In contrast, a study in human ADPKD cystic kidneys reported that ET_A mRNA levels were 5- to 10-

fold higher than non-ADPKD controls, while ET_B mRNA was unchanged [18]. In addition, neo-expression of ET_A receptors was detected overlying cyst epithelia, glomeruli, and medium-sized arteries – regions where immunoreactive ET-1 was also present [18]. The increase in total kidney ET_A mRNA levels in human ADPKD has been confirmed in two subsequent studies using microarray platforms [19, 20].

Activation of the renal ET system in ADPKD could result from stimulation by changes in the circulating and local renin-angiotensin-aldosterone system due to cyst enlargement [21], resulting in focal ischemia and tissue hypoxia, both of which are strong stimuli for ET synthesis [22].

Endothelin-1 and Cyst Initiation in Transgenic Mice

Overexpression of ET-1 in transgenic mice is sufficient to trigger cyst initiation as reported by two independent groups [23, 24]. ET-1 transgenic mice developed multiple cortical kidney cysts in addition to interstitial fibrosis and glomerulosclerosis, thus providing direct evidence that ET-1 alone can contribute to the initiation and progression of cystic kidney diseases [24]. Overexpression of ET-1 in these mice was not associated with systemic hypertension though salt-dependent hypertension could be induced [23]. The mechanism of cyst formation in these models mice was not clarified though intrarenal ischemia and tubular occlusion were proposed as potential explanations [23].

Endothelin-1 and Tubular Cell Proliferation

ET-1 has been shown to be a potent mitogen in vitro for normal human renal tubular cells and interstitial fibroblasts; in addition, it can stimulate collagen I gene expression in fibroblasts [25, 26]. Tubular ET-1 synthesis has been documented in experiments from different species [9]. Tubular cell proliferation is an early feature of precystic tubules in human ADPKD and many rodent PKD models [27]. A role for the ET_B receptor in mediating the mitogenic effect of ET-1 has been demonstrated in human proximal tubular cells as part of a hypoxia-inducible forward feedback autocrine loop involved in renal tubular regeneration [25].

Selective ET_A or ET_B Blockade in Polycystic Kidney Disease Rodent Models

To date, there have only been two published preclinical studies utilizing selective ET_A and ET_B receptor antagonists to delay cystic disease progression in rodent PKD models [22, 28]. Chronic administration of an ET_A selective antagonist

(LU135252 or darusentan) for 4 months led to more severe cystic disease in the Han:SPRD rat model. This was associated with an increase in tubular cell proliferation. The deleterious effect of ET_A blockade was ameliorated by a nonselective ET receptor antagonist (LU224332), presumably by blocking unopposed ET_B receptors [22]. Alternatively, ET_A blockade could directly result in tubular cell proliferation through enhanced tubular expression of hypoxia inducible factor induced by local ischemia [29]. The finding that ET_A blockade induces tubular cell proliferation is different to that reported in other models of CKD or prostate cancer [30, 31], and appears characteristic of PKD (see below).

In a second study, $Pkd2^{WS25/-}$ mice were treated from 5 to 16 weeks of age with the highly selective orally active receptor antagonists atrasentan (or ABT-627, ET_A) or A-192621 (ET_B), singly or in combination [28]. The doses chosen for the study were shown to be effective in blocking the pressor effects of intravenous ET-1 (ET_A and ET_B) and sarafotoxin-6c (ET_B), respectively, in rats after 7 days oral administration. Selective ET_A blockade resulted in a significant increase in tubular cell proliferation, but did not alter the overall cystic phenotype. Unexpected, selective ET_B blockade led to accelerated cystic kidney disease (including interstitial inflammation and fibrosis) without altering tubular cell proliferation (fig. 1). The deleterious effects of ET_B blockade were mediated largely through unopposed ET_A action since these changes were normalized by combined ET_A and ET_B blockade. In addition, ET_B blockade resulted in a decrease in urine volume, reduced urinary sodium excretion, increased urine osmolarity, and renal cAMP concentrations. These features suggest an enhancement of vasopressin action (see below). ET-1 appears to be a major modifying factor for disease progression in this orthologous model. Underlying mechanisms include both vasopressin-dependent (fluid secretion, water, and sodium excretion) and vasopressin-independent pathways (renal fibrosis, proliferation, fibrosis, inflammation, apoptosis, and vasoconstriction) [28]. However, there is a clear difference in the ET receptor subtype responsible for disease progression between the Han:SPRD (ET_A) and $Pkd2^{WS25/-}$ (ET_B) models.

Endothelin-1 and Vasopressin Action

Recent genetic studies utilizing segment-specific knockouts have demonstrated a physiological role for ET-1 acting via ET_B receptors in regulating sodium and water excretion in kidney collecting ducts [32, 33]. ET_B activation inhibits vasopressin action, thereby promoting diuresis and blocking sodium reabsorption by inhibiting Na^+-K^+-ATPase and/or ENaC [34]. ET_B blockade would therefore potentiate vasopressin action in cysts derived from the collecting duct and thus stimulate cyst expansion [28]. A consistent finding in experimental models of ARPKD, ADPKD, and nephronophthisis has been the striking efficacy of vasopressin V2 receptor antagonists in suppressing cyst growth through inhibiting vasopressin-dependent cAMP levels [2].

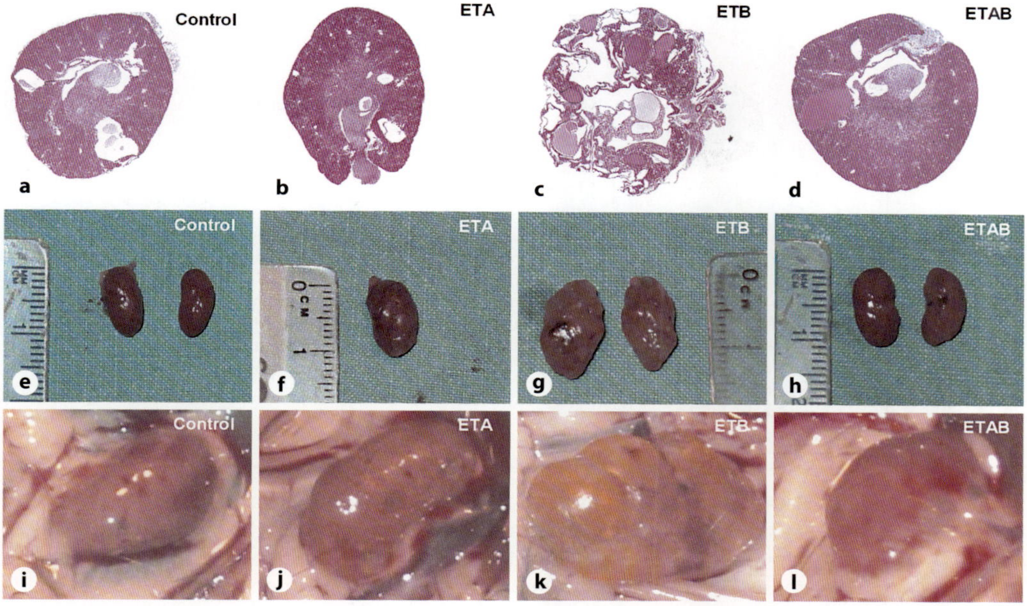

Fig. 1. The effect of ET receptor blockade on cystic disease progression in Pkd2$^{WS25/-}$ mice. Representative kidney cross-sections (hematoxylin and eosin) and gross kidney appearances from affected Pkd2$^{WS25/-}$ mice that were treated from 5 to 16 weeks of age with vehicle alone (**a**, **e**, and **i**), ET_A (**b**, **f**, and **j**), ET_B (**c**, **g**, and **k**), and combined ET_A/ET_B (**d**, **h**, and **l**) receptor antagonists [28].

Clinical Implications

Figure 2 summarizes the potential pathogenic mechanisms by which ET-1 could promote cyst growth in ADPKD. In rodent models, the balance between ET_A and ET_B signaling appears to be critical for maintaining kidney structure and function. ET_B activation may suppress vasopressin action and increase vasodilation, tubular proliferation, and salt and ET-1 excretion in PKD, while ET_A activation may increase vasoconstriction, vasopressin action, inflammation, and fibrosis. In human ADPKD kidney, ET_A expression in several cell types is consistently increased [18], but ET_A blockade in the Pkd2$^{WS25/-}$ model did not alter cystic disease severity [28]. Clearly, there could be species differences in the action of ET-1 between human and rodent kidney, but the increase in ET_A could represent part of a compensatory response to maintain residual renal function.

Although selective ET_A receptor blockade has been found to be effective in reducing proteinuria in CKD patients [35, 36], the possibility of increased tubular cell proliferation through unopposed ET_B action should be considered a potential long-term risk [22, 28]. Selective ET_B blockade is apparently deleterious as in many other pathological conditions and should not be used in

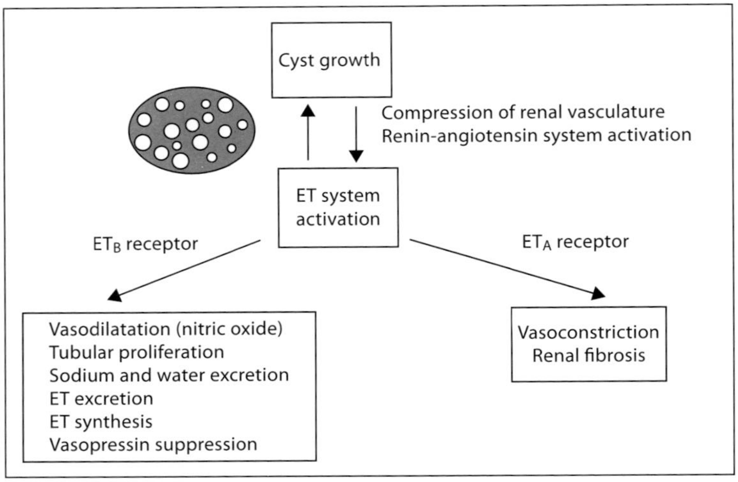

Fig. 2. Pathophysiological pathways activated by ET in the cystic ADPKD kidney. Cysts arise due to germline *PKD1* or *PKD2* mutations. Compression of the renal vasculature leading to ischemia, hypoxia, and activation of the renin-angiotensin system could activate ET-1 synthesis. ET-1 may further stimulate cyst growth and interstitial fibrosis through ET_A and ET_B receptors via vasopressin-dependent and independent pathways.

ADPKD [37–40]. Combined ET_A/ET_B blockade was ineffective in reducing cyst progression in two experimental models. Future studies using different treatment regimens or other PKD models could help confirm these results.

Endothelin-1 and Hypertension in Autosomal Dominant Polycystic Kidney Disease

Increased sympathetic activity and ET-1 levels could result from the stimulation of both circulating and intrarenal renin-angiotensin-aldosterone system due to progressive cyst enlargement, thus leading to systemic hypertension [21]. ET_A blockade with darusentan is effective in patients with resistant hypertension (on 3 or more antihypertensive agents) and can reduce proteinuria in CKD patients, particularly when combined with angiotensin II receptor blockade [41–44]. The major side effect of darusentan is edema, but this may be less of an issue when combined with a diuretic. The treatment strategy of the ongoing HALT PKD trial is to lower BP (110/75 mm Hg) versus standard BP (130/80 mm Hg) targets with combination angiotensin-converting enzyme inhibitor and angiotensin receptor blocker therapy [45]. ET antagonists may still have a role in the treatment of ADPKD patients through their antihypertensive effects, provided that their safety and side effect profile can be clarified in longer term studies.

Conclusions

Numerous studies have confirmed that the ET system plays important roles in the complex pathophysiology associated with cyst formation and disease progression in ADPKD. However, results from selective targeting of both ET receptor subtypes in animal models of PKD have proved disappointing and do not support further extension into clinical trials. Nonetheless, these studies have shed light on the critical balance between ET_A and ET_B action in the cystic kidney, which appears to be necessary to maintain kidney structure and function; inducing an imbalance in receptor signaling clearly leads to more severe disease. These results confirm that ET-1 and its receptors act as major modifying genes for ADPKD. The future challenges will be to translate these findings to alter disease severity or predicting prognosis in man.

References

1 Chang MY, Ong AC: Autosomal dominant polycystic kidney disease: recent advances in pathogenesis and treatment. Nephron Physiol 2008;108:1–7.

2 Torres VE, Harris PC: Autosomal dominant polycystic kidney disease: the last 3 years. Kidney Int 2009;76:149–168.

3 Wilson PD: Polycystic kidney disease. N Engl J Med 2004;350:151–164.

4 Ong AC, Harris PC: Molecular pathogenesis of ADPKD: the polycystin complex gets complex. Kidney Int 2005;67:1234–1247.

5 Grantham JJ: Polycystic kidney disease: neoplasia in disguise. Am J Kidney Dis 1990;15:110–116.

6 Peters DJ, Breuning MH: Autosomal dominant polycystic kidney disease: modification of disease progression. Lancet 2001;358:1439–1444.

7 Chang MY, Kuok CM, Chen YC, Ryu SJ, Tian YC, Wu-Chou YH, Tseng FJ, Yang CW: Comparison of intracerebral hemorrhage and subarachnoid hemorrhage in patients with autosomal-dominant polycystic kidney disease. Nephron Clin Pract 2010;114:c158–c164.

8 Ecder T, Schrier RW: Cardiovascular abnormalities in autosomal-dominant polycystic kidney disease. Nat Rev Nephrol 2009;5:221–228.

9 Ong AC: Perspectives in endothelin research: tubulointerstitial actions of endothelins in the kidney: roles in health and disease. Nephrol Dial Transplant 1996;11:251–257.

10 Fain PR, McFann KK, Taylor MR, Tison M, Johnson AM, Reed B, Schrier RW: Modifier genes play a significant role in the phenotypic expression of PKD1. Kidney Int 2005;67:1256–1267.

11 Barua M, Cil O, Paterson AD, Wang K, He N, Dicks E, Parfrey P, Pei Y: Family history of renal disease severity predicts the mutated gene in ADPKD. J Am Soc Nephrol 2009;20:1833–1838.

12 Reiterova J, Merta M, Stekrova J, Maixnerova D, Obeidova H, Kebrdlova V, Viklicky O, Tesar V: The influence of endothelin-A receptor gene polymorphism on the progression of autosomal dominant polycystic kidney disease and IgA nephropathy. Folia Biol (Praha) 2007;53:134–137.

13 Reiterova J, Merta M, Stekrova J, Cabartova Z, Cibulka R, Maixnerova D, Rysava R, Rihova Z, Tesar V, Motan J: Influence of endothelin-1 gene polymorphisms on the progression of autosomal dominant polycystic kidney disease. Kidney Blood Press Res 2006;29:182–188.

14 Munemura C, Uemasu J, Kawasaki H: Epidermal growth factor and endothelin in cyst fluid from autosomal dominant polycystic kidney disease cases: possible evidence of heterogeneity in cystogenesis. Am J Kidney Dis 1994;24:561–568.
15 Giusti R, Neri M, Angelini D, Carlini A, Fiorini I, Bigongiari P, Antonelli A: Plasma concentration of endothelin and arterial pressure in patients with ADPKD. Contrib Nephrol 1995;115:118–121.
16 Nakamura T, Ebihara I, Fukui M, Osada S, Tomino Y, Masaki T, Goto K, Furuichi Y, Koide H: Increased endothelin and endothelin receptor mRNA expression in polycystic kidneys of cpk mice. J Am Soc Nephrol 1993;4:1064–1072.
17 Hocher B, Zart R, Braun N, Schwarz A, van der Woude F, Rohmeiss P, Koppenhagen K: The endothelin system in polycystic kidneys of Han:SPRD rats. J Cardiovasc Pharmacol 1998;31(suppl 1):S342–S344.
18 Ong AC, Newby LJ, Dashwood MR: Expression and cellular localisation of renal endothelin-1 and endothelin receptor subtypes in autosomal-dominant polycystic kidney disease. Nephron Exp Nephrol 2003;93:e80.
19 Song X, Di Giovanni V, He N, Wang K, Ingram A, Rosenblum ND, Pei Y: Systems biology of autosomal dominant polycystic kidney disease (ADPKD): computational identification of gene expression pathways and integrated regulatory networks. Hum Mol Genet 2009;18:2328–2343.
20 Schieren G, Rumberger B, Klein M, Kreutz C, Wilpert J, Geyer M, Faller D, Timmer J, Quack I, Rump LC, Walz G, Donauer J: Gene profiling of polycystic kidneys. Nephrol Dial Transplant 2006;21:1816–1824.
21 Schrier RW: Renal volume, renin-angiotensin-aldosterone system, hypertension, and left ventricular hypertrophy in patients with autosomal dominant polycystic kidney disease. J Am Soc Nephrol 2009;20:1888–1893.
22 Hocher B, Kalk P, Slowinski T, Godes M, Mach A, Herzfeld S, Wiesner D, Arck PC, Neumayer HH, Nafz B: ET_A receptor blockade induces tubular cell proliferation and cyst growth in rats with polycystic kidney disease. J Am Soc Nephrol 2003;14:367–376.
23 Shindo T, Kurihara H, Maemura K, Kurihara Y, Ueda O, Suzuki H, Kuwaki T, Ju KH, Wang Y, Ebihara A, Nishimatsu H, Moriyama N, Fukuda M, Akimoto Y, Hirano H, Morita H, Kumada M, Yazaki Y, Nagai R, Kimura K: Renal damage and salt-dependent hypertension in aged transgenic mice overexpressing endothelin-1. J Mol Med 2002;80:105–116.
24 Hocher B, Thone-Reineke C, Rohmeiss P, Schmager F, Slowinski T, Burst V, Siegmund F, Quertermous T, Bauer C, Neumayer HH, Schleuning WD, Theuring F: Endothelin-1 transgenic mice develop glomerulosclerosis, interstitial fibrosis, and renal cysts but not hypertension. J Clin Invest 1997;99:1380–1389.
25 Ong AC, Jowett TP, Firth JD, Burton S, Karet FE, Fine LG: An endothelin-1 mediated autocrine growth loop involved in human renal tubular regeneration. Kidney Int 1995;48:390–401.
26 Ong AC, Jowett TP, Firth JD, Burton S, Kitamura M, Fine LG: Human tubular-derived endothelin in the paracrine regulation of renal interstitial fibroblast function. Exp Nephrol 1994;2:134.
27 Chang MY, Parker E, Ibrahim S, Shortland JR, Nahas ME, Haylor JL, Ong AC: Haploinsufficiency of Pkd2 is associated with increased tubular cell proliferation and interstitial fibrosis in two murine Pkd2 models. Nephrol Dial Transplant 2006;21:2078–2084.
28 Chang MY, Parker E, El Nahas M, Haylor JL, Ong AC: Endothelin B receptor blockade accelerates disease progression in a murine model of autosomal dominant polycystic kidney disease. J Am Soc Nephrol 2007;18:560–569.
29 Kakinuma Y, Miyauchi T, Yuki K, Murakoshi N, Goto K, Yamaguchi I: Novel molecular mechanism of increased myocardial endothelin-1 expression in the failing heart involving the transcriptional factor hypoxia-inducible factor-1alpha induced for impaired myocardial energy metabolism. Circulation 2001;103:2387–2394.
30 Growcott JW: Preclinical anticancer activity of the specific endothelin A receptor antagonist ZD4054. Anticancer Drugs 2009;20:83–88.

31 Hocher B, Schwarz A, Reinbacher D, Jacobi J, Lun A, Priem F, Bauer C, Neumayer HH, Raschack M: Effects of endothelin receptor antagonists on the progression of diabetic nephropathy. Nephron 2001;87:161–169.
32 Ge Y, Stricklett PK, Hughes AK, Yanagisawa M, Kohan DE: Collecting duct-specific knockout of the endothelin A receptor alters renal vasopressin responsiveness, but not sodium excretion or blood pressure. Am J Physiol Renal Physiol 2005;289:F692–F698.
33 Ge Y, Ahn D, Stricklett PK, Hughes AK, Yanagisawa M, Verbalis JG, Kohan DE: Collecting duct-specific knockout of endothelin-1 alters vasopressin regulation of urine osmolality. Am J Physiol Renal Physiol 2005;288:F912–F920.
34 Kohan DE: The renal medullary endothelin system in control of sodium and water excretion and systemic blood pressure. Curr Opin Nephrol Hypertens 2006;15:34–40.
35 Wenzel RR, Littke T, Kuranoff S, Jurgens C, Bruck H, Ritz E, Philipp T, Mitchell A: Avosentan reduces albumin excretion in diabetics with macroalbuminuria. J Am Soc Nephrol 2009;20:655–664.
36 Dhaun N, Macintyre IM, Melville V, Lilitkarntakul P, Johnston NR, Goddard J, Webb DJ: Blood pressure-independent reduction in proteinuria and arterial stiffness after acute endothelin-A receptor antagonism in chronic kidney disease. Hypertension 2009;54:113–119.
37 Kitada K, Yui N, Matsumoto C, Mori T, Ohkita M, Matsumura Y: Inhibition of endothelin ET_B receptor system aggravates neointimal hyperplasia after balloon injury of rat carotid artery. J Pharmacol Exp Ther 2009;331:998–1004.
38 Pfab T, Thone-Reineke C, Theilig F, Lange I, Witt H, Maser-Gluth C, Bader M, Stasch JP, Ruiz P, Bachmann S, Yanagisawa M, Hocher B: Diabetic endothelin B receptor-deficient rats develop severe hypertension and progressive renal failure. J Am Soc Nephrol 2006;17:1082–1089.
39 Okada Y, Nakata M, Izumoto H, Takasu M, Tazawa N, Takaoka M, Gariepy CE, Yanagisawa M, Matsumura Y: Role of endothelin ET_B receptor in partial ablation-induced chronic renal failure in rats. Eur J Pharmacol 2004;494:63–71.
40 Nishida M, Eshiro K, Okada Y, Takaoka M, Matsumura Y: Roles of endothelin ET_A and ET_B receptors in the pathogenesis of monocrotaline-induced pulmonary hypertension. J Cardiovasc Pharmacol 2004;44:187–191.
41 Kohan DE: Endothelin, hypertension and chronic kidney disease: new insights. Curr Opin Nephrol Hypertens 2010;19:134–139.
42 Barton M: Reversal of proteinuric renal disease and the emerging role of endothelin. Nat Clin Pract Nephrol 2008;4:490–501.
43 Bakris GL, Lindholm LH, Black HR, Krum H, Linas S, Linseman JV, Arterburn S, Sager P, Weber M: Divergent results using clinic and ambulatory blood pressures: report of a darusentan-resistant hypertension trial. Hypertension 2010;56:824–830.
44 Weber MA, Black H, Bakris G, Krum H, Linas S, Weiss R, Linseman JV, Wiens BL, Warren MS, Lindholm LH: A selective endothelin-receptor antagonist to reduce blood pressure in patients with treatment-resistant hypertension: a randomised, double-blind, placebo-controlled trial. Lancet 2009;374:1423–1431.
45 Chapman AB, Torres VE, Perrone RD, Steinman TI, Bae KT, Miller JP, Miskulin DC, Rahbari Oskoui F, Masoumi A, Hogan MC, Winklhofer FT, Braun W, Thompson PA, Meyers CM, Kelleher C, Schrier RW: The HALT polycystic kidney disease trials: design and implementation. Clin J Am Soc Nephrol 2010;5:102–109.

Prof. Albert C.M. Ong
Kidney Genetics Group, Academic Unit of Nephrology
The Henry Wellcome Laboratories for Medical Research, University of Sheffield Medical School
Beech Hill Road
Sheffield S10 2RX (UK)
Tel. +44 114 271 3402, E-Mail a.ong@sheffield.ac.uk

Barton M, Kohan DE (eds): Endothelin in Renal Physiology and Disease.
Contrib Nephrol. Basel, Karger, 2011, vol 172, pp 210–222

Endothelin Antagonism and Reversal of Proteinuric Renal Disease in Humans

Matthias Barton

Molecular Internal Medicine, University of Zürich, Zürich, Switzerland

Abstract

Endothelin (ET)-1 is a powerful mitogen and vasoconstrictor that contributes to cardiovascular and renal pathologies. In the kidney, ET causes vasoconstriction, sodium retention, mesangial cell inflammation and proliferation, hypertrophy of glomerular capillaries, and podocyte injury. The latter, due to destruction of the glomerular filtration barrier, is a key factor for renal protein loss. Experimental and recent clinical studies suggest that orally active drugs inhibiting ET_A receptors are capable of not only inhibiting the progression, but also reversing glomerulosclerosis-related renal injury. Clinical studies using ET receptor antagonists (ERAs) have found regression of proteinuria which serves as a functional indicator of glomerular filtration barrier injury. The effects of ERA therapy can be observed in the presence of inhibition of the renin-angiotensin system, suggesting a renin-angiotensin system-independent therapeutic effect of ERAs. Thus, ET blockade is not an 'add-on' treatment, but represents an independent therapeutic principle. This article will discuss the underlying mechanisms of the antiproteinuric effects of ET antagonists, and summarize recent clinical trials in the field and the therapeutic potential of the ERA class of drugs for renal medicine.

Endothelin in Renal Physiology and Disease

Endogenous endothelin (ET) controls renal water and sodium excretion and acid-base balance, and maintains renal cell growth and tonic vasoconstriction [1, 2]. In diseases such as diabetes or arterial hypertension, ET-dependent mechanisms contribute to proteinuria and glomerular injury, which involves mesangial activation of fibrogenic pathways [Gagliardini et al., this vol., pp. 171–184; 3–5].

Inflammation is a common denominator of all fibroproliferative diseases [5–7]. Excessive formation of proinflammatory and fibrogenic peptides such as ET appears to play an essential role in this process [5]. Renal ET also activates pathways that have been independently associated with the physiopathology of renal disease progression, including formation of reactive oxygen species production [8]. Since angiotensin receptor antagonism inhibits ET-1 production [9] but the protective effects of angiotensin receptor blockade (ARB) on podocyte injury occur even in the absence of functional angiotensin receptors [5, 10], it is likely that secondary effects via alternative pathways such as ET are involved. Trophic effects of ET promote mesangial cell proliferation as well as hypertrophy of the glomerular basement membrane [5]. Glomerular basement membrane hypertrophy is indirectly mediated via injury of glomerular epithelial cells (podocytes) and is an important indicator of injury of the glomerulus [Fligny et al., this vol., pp. 120–138].

Other important factors that aggravate glomerular injury via ET are hyperglycemia, high levels of sodium, and aldosterone [5]. Overexpression of ET is associated with inflammatory changes in the kidney [11]. These changes are essential for glomerulosclerosis to develop and are sensitive to inhibition by ET_A receptor antagonists [6]. The groups of David and Jennifer Pollock, using the highly selective ET_A receptor antagonist atrasentan, have shown that ET blockade exerts anti-inflammatory effects which contribute at least in part to the ET receptor antagonist (ERA)-mediated inhibition of experimental diabetic nephropathy [Saleh and Pollock, this vol., pp. 160–170; 8, 12]. Preclinical prevention studies using transgenic animal models [11] support the notion that the ET pathway is critical for the pathogenesis of proteinuria and glomerulosclerosis and that this effect is at least in part due to the proinflammatory activity of ET [Saleh and Pollock, this vol., pp. 160–170; 6, 11].

Proteinuria Involves Podocyte Injury via Endothelin A Receptors

To date, no information regarding the effects of ERA treatment on glomerular epithelial cell (podocyte) injury or the severity of glomerulosclerosis in humans are available from the clinical studies which have been completed. Instead, the most often used indirect read-outs of podocyte injury in clinical trials are albuminuria or the albuminuria/creatinine ratio [13–15]. According to Wiggins [16], podocytopathy is a general element of all glomerular diseases. In particular, focal segmental glomerulosclerosis due to aging has been characterized as a podocyte disease [17].

Preclinical studies have shown that podocytes are involved in the ET-mediated disease processes of glomerulosclerosis and the proteinuria associated with it [Gagliardini et al., this vol., pp. 171–184; Fligny et al., this vol., pp. 120–138; 18–22]. Podocytes are terminally differentiated glomerular epithelial cells that regulate the protein filter function of the glomerulus and its slit diaphragm [4]. Podocytes contain ET-binding sites and human podocytes express a fully functional ET system, including ET receptors, preproET-1, and ET-converting enzyme-1 (reviewed in [4, 5]).

The actin cytoskeleton ensures intact podocyte function and structural integrity of the cell [Gagliardini et al., this vol., pp. 171–184; 4]. Excessive levels of ET-1 or angiotensin II causes podocyte actin cytoskeleton disruption and increases albumin permeability [18, 19]. The first evidence demonstrating that ET_A receptors directly contribute to podocyte injury came from cell experiments and preclinical in vivo studies [4, 20, Fligny et al., this vol., pp. 120–138]. Using human podocyte injury, we found that podocyte injury induced by puromycin aminonucleoside (which, due to its toxicity, causes glomerulosclerosis in vivo [5]) is completely prevented by the selective ET_A receptor antagonists, BQ-123 or darusentan, or by gene silencing of the ET_A receptor [Fligny et al., this vol., pp. 120–138; 20]. Protein loading ex vivo stimulates podocytes to produce ET [18], which, in turn, damages the podocyte architecture with consecutive increases in glomerular permselectivity for proteins. This deleterious effect is also regulated in an ET_A-dependent manner [18]. More recently, also acute ET-mediated direct effects of glomerular permeability in isolated glomeruli have also been reported [23].

Reversal of Proteinuria: Underlying Mechanisms

Numerous prevention-type preclinical studies using ET antagonists have demonstrated inhibition of glomerulosclerosis and the proteinuria associated with it [4, 5]; however, only few studies have investigated the effects of ERA treatment in established renal disease (i.e. in regression-type studies). In aging-associated focal segmental glomerulosclerosis [17], ET_A blockade results in structural and functional reversal of proteinuric renal disease [20]. The antisclerotic and anti-proteinuric effects were independent of systemic blood pressure, renal blood flow, and glomerular filtration rate.

Electron microscopy studies revealed that treatment also reversed podocyte injury [20], which was characterized by vacuolar degeneration of podocyte cell bodies and foot process fusion (fig. 1). These vacuolar changes in the aging kidney, which were recently identified to be at least in part mediated by podocyte autophagy [24], disappeared after darusentan treatment, suggesting that ET blockade affects podocyte autophagy mechanisms. This hypothesis is further supported by the notion that angiotensin II, which increases renal ET production in vivo [9, 25], similarly stimulates podocyte autophagy [26, 27]. Electron microscopy studies revealed normalization of the thickness of the previously hypertrophied glomerular basement membrane after ET_A blockade (fig. 1); the thickness of the basement membrane is proportional to the degree of disease and is dependent on podocyte-derived matrix proteins (reviewed in [4]).

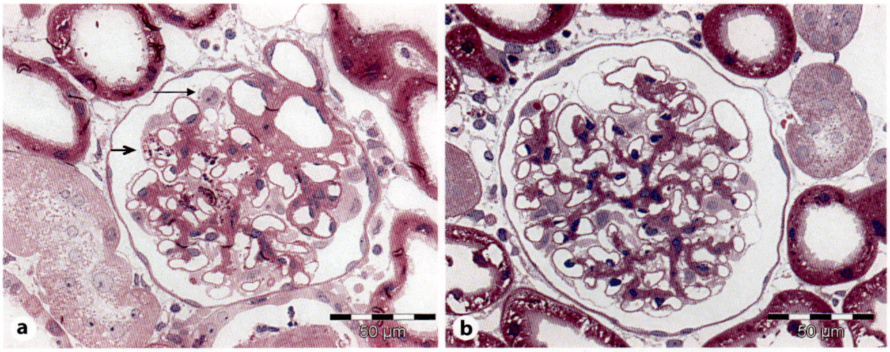

Fig. 1. Results from preclinical studies investigating the effects of ET_A receptor blockade on pre-existing spontaneous focal segmental glomerulosclerosis due to aging. Representative semithin sections stained with hematoxylin and eosin are shown. **a** Representative glomerular injury in the absence of treatment in aged rats: moderate mesangial matrix expansion and hypertrophy of podocytes with enlarged nuclei, prominent nucleoles (small arrow), and increased number of intracytoplasmatic vacuoles (large arrow) indicative of podocyte injury and cell activation. **b** After 4 weeks of treatment with the ET_A antagonist darusentan, regression of glomerular injury is noted, which is associated with a reduction in mesangial matrix expansion and podocyte hypertrophy. Bar, 50 µm. Figure in part reproduced from Ortmann et al. [20] with permission of the publisher.

Confirming the seminal work by Hocher et al. [11] demonstrating a role for ET in glomerulosclerosis and renal inflammation, these data were the first in vivo observations showing that ET represents a previously unidentified factor regulating the architecture of the glomerular capillary, and that capillary and glomerular injury can be reversed by ERA treatment. Established glomerulosclerosis can also be reversed in animals with chronic nitric oxide (NO) deficiency (also a feature present in aging), again effects that were independent of blood pressure [28, 29].

In malignant hypertension, ET_A blockade using atrasentan, but not combined $ET_{A/B}$ receptor blockade, was able to afford protection of podocytes, and effects were observed despite continued hypertension [21, 22, 30]. These important findings suggest that (at least in malignant hypertension) the unblocked ET_B receptor is essential in mediating the protective effects on podocyte injury. A central role of ET in controlling podocyte function is also evident from the work by Ariela Benigni's laboratory who, using a different ET antagonist [Gagliardini et al., this vol., pp. 171–184; 31], also reported ET_A-mediated nephroprotection.

There are several mechanisms involved in disease regression following ERA treatment [4]. They involve changes in podocyte function and structure [4, 5] as well as capillary restructuring/angiogenesis [Gagliardini et al., this vol., pp. 171–184; 31], which has been shown to be involved in renal repair (fig. 2). These effects possibly occur via signaling through vascular endothelial growth factor-mediated angiogenesis which mediates its effects through NO and/or via

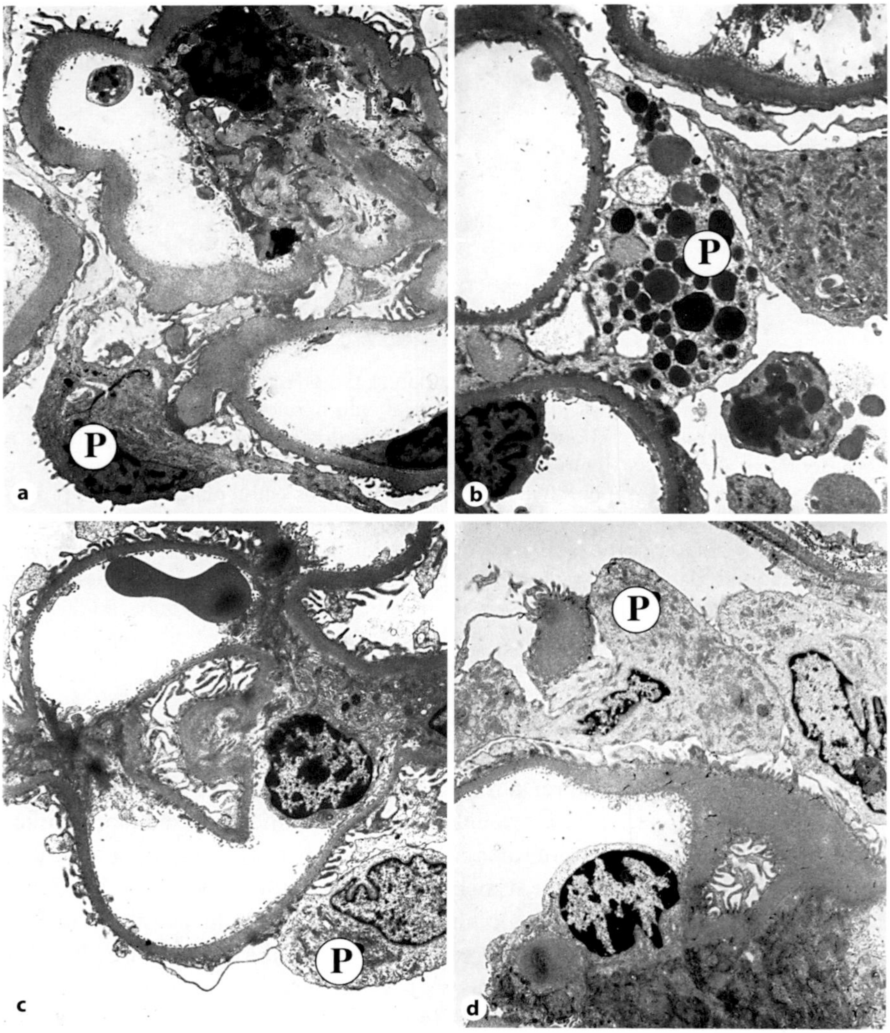

Fig. 2. Results from preclinical studies investigating the effects of ET_A receptor blockade on ultrastructural glomerular changes in pre-existing spontaneous focal segmental glomerulosclerosis due to aging. Representative transmission electron micrographs of podocytes and glomerular basement membranes from untreated (upper panels) and darusentan-treated aged kidneys (lower panels). **a** Untreated, showing glomerular basement membrane hypertrophy and podocyte injury and detachment. **b** Untreated, high-power micrograph showing thickening of glomerular capillary basement membrane with podocyte detachment. Injury of podocytes is characterized by hypertrophy, vacuolar degeneration, and diffuse effacement of foot processes. **c** Darusentan-treatment resulting in attachment of the podocyte to the basement membrane and reversal of glomerular capillary hypertrophy. **d** High-power micrograph showing podocyte recovery after darusentan treatment. ET blockade resulted in regression of podocyte injury and disappearance of vacuolar degeneration. P = Podocyte. Figure modified and in part reproduced from Ortmann et al. [20] with permission of the publisher.

accelerated formation of neovessels [Gagliardini et al., this vol., pp. 171–184]. Since ET_A receptor blockade not only increases the bioactivity of NO [32], NO being a strong stimulus for angiogenesis to occur, the L-arginine/NO pathway may also play a role in glomerular repair following ET blockade [5]. Whether and to what extent inhibition of ET production also underlies the antiproteinuric effects of drugs other than ERAs [angiotensin-converting enzyme (ACE) inhibitors, ARBs] is not fully understood, but inhibition of ET formation by ARBs has been reported in experimental studies [9]. However, recent work from Ariela Benigni's group suggests that ET antagonists indeed promote capillary (re-) growth in a nonaging model of glomerulosclerosis, and that treatment with antiproteinuric drugs actually leads to repopulation of podocytes in the aging kidney [Gagliardini et al., this vol., pp. 171–184; 31]. Work from Marcus Moeller's laboratory also suggest that parietal epithelial cells repopulate the glomerular epithelial cells (i.e. podocytes) [33, 34], the latter thought to be terminally differentiated. Such cellular repopulation of the glomerulus may represent yet another mechanism by which ERAs could contribute to the disease regression.

Reversal of Proteinuria by Endothelin Antagonism: Clinical Studies

A still much debated question is whether and how reversal or regression of renal disease can be defined from a clinical standpoint. Currently, no data on structural changes and possible regression of glomerular disease or podocyte injury after ERA treatment are available from clinical studies. Essentially, these would require histology from repeated kidney biopsies at identical biopsy locations. Because such interventions can hardly be justified from an ethical standpoint and are technically extremely demanding, these questions will remain unanswered. However, by studying functional injury of the glomerulus (i.e. damage of the slit diaphragm and consecutive proteinuria), reversal/regression of functional injury can readily be assessed indirectly.

Endothelin Antagonists in Clinical Development for Proteinuric Renal Disease
Atrasentan (ABT-627, Xinlay^{TM}) is a nonsulfonamide, highly selective ET_A selective antagonist. Atrasentan was the first ERA for which first clinical evidence for antiproteinuric activity was reported more than a decade ago by Rabelink's group [Rabelink and Kohan, this vol., pp. 235–242; 35]. These investigators reported different degrees of reversal of proteinuria in 10 patients with type 1 diabetes treated with the ET_A antagonist atrasentan for 12 weeks [Rabelink and Kohan, this vol., pp. 235–242; 35]. The effects were seen on top of standard renin-angiotensin system blockade with an angiotensin receptor antagonist or an ACE inhibitor, and in some patients blood pressure-lowering effects of ERA treatment were observed [Rabelink and Kohan, this

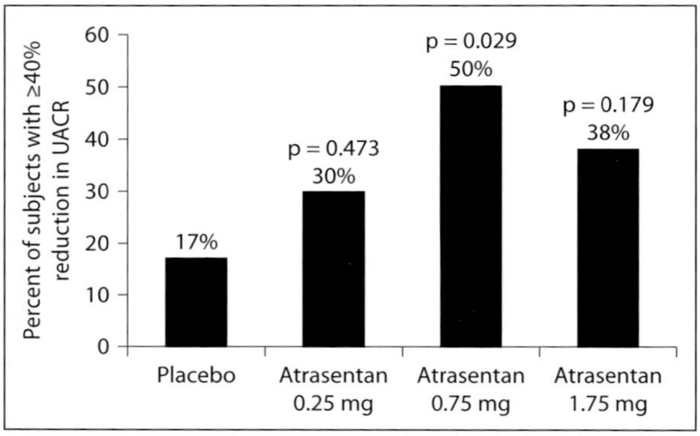

Fig. 3. Effects of 8-week treatment with placebo or different doses of the ET_A receptor antagonist atrasentan (Xinlay^{TM}) in obese patients with type 2 diabetes (n = 89). Data shown are given as percent of patients experiencing regression of proteinuria [measured as urinary albumin-creatinine ratio (UACR)]. Treatment significantly increased the percentage of subjects achieving at least 40% reduction in UACR compared to placebo. Figure modified and in part reproduced from Kohan et al. [13] with permission of the publisher.

vol., pp. 235–244; 35]. The antiproteinuric effects of atrasentan have recently been confirmed in a larger trial by Kohan et al. [13] in patients with proteinuria due to diabetic nephropathy. In this study, like in all other ERA studies in diabetics, patients were obese and frequently on insulin therapy. ERA therapy had pronounced antiproteinuric effects (fig. 3) and moderate effects on blood pressure were observed [13]. Although this has not been directly assessed, mechanisms involved in the regression of proteinuria are likely to involve structural and, thus, functional improvements of podocytes (fig. 2) [4]. Indeed, in preclinical studies atrasentan has been shown to restore podocyte structure in association with regression of proteinuria, and the effects were largely pressure independent [21, 22].

Endothelin Antagonists for which Clinical Development Has Been Discontinued
In the past decade, another four ERAs have been investigated for the potential treatment of proteinuria in diabetics and patients with resistant hypertension.

Bosentan (Ro 47–0203, Tracleer^{TM}) is a sulfonamide-based receptor unselective ERA, which Roche Pharmaceuticals (the original developer) licensed to Actelion Ltd. Bosentan was investigated for the potential indication of antiproteinuric therapy in patients with diabetic nephropathy. Bosentan was studied in a small trial involving 35 diabetics that were on ACE inhibition or ARB treatment and continued for 16 weeks. According to information provided by the investigators, the study was terminated in 2010 because of problems with patient recruiting and no results from the patients studied were published [36].

Avosentan (SPP301/Ro-67-0656), is a sulfonamide-based nonselective ERA [Davenport and Maguire, this vol., pp. 1–17] that was licensed from Roche to Speedel Pharmaceuticals (now Novartis). A first dose-finding trial found substantial reductions in albuminuria in 286 obese patients with mild-to-moderate type 2 diabetic nephropathy after 12 weeks of avosentan treatment. As with atrasentan, antiproteinuric effects were obtained on top of standard antiproteinuric therapy with ACE inhibition/angiotensin receptor blockade. This indicates an additional and independent beneficial effect of ET blockade. Like with atrasentan [37], reductions in LDL-cholesterol levels were also observed with avosentan. In the second study, the phase 3 ASCEND trial (Avosentan on Doubling of Serum Creatinine, End-Stage Renal Disease and Death in Diabetic Nephropathy) [Gagliardini et al., this vol., pp. 171–184; Rabelink and Kohan, this vol., pp. 235–244; Benz and Amann, this vol., pp. 139–148], patients with advanced renal failure with average estimated glomerular filtration rates of 33 ml/min were followed for 3 and 6 months, and again substantial reductions in albuminuria were noted that were similar at both doses used. The antiproteinuric effects were independent of systemic blood pressure, suggesting that the structural effects of ERA treatment on the kidney predominate. The study was not completed, however, as it was terminated early in 2007 after numerous patients suffered drug-related side effects, including severe heart failure, and because deaths had occurred. Following these disappointing results, which were related to both the severe morbidity of the study patients and drug toxicity of avosentan, drug development for avosentan for proteinuric renal disease was abandoned after Speedel Pharmaceuticals was acquired by Novartis in July 2008.

Darusentan (LU135252/BSF135252), a propanoic acid-derived ET_A-selective ERA, was previously studied with regard to renal protection in resistant arterial hypertension. Darusentan, originally developed by Knoll AG/BASF (Germany) was effective in reducing blood pressure in a phase 2 dose-finding trial in patients with arterial hypertension [38] and in patients with therapy-resistant hypertension already on treatment with two antihypertensive drugs and a diuretic [30, 39]. Despite results from preclinical studies demonstrating reversal of proteinuria and pre-existing glomerulosclerosis after darusentan treatment (fig. 1, 2), results reporting antiproteinuric effects in patients were not available until 2009 [40]. As with most ERA studies in renal patients, the most frequent side effects were fluid retention and headaches [41]. Since then, limitations in the study design regarding the type of blood pressure measurements have become evident [41]. Although darusentan reduced blood pressure and treatment reached several of the secondary endpoints (including a marked reduction in urinary albumin excretion in patients with proteinuric hypertensive nephropathy [40]) and despite good results for drug safety, Gilead Sciences decided to abandon further clinical development of darusentan shortly after publication of the study results in December 2009.

Sitaxentan (TBC11251/Thelin™), a sulfonamide-based ET_A receptor antagonist with very high receptor selectivity developed by Texas Biotechnology Corporation/Encysive, was acquired by Pfizer in early 2008. Sitaxentan was under clinical development as a renal drug in patients with proteinuric nondiabetic renal disease of different etiologies, including focal segmental glomerulosclerosis and autoimmune disease [42]. Using sitaxentan, the group of David Webb found impressive reductions in proteinuria that were largely independent of blood pressure [Dhaun et al., this vol., pp. 243–254; 42–44]. Sitaxentan was found to have only mild or no edema occurring in these patients, which – unlike all other clinical studies in renal patients [13–15, 40] – were not obese. Although the drug had been clinically approved since 2006 for the treatment of pulmonary arterial hypertension by federal health agencies in three continents (Canada, European Union, and Australia), several cases of fatal liver failure were reported in patients on sitaxentan therapy. Pfizer decided to withdraw sitaxentan from the market in December 2010.

Will Endothelin Antagonists Become Novel Therapeutics in Renal Medicine?

Over the past decade, the pharmaceutical industry has made considerable efforts to test the potential of ERAs as new drugs in the treatment of proteinuric renal diseases [Barton and Kohan, this vol., pp. 255–260]. Despite initial highly promising results, four drug candidates failed, three of which were sulfonamide-derived compounds with potential serious side effect profiles. I have previously pointed out the importance of ensuring drug safety for ERAs, particularly in renal patients [5]. Also, it should be noted that ERAs are not to be considered as a so-called 'add-on' therapy as recently proposed by some authors [45]. In fact, the demonstration that ERAs also act independently of renin-angiotensin system inhibitors, and that effects are largely independent from blood pressure are indeed compatible with the concept that ET antagonism indeed represents an independent therapeutic principle. Clinical development of ERAs needs to be continued, with particular attention on issues related to drug safety. If successful, ET receptor blockade might become a novel therapeutic paradigm to interfere with and possibly even reverse human proteinuric renal disease (fig. 4), even on top of currently approved antiproteinuric medications. The clinical data currently available clearly show that this class of drugs has therapeutic potential. Ultimately, ERAs might still evolve as one of the urgently needed next steps to treat chronic proteinuric renal disease in humans.

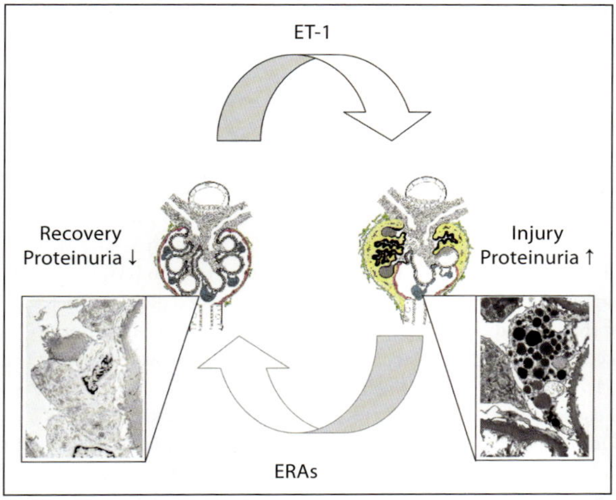

Fig. 4. Proposed mechanisms of effects of ET antagonism on regression of glomerular injury and proteinuria. Systemic or local activation of the ET systems results in inflammatory activation, resulting in mesangial proliferation and subsequent glomerular filtration barrier defects with progressive injury of podocytes and glomerular basement membranes (right). Pharmacological intervention blocking the effects of ET-1 via ET_A receptors can at least in part re-establish an intact filtration barrier leading to partial reversal of proteinuria. Treatment also results in repair of podocyte and glomerular basement membrane injury (left). Figure in part reproduced from Barton et al. [30] with permission of the publisher.

Acknowledgements

This work was supported by the Swiss National Science Foundation (grants No. 3200–108258 and K33KO-122504). I thank all present and previous colleagues who have participated in the work discussed here for their contribution.

References

1 Kohan DE: Endothelin-1 and hypertension: from bench to bedside. Curr Hypertens Rep 2008;10:65–69.
2 Barton M, Yanagisawa M: Endothelin: 20 years from discovery to therapy. Can J Physiol Pharmacol 2008;86:485–498.
3 Benigni A, Perico N, Remuzzi G: Research on renal endothelin in proteinuric nephropathies dictates novel strategies to prevent progression. Curr Opin Nephrol Hypertens 2001;10:1–6.
4 Barton M: Reversal of proteinuric renal disease and the emerging role of endothelin. Nat Clin Pract Nephrol 2008;4:490–501.
5 Barton M: Therapeutic potential of endothelin receptor antagonists for chronic proteinuric renal disease in humans. Biochim Biophys Acta 2010;1802:1203–1213.

6 Nett PC, Teixeira MM, Candinas D, Barton M: Recent developments on endothelin antagonists as immunomodulatory drugs – from infection to transplantation medicine. Recent Pat Cardiovasc Drug Discov 2006;1:265–276.
7 Barton M, Minotti R, Haas E: Inflammation and atherosclerosis. Circ Res 2007;101:750–751.
8 Saleh MA, Boesen EI, Pollock JS, Savin VJ, Pollock DM: Endothelin receptor A-specific stimulation of glomerular inflammation and injury in a streptozotocin-induced rat model of diabetes. Diabetologia 2011;54:979–988.
9 d'Uscio LV, Shaw S, Barton M, Luscher TF: Losartan but not verapamil inhibits angiotensin II-induced tissue endothelin-1 increase: role of blood pressure and endothelial function. Hypertension 1998;31:1305–1310.
10 Matsusaka T, Asano T, Niimura F, Kinomura M, Shimizu A, Shintani A, Pastan I, Fogo AB, Ichikawa I: Angiotensin receptor blocker protection against podocyte-induced sclerosis is podocyte angiotensin II type 1 receptor-independent. Hypertension 2010;55:967–973.
11 Hocher B, Thone-Reineke C, Rohmeiss P, Schmager F, Slowinski T, Burst V, Siegmund F, Quertermous T, Bauer C, Neumayer HH, Schleuning WD, Theuring F: Endothelin-1 transgenic mice develop glomerulosclerosis, interstitial fibrosis, and renal cysts but not hypertension. J Clin Invest 1997;99:1380–1389.
12 Sasser JM, Sullivan JC, Hobbs JL, Yamamoto T, Pollock DM, Carmines PK, Pollock JS: Endothelin a receptor blockade reduces diabetic renal injury via an anti-inflammatory mechanism. J Am Soc Nephrol 2007;18:143–154.
13 Kohan DE, Pritchett Y, Molitch M, Wen S, Garimella T, Audhya P, Andress DL: Addition of atrasentan to renin-angiotensin system blockade reduces albuminuria in diabetic nephropathy. J Am Soc Nephrol 2011;22:763–772.
14 Mann JF, Green D, Jamerson K, Ruilope LM, Kuranoff SJ, Littke T, Viberti G: Avosentan for overt diabetic nephropathy. J Am Soc Nephrol 2010;21:527–535.
15 Wenzel RR, Littke T, Kuranoff S, Jurgens C, Bruck H, Ritz E, Philipp T, Mitchell A: Avosentan reduces albumin excretion in diabetics with macroalbuminuria. J Am Soc Nephrol 2009;20:655–664.
16 Wiggins RC: The spectrum of podocytopathies: a unifying view of glomerular diseases. Kidney Int 2007;71:1205–1214.
17 Floege J, Hackmann B, Kliem V, Kriz W, Alpers CE, Johnson RJ, Kuhn KW, Koch KM, Brunkhorst R: Age-related glomerulosclerosis and interstitial fibrosis in Milan normotensive rats: a podocyte disease. Kidney Int 1997;51:230–243.
18 Morigi M, Buelli S, Angioletti S, Zanchi C, Longaretti L, Zoja C, Galbusera M, Gastoldi S, Mundel P, Remuzzi G, Benigni A: In response to protein load podocytes reorganize cytoskeleton and modulate endothelin-1 gene: implication for permselective dysfunction of chronic nephropathies. Am J Pathol 2005;166:1309–1320.
19 Morigi M, Buelli S, Zanchi C, Longaretti L, Macconi D, Benigni A, Moioli D, Remuzzi G, Zoja C: Shigatoxin-induced endothelin-1 expression in cultured podocytes autocrinally mediates actin remodeling. Am J Pathol 2006;169:1965–1975.
20 Ortmann J, Amann K, Brandes RP, Kretzler M, Munter K, Parekh N, Traupe T, Lange M, Lattmann T, Barton M: Role of podocytes for reversal of glomerulosclerosis and proteinuria in the aging kidney after endothelin inhibition. Hypertension 2004;44:974–981.
21 Opocensky M, Kramer HJ, Backer A, Vernerova Z, Eis V, Cervenka L, Certikova Chabova V, Tesar V, Vaneckova I: Late-onset endothelin-A receptor blockade reduces podocyte injury in homozygous ren-2 rats despite severe hypertension. Hypertension 2006;48:965–971.
22 Vernerova Z, Kramer HJ, Backer A, Cervenka L, Opocensky M, Huskova Z, Vanourkova Z, Eis V, Chabova VC, Tesar V, Maly J, Vaneckova I: Late-onset endothelin receptor blockade in hypertensive heterozygous ren-2 transgenic rats. Vascul Pharmacol 2008;48:165–173.
23 Saleh MA, Boesen EI, Pollock JS, Savin VJ, Pollock DM: Endothelin-1 increases glomerular permeability and inflammation independent of blood pressure in the rat. Hypertension 2010;56:942–949.

24 Hartleben B, Godel M, Meyer-Schwesinger C, Liu S, Ulrich T, Kobler S, Wiech T, Grahammer F, Arnold SJ, Lindenmeyer MT, Cohen CD, Pavenstadt H, Kerjaschki D, Mizushima N, Shaw AS, Walz G, Huber TB: Autophagy influences glomerular disease susceptibility and maintains podocyte homeostasis in aging mice. J Clin Invest 2010;120:1084–1096.
25 Barton M, Shaw S, d'Uscio LV, Moreau P, Luscher TF: Angiotensin II increases vascular and renal endothelin-1 and functional endothelin converting enzyme activity in vivo: role of ET_A receptors for endothelin regulation. Biochem Biophys Res Commun 1997;238:861–865.
26 Weide T, Huber TB: Implications of autophagy for glomerular aging and disease. Cell Tissue Res 2011;343:467–473.
27 Yadav A, Vallabu S, Arora S, Tandon P, Slahan D, Teichberg S, Singhal PC: Ang II promotes autophagy in podocytes. Am J Physiol Cell Physiol 2010;299:C488–C496.
28 Dussaule JC, Chatziantoniou C: Reversal of renal disease: is it enough to inhibit the action of angiotensin II? Cell Death Differ 2007;14:1343–1349.
29 Placier S, Boffa JJ, Dussaule JC, Chatziantoniou C: Reversal of renal lesions following interruption of nitric oxide synthesis inhibition in transgenic mice. Nephrol Dial Transplant 2006;21:881–888.
30 Barton M, Mullins JJ, Bailey MA, Kretzler M: Role of endothelin receptors for renal protection and survival in hypertension: waiting for clinical trials. Hypertension 2006;48:834–837.
31 Gagliardini E, Corna D, Zoja C, Sangalli F, Carrara F, Rossi M, Conti S, Rottoli D, Longaretti L, Remuzzi A, Remuzzi G, Benigni A: Unlike each drug alone, lisinopril if combined with avosentan promotes regression of renal lesions in experimental diabetes. Am J Physiol Renal Physiol 2009;297:F1448–F1456.
32 Barton M, Haudenschild CC, d'Uscio LV, Shaw S, Munter K, Luscher TF: Endothelin ET_A receptor blockade restores NO-mediated endothelial function and inhibits atherosclerosis in apolipoprotein E-deficient mice. Proc Natl Acad Sci USA 1998;95:14367–14372.
33 Appel D, Kershaw DB, Smeets B, Yuan G, Fuss A, Frye B, Elger M, Kriz W, Floege J, Moeller MJ: Recruitment of podocytes from glomerular parietal epithelial cells. J Am Soc Nephrol 2009;20:333–343.
34 Smeets B, Uhlig S, Fuss A, Mooren F, Wetzels JF, Floege J, Moeller MJ: Tracing the origin of glomerular extracapillary lesions from parietal epithelial cells. J Am Soc Nephrol 2009;20:2604–2615.
35 Honing MLH, Bouter PK, Ballard DE, Padley P, Morrison PJ, Rabelink TJ: Abt-627, a selective ET_A-receptor antagonist, reduces proteinuria in patients with diabetes mellitus; in: Regulation of Vascular Tone in Humans by Endothelium-Derived Mediators (doctoral thesis). University of Utrecht, 2000, pp 89–102.
36 Bosentan use in patients with diabetic nephropathy. 2010. http://clinicaltrials.Gov/ct2/show/nct00638131 (accessed March 26, 2011).
37 Reriani M, Raichlin E, Prasad A, Mathew V, Pumper GM, Nelson RE, Lennon R, Rihal C, Lerman LO, Lerman A: Long-term administration of endothelin receptor antagonist improves coronary endothelial function in patients with early atherosclerosis. Circulation 2010;122:958–966.
38 Nakov R, Pfarr E, Eberle S: Darusentan: an effective endothelinA receptor antagonist for treatment of hypertension. Am J Hypertens 2002;15:583–589.
39 Black HR, Bakris GL, Weber MA, Weiss R, Shahawy ME, Marple R, Tannoury G, Linas S, Wiens BL, Linseman JV, Roden R, Gerber MJ: Efficacy and safety of darusentan in patients with resistant hypertension: results from a randomized, double-blind, placebo-controlled dose-ranging study. J Clin Hypertens (Greenwich) 2007;9:760–769.
40 Weber MA, Black H, Bakris G, Krum H, Linas S, Weiss R, Linseman JV, Wiens BL, Warren MS, Lindholm LH: A selective endothelin-receptor antagonist to reduce blood pressure in patients with treatment-resistant hypertension: a randomised, double-blind, placebo-controlled trial. Lancet 2009;374:1423–1431.

41 Bakris GL, Lindholm LH, Black HR, Krum H, Linas S, Linseman JV, Arterburn S, Sager P, Weber M: Divergent results using clinic and ambulatory blood pressures: report of a darusentan-resistant hypertension trial. Hypertension 2010;56:824–830.
42 Dhaun N, MacIntyre IM, Bellamy CO, Kluth DC: Endothelin receptor antagonism and renin inhibition as treatment options for scleroderma kidney. Am J Kidney Dis 2009;54:726–731.
43 Dhaun N, Macintyre IM, Melville V, Lilitkarntakul P, Johnston NR, Goddard J, Webb DJ: Effects of endothelin receptor antagonism relate to the degree of renin-angiotensin system blockade in chronic proteinuric kidney disease. Hypertension 2009;54:e19–e20.
44 Dhaun N, Macintyre IM, Melville V, Lilitkarntakul P, Johnston NR, Goddard J, Webb DJ: Blood pressure-independent reduction in proteinuria and arterial stiffness after acute endothelin-A receptor antagonism in chronic kidney disease. Hypertension 2009;54:113–119.
45 Ritz E, Wenzel RR: Endothelin antagonist as add-on treatment for proteinuria in diabetic nephropathy: is there light at the end of the tunnel? J Am Soc Nephrol 2011;22:593–595.

Matthias Barton, MD
Molecular Internal Medicine, University of Zürich, LTK Y44 G22
Winterthurerstrasse 190
CH–8057 Zürich (Switzerland)
Tel. +41 77 439 5554, E-Mail barton@access.uzh.ch

Barton M, Kohan DE (eds): Endothelin in Renal Physiology and Disease.
Contrib Nephrol. Basel, Karger, 2011, vol 172, pp 223–234

Endothelin Antagonism in Patients with Resistant Hypertension and Hypertension Nephropathy

Ivana Lazich · George L. Bakris

Hypertensive Diseases Unit, University of Chicago, Pritzker School of Medicine, Chicago, Ill., USA

Abstract

Resistant hypertension is a failure to achieve a blood pressure (BP) goal of <140/90 mm Hg despite treatment with at least three different antihypertensive medications classes at a maximally tolerated dose and including a diuretic. The most important systems that require alteration include the renin-angiotensin-aldosterone system, sympathetic nervous system, and (more recently) the endothelin (ET) system. To date, several clinical trials have assessed the effects of ET antagonism, both selective and nonselective, on BP control in humans. The nonselective ET antagonist bosentan was evaluated in patients with mild-to-moderate hypertension. Bosentan was able to lower diastolic BP when compared to placebo and, similarly, to enalapril. Similar findings have been published for darusentan, an ET receptor antagonist with higher affinity for the type A receptor. More recent research has focused on the impact of ET in the setting of resistant hypertension. Studies with darusentan as an add-on therapy in patients with resistant hypertension found a significant BP-lowering effect of 17/10 mm Hg compared to placebo. This BP-lowering effect was similar at higher doses. In a similar patient cohort, darusentan also lowered mean 24-hour BP to a greater extent than the central α_2-agonist guanfacine. Another selective ET_A antagonist, atrasentan, provided other benefits on metabolism in addition to its antihypertensive effect. Atrasentan significantly decreased glucose in diabetes and improved lipid profiles while slowing coronary artery disease progression. Selective ET receptor blockade also has dose-dependent side effects. In a large number of trials, almost one third of the patients suffered excessive fluid retention and edema that was significantly higher than in the placebo groups. One trial, ASCEND, was terminated early due to an increased incidence of fluid retention and increased episodes of heart failure. Thus, this class of agents is effective in resistant hypertension, but lower doses with fewer side effects need to be developed.

Hypertension is a chronic disorder associated with a high risk for cardiovascular morbidity and chronic kidney disease (CKD) development. The prevalence of hypertension is estimated to be 47% in individuals older than 55 years and 56% in those 75–85 years, as estimated from the data collected in the National Health Examination Survey (NHANES) for the time period between 2004 and 2007 [1]. Despite well-defined blood pressure (BP) goals associated with improved health outcomes, large portions of these individuals remain uncontrolled. To date, several trials have demonstrated that between 20 and 35% of the participants remain above the set BP goal of <140/90 mm Hg [2, 3]. Considering the risk of cardiovascular mortality doubles for every 20-mm Hg increase in systolic BP and 10-mm Hg diastolic BP above 115/75 mm Hg, achieving the BP target has an impact on health outcomes [4]. Moreover, hypertension is estimated to be the second most common cause of CKD in some cases progressing to end-stage kidney disease [5].

The presence of CKD plays a prominent role in hypertension control. Over 80% of patients with CKD have hypertension, but their control rate is <40% [6]. With the ongoing epidemic of obesity and diabetes mellitus as well as the increasing number of elderly, the incidence of uncontrolled hypertension as well as CKD will increase. The significance of its impact on patient health outcomes as well as healthcare maintenance expenditures demands effective prevention, timely detection, and treatment of the above conditions.

Diagnosis of Resistant Hypertension

Uncontrolled hypertension includes all hypertensive patients lacking adequate BP control despite seemingly appropriate medical treatment. Not all patients in this category, however, have true resistant hypertension. The definition of resistant hypertension is a failure to achieve a BP goal of <140/90 mm Hg despite treatment with at least three different medications classes at a maximally tolerated dose and including a diuretic [7]. The true prevalence of resistant hypertension is unknown. Data from small studies suggest about 5% of patients in general medical practice and up to 50% of patients in clinical hypertension specialty clinics have resistant hypertension [8]. Most recently, evaluation of the NHANES for 2003–2008 has found that the prevalence of resistant hypertension, as defined by JNC 7, is 8.9% among adults with hypertension and 12.8% among the patients who are treated. Not surprisingly, individuals with resistant hypertension were more likely to be older and African American, and have a high BMI, kidney disease and albuminuria, history of cardiovascular morbidity, and diabetes mellitus ($p < 0.001$) [9].

Therefore, identifying individuals that need more focused management of BP is a priority. The approach to achieve BP targets in these patients involves ruling out causes of so-called 'pseudo-resistant' hypertension, such as inadequate

Table 1. Factors that contribute to resistant hypertension

Physiological changes throughout life
High age (>70 years)
Menopause
Ovariectomy (surgical menopause)
Lifestyle factors contributing to resistant hypertension
Obesity or overweight
High-salt diet
Physical inactivity
Ingestion of low-fiber, high-fat diet
Heavy alcohol ingestion
Substances that can interfere with blood pressure control
Salt: increases volume and decreases NO responsiveness by increasing oxidant stress
Alcohol: increased catecholamines
Oral contraceptives: volume retention
Cyclosporine: volume/endothelin increases
Erythropoietin: increased blood volume
Natural licorice: increased aldosterone
Herbal compounds (ephedra, ma huang): increased catecholamines

medication dosing, confounders that prevent achievement of BP goals, cost of medication, patient nonadherence, white coat hypertension, and/or errors in BP measuring techniques (table 1) [10, 11]. In this regard, important lifestyle contributors such as excess dietary salt intake are a common and important etiology of resistant hypertension. Salt sensitivity is associated with abnormal endothelial cell function, hyperlipidemia, microalbuminuria, and insulin resistance [12]. In patients diagnosed with 'essential hypertension,' the presence of salt sensitivity may aggravate hypertension and require additional antihypertensive therapy [13]. Salt sensitivity is increasingly prevalent with advanced aging and among Black hypertensive patients [14]. An average individual exceeds approximately 4 times the recommended dose of 2.5 mg per day [15]. In this respect, higher sensitivity to increased salt load contributes to a lower rate of BP control in this patient subgroup. Additionally, larger amounts of alcohol intake that should not exceed 1 oz. for men and 0.5 oz. for women per day are commonly a source of hypertension resistance [6]. Evaluation of over-the-counter medications, including vitamin and herbal supplements, could potentially reveal agents that may increase BP (table 1) [16, 17].

Obesity and age, as well as cessation of estrogen production after menopause in women, contribute to the development of resistant hypertension [12]. The exact mechanisms of obesity-induced resistant hypertension have not been completely elucidated, but insulin resistance, increased production of aldosterone facilitated by adipocytes, and higher incidence of sleep apnea with increased sympathetic drive all contribute [18, 19]. Several mechanisms connect insulin resistance and compensatory hyperinsulinemia with BP elevation in the context of the metabolic syndrome, i.e. sodium retention, sympathetic activation, and impairment of endothelial nitric oxide (NO) production. Accumulating evidence suggests that activation of the endothelin (ET) system seems to be another important mechanism [20]. In vitro studies have shown that insulin stimulates both ET-1 production and action on the vascular wall. In vivo, high levels of insulin result in increased circulating ET-1 levels in healthy and insulin-resistant subjects, a relationship not observed with NO production. Moreover, ET receptor antagonism effectively reduces BP in animal models of insulin resistance and hypertension [20], and successfully improves insulin sensitivity [21, 22]. Conversely, elevation of ET-1 levels can further increase insulin resistance. The relationship between ET and insulin-resistant states is important since the incidence of obesity around the world is increasing and insulin resistance is more prevalent.

Increased arterial stiffness with advanced aging is also associated with high systolic BPs that are difficult to control [14]. Gender/hormonal differences also affect arterial stiffness. Indeed, in menopausal women, prolactin serum levels correlate with increasing central aortic systolic and diastolic BPs as well as arterial stiffness with a prolactin level of >9 ng/ml predicting high peripheral BP [23]. To help understand resistant hypertension better and gain better BP control rates, work on the ET system has been ongoing. Recently, new advances in this system may help in understanding the mechanisms of resistant hypertension and how to achieve better BP goals.

The Endothelin System in Hypertension and Kidney Disease

ETs represent a group of three distinctive 21-amino acid peptides: ET-1, ET-2, and ET-3, with ET-1 being one of the most potent vasoconstrictor substances found in humans. ET exerts its actions in an autocrine or paracrine manner [24, 25]. The regulation of ET production is covered elsewhere in this book, but in the context of elevated BP it is stimulated by physical stress [26], angiotensin II [27], vasopressin [28], insulin [29], and TGFβ [30], among others. Conversely, NO [31] and natriuretic peptides [32] inhibit ET synthesis. Additionally, ET-1 likely plays an important role in the development of hypertension in insulin-resistant states [33].

ET exerts its action through two pharmacologically distinct G protein-coupled receptors: ET_A and ET_B. The ET_A receptor primarily exerts the vasoconstrictor and proliferative effects of ETs [34], while the ET_B receptor induces endothelial dependent vasodilatation through NO and prostaglandin production [35, 36]. Moreover, this receptor subtype also modulates sodium and water handling in the kidney [37].

Considering the role of ET in general, its impact on the kidney is reflective in the control of BP and/or occurrence and progression of kidney disease. This is evidenced by the effect of ET on the afferent and efferent arterioles of the glomerular, as well as the arcuate and interlobular arteries regulating renal blood flow. Long-term vasoconstriction through ET_A receptor stimulation decreases renal blood flow and induces decline in glomerular filtration rate (GFR), but increases filtration fraction [38]. This subsequently, enhances sodium and water reabsorption possibly leading to volume expansion and BP elevations. The role of the ET_B receptor, however, induces transitory vasodilatation likely related to NO and prostaglandins [39]. In addition, this receptor subtype plays an important role in medullary tubular handling of sodium, water, and acids. A number of animal experiments have determined that ET-1 modulates permeability and electrolyte transport activity in different parts of the nephron. Moreover, it is induced by the high-sodium load. The ascending nephron along with the collecting duct responds to ET-1 via ET_B receptors with powerful natriuresis and diuresis that antagonize the work of vasopressin and angiotensin II. These effects are most likely secondary to an increase in NO production [40]. The disruption of this system in vivo results in salt-sensitive hypertension as evidenced in ET_B receptor gene knockout mice [41]. More recent evidence also suggests that the ET_A receptor might be contributing to ET-induced sodium excretion; however, the exact mechanism remains to be elucidated [42, 43]. As detailed further in this chapter, ET antagonists tend to reverse these effects leading to the side effects of fluid retention and edema formation.

The ET system also plays important role in proliferative and inflammatory cascades in the kidney. Evaluation of rodent models with overexpression of ET-1 revealed significant development of glomerulosclerosis and interstitial fibrosis [44]. Moreover, rats with experimentally decreased renal mass have increased expression of ET-1 gene paralleling the progression of injury [45]. The system operates in close connection with the renin-angiotensin-aldosterone system. Evidence suggests that ET activates the angiotensin-converting enzyme function and, more importantly, represents a downstream mediator of angiotensin II [46, 47]. In this setting, the likely mechanisms responsible for the renal injury are the activation of inflammatory cascades, increase in extracellular matrix components, and thickening of the glomerular basement membrane – hallmarks of kidney disease. Additionally, podocyte damage ensues by the way of its actin cytoskeleton disruption, allowing for glomerular protein leakage [48]. Therefore, it is not surprising to find increased availability of ET-1 in renal tissue of proteinuric nephropathies [49]. This increased ET production in renal tissue can be measured in urine and actually correlates with the level of renal injury and proteinuria [50]. Interestingly, podocyte injury and ET-1 accumulation is further exacerbated by the increased dietary protein load [51]. It is likely that the majority of the above effects, exerted by ETs, stem from the activation of the ET_A receptor [48, 52].

Kidney Disease and Resistant Hypertension

Longstanding elevated pressures are deleterious to the kidney. The underlying process responsible for the damage is multifactorial and based on vascular injury related to the elevated pressure in kidney vessels. This induces two separate compensatory processes. The first is an intimal thickening resulting ultimately in the loss of blood flow, i.e. autoregulation. This results in a failure of the kidney to dampen higher systemic pressures from further injuring the kidney [53]. The second process involves direct injury to the endothelium, resulting in decreased NO production, poor vasodilation, and tissue ischemia. This ultimately leads to the development of albuminuria through podocyte injury and finally glomerulosclerosis, the hallmark of hypertension-induced kidney injury. The maladaptive changes in the hemodynamics as well as metabolism further activate and stimulate regulatory mechanisms, including the renin-angiotensin-aldosterone system and ET system. These maladaptive changes contribute to the kidney's inability to reduce BP through the common result of pressure-induced natriuresis, thus contributing to vascular injury and fibrosis of the renal vessels [54].

Endothelin System Blockade in the Management of Resistant Hypertension

Patients diagnosed with resistant hypertension are usually on more than three different medications. The general approach in patients with resistant hypertension demands simultaneous pharmacological blockade of mechanistically different pathways involved in hypertensive control [8]. The most important systems that require alteration include the renin-angiotensin-aldosterone system and sympathetic nervous system. Additionally, lifestyle modification is imperative. Specifically, the decrease in dietary sodium intake to <2.4 g/day can reduce the number of antihypertensive medications needed and enhance their BP-lowering effect as recently evidenced in individuals with resistant hypertension [55].

In recent decades, additional targets have emerged, one of them being the ET system. Successful blockade of the ET system is exerted through the ET_A receptor and ET_B receptor. For more than 20 years, many experimental compounds with the ability to antagonize them have been evaluated and used in animal studies as well as human trials [56]. Initial application utilized a nonspecific ET receptor antagonist and was limited to the management of heart failure and pulmonary hypertension, with resistant hypertension and CKD being a more recent focus as selective ET_A receptor antagonists were developed [57].

ET involvement in hypertension was initially observed in patients with hemangioendothelioma and elevated plasma ET-1 whose hypertension resolved after tumor removal along with normalization of ET levels [58]. In addition, ET antagonists were evaluated extensively in hypertensive animal models with only moderate impact on BP lowering. Interestingly, the most pronounced BP lowering was among those who were salt-sensitive [59, 60]. However, ET_A and ET_B blockade was able to reverse the vascular remodeling and vasoconstriction seen in hypertensive animals, partly because of enhancement of endothelial dependent NO vasodilatation [56, 61].

To date, several clinical trials have assessed the effects of ET antagonism on hypertension in humans. Krum et al. [62] initially evaluated the ET antagonist bosentan in 293 individuals with mild-to-moderate hypertension. Bosentan was able to lower diastolic BP for 5.7 mm Hg when compared to placebo and similarly as enalapril. The effect was observed for the bosentan doses ranging from 500 mg to 2 g per day. These doses were 10 times higher than what is currently used in the treatment of pulmonary artery hypertension [62]. Similar findings were published for darusentan, an ET receptor antagonist with a higher affinity for the type A receptor [63]. More recent research has focused on the impact of selective ET receptor blockade in the setting of the resistant hypertension. Weber et al. [64] evaluated darusentan as an add-on therapy in 379 patients with resistant hypertension. The significant BP-lowering effect of 17/10 mm Hg was observed when compared to placebo with the lowest dose of 50 mg. This BP-lowering effect was similar for all higher doses as well. Of note is that a small subgroup of patients with proteinuria experienced considerable reduction at the end of the short study [64]. Additionally, in a similar patient population, darusentan was found to lower mean 24-hour BP to a greater extent than the central α_2-agonist guanfacine [65]. In this investigator driven and company sponsored study, office and not 24-hour BP was the primary endpoint. For reasons no one can explain, darusentan had a statistically better BP throughout the study except for the final endpoint when all data converged. We suspect there was a problem with a mix-up between the drug and placebo groups; however, this was never confirmed as all other explanations failed to answer the question. Moreover, the 24-hour ambulatory BP analysis confirmed darusentan was clearly superior to the other groups when BP was assessed as a continuous variable over time [65].

In a trial of another selective ET_A receptor blocker, atrasentan, the antihypertensive effect was not the only benefit observed. The significant decrease in glycemia and improvement in lipid metabolism was noted with a slowing down of coronary artery disease progression [66].

The successful antihypertensive effect might also be partly due to the beneficial nephroprotective functions of ET receptor antagonists. In hypertensive rats (primarily salt-sensitive), administration of ET antagonists exerts renoprotection regardless of the level of BP control [25]. Moreover, in the work of Vaneckova and colleagues [67], atrasentan (a selective ET_A blocker), but not nonselective antagonism, prevented podocyte injury, which is known to precede proteinuria and glomerulosclerosis, in uncontrolled hypertensive animals. The reversal of podocyte injury was achieved independently of hypertensive control and, more importantly, was correlated with improved survival [67]. Goddard et al. [68] evaluated ET blockade of ET_A and ET_B receptors in combination or separately in 8 hypertensive patients with CKD. The ET_A- blocked subjects' mean arterial BP fell 12.9 mm Hg ($p < 0.01$ when compared to placebo), which was superior to combined ET_A and ET_B blockade. Moreover, the significant increase in renal blood flow and decrease in renal vascular resistance was observed when compared to placebo or addition of ET_B receptor blocker [68]. Pharmacotherapy with a newer ET_A antagonist, sitaxentan has also shown promising antiproteinuric effects along with reduction in arterial stiffness, which could definitely translate into more apprehensive cardiovascular benefits and better hypertensive control [69].

Despite the promising effects on BP, the side effects of ET receptor antagonists, selective and nonselective, remain a concern. In a large number of trials, almost one third of the patients suffered excessive fluid retention and edema that was significantly higher than in the placebo groups. In fact, the most robust study to date, the ASCEND study, which was designed to evaluate ET antagonism effects on hard renal endpoints (doubling of the creatinine, end-stage renal disease development, or death) in patients with diabetic nephropathy, was terminated early due to the increased incidence of fluid retention and secondary weight gain as well as episodes of heart failure which raised safety concerns. In addition, there was a trend of higher mortality in the experimental group [70]. This fact alone might preclude the implementation of these drugs as a standard of therapy. It is, however, important to note, as evidenced in the above trials, that the lower doses of darusentan do not cause significant fluid retention, but do have a positive impact on BP control. The selectivity of the different antagonists available might also improve the side effect profile as well as performance. This stems from the fact that the ET_B receptor exerts protective diuretic and natriuretic actions in the kidney and promotes NO-dependent vasodilatation [71]. This, however, is still a matter of debate. Additional concern for this medication class is the teratogenicity that implies it must be avoided in women that are of child-bearing potential or pregnant [57].

In light of the data from the aforementioned studies, it is not likely that ET receptor blockers will be commonly used for hypertension management. The use of selective ET_A receptor blockers, though, will have better prospects of being accepted as a reasonable choice once they have a better side effect profile.

Conclusion

Given that ET_A receptor blockade clearly improves BP in patients with resistant hypertension but has dose-related edema as a major consequence, its tolerability may limit its effectiveness for routine use. Most of the data gathered through extensive animal experiments suggest the significant therapeutic efficacy for the treatment of hypertension as well as chronic renal impairment and proteinuria. Smaller studies have also confirmed these outcomes in human subjects with the above conditions. While large outcome trials clearly confirm the benefits of using ET_A blockers to achieve BP control, there has not been much experience in people with stage 3 or higher CKD. One exception is atrasentan when used at lower doses to reduce albuminuria in diabetic nephropathy; it was better tolerated than other ERAs and, thus, can be given safely and effectively to this cohort. The increases in edema seen with ET receptor antagonists are dose-dependent and not seen at lower doses, yet BP lowering, albeit of a lower magnitude, is still present. Further work with selective agents at lower doses is needed to retain this valuable class of drugs and implement them into our armamentarium for treatment of resistant hypertension. It is conceivable that the lower dosing and pairing with appropriate diuretics might alleviate the above adverse effects while still maintaining the protection of the vasculature and kidney.

References

1 Schoenborn C, Heyman K: Health characteristics of adults aged 55 years and over: United States, 2004 - 2007. National Health Statistics Reports, No.16. Centers for Disease Control and Prevention. 2009. www.cdc.gov (accessed April 2011).

2 Lindholm LH, Ibsen H, Dahlof B. et al: Cardiovascular morbidity and mortality in patients with diabetes in the Losartan Intervention For Endpoint Reduction in Hypertension study (LIFE): a randomised trial against atenolol. Lancet 2002;359:1004–1010.

3 Major cardiovascular events in hypertensive patients randomized to doxazosin vs chlorthalidone: the antihypertensive and lipid-lowering treatment to prevent heart attack trial (ALLHAT). ALLHAT Collaborative Research Group. JAMA 2000;283:1967–1975.

4 Lewington S, Clarke R, Qizilbash N, Peto R, Collins R: Age-specific relevance of usual blood pressure to vascular mortality: a meta-analysis of individual data for one million adults in 61 prospective studies. Lancet 2002;360:1903–1913.

5 2010 atlas of chronic kidney disease and end-stage renal disease in the United States. United States Renal Data System. 2010. www.USRDS.org (accessed April 2011).

6 Chobanian AV, Bakris GL, Black HR, et al: Seventh report of the Joint National Committee on Prevention, Detection, Evaluation, and Treatment of High Blood Pressure. Hypertension 2003;42:1206–1252.

7 Kaplan NM: Resistant hypertension. J Hypertens 2005;23:1441–1444.

8 Sarafidis PA, Bakris GL: Resistant hypertension: an overview of evaluation and treatment. J Am Coll Cardiol 2008;52:1749–1757.

9 Persell SD: Prevalence of Resistant Hypertension in the United States, 2003–2008. Hypertension 2011, E-pub ahead of print.

10 Moser M, Setaro JF: Clinical practice. Resistant or difficult-to-control hypertension. N Engl J Med 2006;355:385–392.

11 Calhoun DA, Jones D, Textor S, et al: Resistant hypertension: diagnosis, evaluation, and treatment: a scientific statement from the American Heart Association Professional Education Committee of the Council for High Blood Pressure Research. Circulation 2008;117:e510–e526.

12 Barton M, Meyer MR: Postmenopausal hypertension: mechanisms and therapy. Hypertension 2009;54:11–18.

13 Mishra SI, Jones-Burton C, Fink JC, Brown J, Bakris GL, Weir MR: Does dietary salt increase the risk for progression of kidney disease? Curr Hypertens Rep 2005;7:385–391.
14 Sarafidis PA, Bakris GL: State of hypertension management in the United States: confluence of risk factors and the prevalence of resistant hypertension. J Clin Hypertens (Greenwich) 2008;10:130–139.
15 Nishizaka MK, Pratt-Ubunama M, Zaman MA, Cofield S, Calhoun DA: Validity of plasma aldosterone-to-renin activity ratio in African American and white subjects with resistant hypertension. Am J Hypertens 2005;18:805–812.
16 Johnson AG, Nguyen TV, Day RO. Do nonsteroidal anti-inflammatory drugs affect blood pressure? A meta-analysis. Ann Intern Med 1994;121:289–300.
17 Grossman E, Messerli FH: Secondary hypertension: interfering substances. J Clin Hypertens (Greenwich) 2008;10:556–566.
18 Morris MJ: Cardiovascular and metabolic effects of obesity. Clin Exp Pharmacol Physiol 2008;35:416–419.
19 Sarafidis PA: Obesity, insulin resistance and kidney disease risk: insights into the relationship. Curr Opin Nephrol Hypertens 2008;17:450–456.
20 Sarafidis PA, Bakris GL: Review: insulin and endothelin: an interplay contributing to hypertension development? J Clin Endocrinol Metab 2007;92:379–385.
21 Shemyakin A, Salehzadeh F, Bohm F, et al: Regulation of glucose uptake by endothelin-1 in human skeletal muscle in vivo and in vitro. J Clin Endocrinol Metab 2010;95:2359–2366.
22 Ahlborg G, Shemyakin A, Bohm F, Gonon A, Pernow J: Dual endothelin receptor blockade acutely improves insulin sensitivity in obese patients with insulin resistance and coronary artery disease. Diabetes Care 2007;30:591–596.
23 Georgiopoulos GA, Stamatelopoulos KS, Lambrinoudaki I, et al: Prolactin and preclinical atherosclerosis in menopausal women with cardiovascular risk factors. Hypertension 2009;54:98–105.
24 Yanagisawa M, Kurihara H, Kimura S, et al: A novel potent vasoconstrictor peptide produced by vascular endothelial cells. Nature 1988;332:411–415.
25 Luscher TF, Barton M: Endothelins and endothelin receptor antagonists: therapeutic considerations for a novel class of cardiovascular drugs. Circulation 2000;102:2434–2440.
26 Yoshizumi M, Kurihara H, Sugiyama T, et al: Hemodynamic shear stress stimulates endothelin production by cultured endothelial cells. Biochem Biophys Res Commun 1989;161:859–864.
27 Bakris GL, Re RN: Endothelin modulates angiotensin II-induced mitogenesis of human mesangial cells. Am J Physiol 1993;264:F937–F942.
28 Bakris GL, Fairbanks R, Traish AM: Arginine vasopressin stimulates human mesangial cell production of endothelin. J Clin Invest 1991;87:1158–1164.
29 Hu RM, Levin ER, Pedram A, Frank HJ: Insulin stimulates production and secretion of endothelin from bovine endothelial cells. Diabetes 1993;42:351–358.
30 Zoja C, Orisio S, Perico N, et al: Constitutive expression of endothelin gene in cultured human mesangial cells and its modulation by transforming growth factor-beta, thrombin, and a thromboxane A2 analogue. Lab Invest 1991;64:16–20.
31 Boulanger C, Luscher TF: Release of endothelin from the porcine aorta. Inhibition by endothelium-derived nitric oxide. J Clin Invest 1990;85:587–590.
32 Emori T, Hirata Y, Imai T, Eguchi S, Kanno K, Marumo F: Cellular mechanism of natriuretic peptides-induced inhibition of endothelin-1 biosynthesis in rat endothelial cells. Endocrinology 1993;133:2474–2480.
33 Sarafidis PA, Bakris GL: Insulin and endothelin: an interplay contributing to hypertension development? J Clin Endocrinol Metab 2007;92:379–385.
34 Ihara M, Noguchi K, Saeki T, et al: Biological profiles of highly potent novel endothelin antagonists selective for the ET_A receptor. Life Sci 1992;50:247–255.
35 Hirata Y, Emori T, Eguchi S, et al: Endothelin receptor subtype B mediates synthesis of nitric oxide by cultured bovine endothelial cells. J Clin Invest 1993;91:1367–1373.
36 Lin H, Smith MJ Jr, Young DB: Roles of prostaglandins and nitric oxide in the effect of endothelin-1 on renal hemodynamics. Hypertension 1996;28:372–378.

37 Kohan DE: The renal medullary endothelin system in control of sodium and water excretion and systemic blood pressure. Curr Opin Nephrol Hypertens 2006;15:34–40.
38 Tomobe Y, Miyauchi T, Saito A, et al: Effects of endothelin on the renal artery from spontaneously hypertensive and Wistar Kyoto rats. Eur J Pharmacol 1988;152:373–374.
39 Matsuura T, Miura K, Ebara T, et al: Renal vascular effects of the selective endothelin receptor antagonists in anaesthetized rats. Br J Pharmacol 1997;122:81–86.
40 Plato CF, Pollock DM, Garvin JL: Endothelin inhibits thick ascending limb chloride flux via ET(B) receptor-mediated NO release. Am J Physiol Renal Physiol 2000;279:F326–F333.
41 Ge Y, Bagnall A, Stricklett PK, et al: Collecting duct-specific knockout of the endothelin B receptor causes hypertension and sodium retention. Am J Physiol Renal Physiol 2006;291:F1274–F1280.
42 Ge Y, Bagnall A, Stricklett PK, Webb D, Kotelevtsev Y, Kohan DE: Combined knockout of collecting duct endothelin A and B receptors causes hypertension and sodium retention. Am J Physiol Renal Physiol 2008;295:F1635–F1640.
43 Nakano D, Pollock DM: Contribution of endothelin A receptors in endothelin 1-dependent natriuresis in female rats. Hypertension 2009;53:324–330.
44 Hocher B, Thone-Reineke C, Rohmeiss P, et al: Endothelin-1 transgenic mice develop glomerulosclerosis, interstitial fibrosis, and renal cysts but not hypertension. J Clin Invest 1997;99:1380–1389.
45 Bruzzi I, Corna D, Zoja C, et al: Time course and localization of endothelin-1 gene expression in a model of renal disease progression. Am J Pathol 1997;151:1241–1247.
46 Kawaguchi H, Sawa H, Yasuda H: Effect of endothelin on angiotensin converting enzyme activity in cultured pulmonary artery endothelial cells. J Hypertens 1991;9:171–174.
47 Kawaguchi H, Sawa H, Yasuda H: Endothelin stimulates angiotensin I to angiotensin II conversion in cultured pulmonary artery endothelial cells. J Mol Cell Cardiol 1990;22:839–842.
48 Barton M: Therapeutic potential of endothelin receptor antagonists for chronic proteinuric renal disease in humans. Biochim Biophys Acta 2010;1802:1203–1213.
49 Lehrke I, Waldherr R, Ritz E, Wagner J: Renal endothelin-1 and endothelin receptor type B expression in glomerular diseases with proteinuria. J Am Soc Nephrol 2001;12:2321–2329.
50 Ohta K, Hirata Y, Shichiri M, et al: Urinary excretion of endothelin-1 in normal subjects and patients with renal disease. Kidney Int 1991;39:307–311.
51 Wesson DE, Nathan T, Rose T, Simoni J, Tran RM: Dietary protein induces endothelin-mediated kidney injury through enhanced intrinsic acid production. Kidney Int 2007;71:210–217.
52 Sorokin A, Kohan DE: Physiology and pathology of endothelin-1 in renal mesangium. Am J Physiol Renal Physiol 2003;285:F579–F589.
53 Hill GS: Hypertensive nephrosclerosis. Curr Opin Nephrol Hypertens 2008;17:266–270.
54 Johnson RJ, Herrera-Acosta J, Schreiner GF, Rodriguez-Iturbe B: Subtle acquired renal injury as a mechanism of salt-sensitive hypertension. N Engl J Med 2002;346:913–923.
55 Pimenta E, Gaddam KK, Oparil S, et al: Effects of dietary sodium reduction on blood pressure in subjects with resistant hypertension: results from a randomized trial. Hypertension 2009;54:475–481.
56 Barton M, Yanagisawa M: Endothelin: 20 years from discovery to therapy. Can J Physiol Pharmacol 2008;86:485–498.
57 Kohan DE: Endothelin, hypertension and chronic kidney disease: new insights. Curr Opin Nephrol Hypertens 2010;19:134–139.
58 Yokokawa K, Tahara H, Kohno M, et al: Hypertension associated with endothelin-secreting malignant hemangioendothelioma. Ann Intern Med 1991;114:213–215.
59 Kirchengast M, Luz M: Endothelin receptor antagonists: clinical realities and future directions. J Cardiovasc Pharmacol 2005;45:182–191.
60 Schneider MP, Boesen EI, Pollock DM: Contrasting actions of endothelin ET(A) and ET(B) receptors in cardiovascular disease. Annu Rev Pharmacol Toxicol 2007;47:731–759.

61 Dhaun N, Goddard J, Kohan DE, Pollock DM, Schiffrin EL, Webb DJ: Role of endothelin-1 in clinical hypertension: 20 years on. Hypertension 2008;52:452–459.
62 Krum H, Viskoper RJ, Lacourciere Y, Budde M, Charlon V: The effect of an endothelin-receptor antagonist, bosentan, on blood pressure in patients with essential hypertension. Bosentan Hypertension Investigators. N Engl J Med 1998;338:784–790.
63 Nakov R, Pfarr E, Eberle S: Darusentan: an effective endothelinA receptor antagonist for treatment of hypertension. Am J Hypertens 2002;15:583–589.
64 Weber MA, Black H, Bakris G, et al: A selective endothelin-receptor antagonist to reduce blood pressure in patients with treatment-resistant hypertension: a randomised, double-blind, placebo-controlled trial. Lancet 2009;374:1423–1431.
65 Bakris GL, Lindholm LH, Black HR, et al: Divergent results using clinic and ambulatory blood pressures: report of a darusentan-resistant hypertension trial. Hypertension 2010;56:824–830.
66 Raichlin E, Prasad A, Mathew V, et al: Efficacy and safety of atrasentan in patients with cardiovascular risk and early atherosclerosis. Hypertension 2008;52:522–528.
67 Opocensky M, Kramer HJ, Backer A, et al: Late-onset endothelin-A receptor blockade reduces podocyte injury in homozygous Ren-2 rats despite severe hypertension. Hypertension 2006;48:965–971.
68 Goddard J, Johnston NR, Hand MF, et al: Endothelin-A receptor antagonism reduces blood pressure and increases renal blood flow in hypertensive patients with chronic renal failure: a comparison of selective and combined endothelin receptor blockade. Circulation 2004;109:1186–1193.
69 Dhaun N, Macintyre IM, Kerr D, et al: Selective endothelin-a receptor antagonism reduces proteinuria, blood pressure, and arterial stiffness in chronic proteinuric kidney disease. Hypertension 2011;57:772–779.
70 Mann JF, Green D, Jamerson K, et al: Avosentan for overt diabetic nephropathy. J Am Soc Nephrol 2010;21:527–535.
71 Verhaar MC, Strachan FE, Newby DE, et al: Endothelin-A receptor antagonist-mediated vasodilatation is attenuated by inhibition of nitric oxide synthesis and by endothelin-B receptor blockade. Circulation 1998;97:752–756.

George L. Bakris, MD, Professor of Medicine, Director
Hypertensive Diseases Unit, The University of Chicago Medical Center
5841 S. Maryland Avenue, MC1027
Chicago, IL 60637 (USA)
Tel. +1 773 702 7936, E-Mail gbakris@gmail.com

Barton M, Kohan DE (eds): Endothelin in Renal Physiology and Disease.
Contrib Nephrol. Basel, Karger, 2011, vol 172, pp 235–242

Endothelin Receptor Blockade in Patients with Diabetic Nephropathy

Ton J. Rabelink[a] · Donald E. Kohan[b]

[a]LUMC Department of Nephrology, Leiden, The Netherlands; [b]Division of Nephrology, University of Utah, Health Sciences Center, Salt Lake City, Utah, USA

Abstract

Diabetic nephropathy constitutes a major health care challenge. In the current review we summarize the rationale and preclinical data that suggest involvement of the endothelin (ET) system in the pathogenesis of this complication of diabetes. Early clinical studies suggest that blockade of the ET system indeed may have renoprotective effects, as reflected by a strong reduction in albumin excretion. A major challenge in the clinical development of ET receptor blockades for this indication will be to tease out the protective effects from potential off-target effects. Of particular concern is the edema formation during ET receptor blockade.

Diabetic nephropathy is the major cause of end-stage renal disease (ESRD). While optimal blood pressure (BP) treatment using drugs that interfere with the renin-angiotensin system (RAS) has been shown to slow progression of diabetic nephropathy, many patients still progress to ESRD [1]. Therefore, there is a great need for new strategies to slow the progression of chronic kidney disease in patients with diabetic nephropathy. One such strategy involves targeting the endothelin (ET) system. Numerous studies involving experimental animals have strongly implicated ET in the pathogenesis of diabetic nephropathy, while emerging clinical trials show promise for ET receptor antagonists in the treatment of diabetic nephropathy. In this chapter we will review the pathophysiology of diabetic nephropathy and its relationship to the ET system, as well as the early clinical experience with blockade of ET receptors in this disorder.

Pathophysiology of Diabetic Nephropathy

The histologic features of nephropathy in patients with type 2 diabetes are heterogeneous. Mesangial expansion, mesangiolysis, and disruption of the endothelial attachment to the mesangium can be observed, just like in type 1 diabetes patients, which can then lead to formation of microaneurysms that can evolve into nodular lesions. However, the large majority of type 2 patients with nephropathy also exhibit glomerulosclerosis and ischemia due to secondary capillary rarefaction, similar to that seen in hypertension and atherosclerotic renal disease [2].

Diabetic nephropathy involves virtually every renal cell type. Endothelial cell involvement is important, leading to remodeling of the vessel wall. The endothelial phenotype is determined by its interaction with perivascular cells such as pericytes, mesangial cells, and fibroblasts. In vascular disease and particularly in hyperglycemia, the stabilizing interaction between endothelial cells and perivascular cells becomes disrupted [3]. Endothelial cells proliferate, perivascular cells may transform into myofibroblasts, and inflammatory cells are recruited to the vessel wall. While such phenomena may be an adaptation to ischemia, destabilization of the microvascular network ultimately leads to rarefaction and ischemic organ failure. Glomerular endothelial cell involvement, due to redox activation, hyperglycemia, or other factors, causes disruption of the underlying glycocalyx [4]. Consequently, podocytes are exposed to filtered albumin that is often chemically modified (e.g. glycosylated). The podocytes express a wide range of scavenger receptors that respond to injury, including cytoskleletal rearrangements and mesenchymal transformation. Interestingly, the ET system, through the ET_A receptor, appears to be involved in this response. For example, blockade of the ET_A receptor modifies podocyte remodeling after exposure to puromycin aminonucleoside as well as shigatoxin [5, 6]. Subsequent loss of podocyte and slit diaphragm structure may further contribute to filtration of macromolecules and albumin. Ensuing tubule cell exposure to albumin and other macromolecules may elicit additional inflammation and injury [7].

Preclinical Studies in Models of Diabetic Nephropathy

The ET system seems to be involved in all of the above pathophysiological processes. ET-1, via activation of the endothelial ET_B receptor and production of nitric oxide, is important in maintaining quiescence of the endothelium [8]. In contrast, activation of the ET_A receptor causes endothelial cell inflammatory activation [9]. Hence, selective blockade of the ET_A receptor would be a plausible strategy to facilitate endothelial quiescence. At the glomerular level, podocytes produce ET-1 in response to albumin. Blockade of the ET_A

receptor prevents cytoskeletal changes in podocytes upon albumin exposure in models that primarily injure the podocytes (e.g. puromycin aminonucleoside nephrosis) [10]. Podocyte structure could even be preserved in a model of age-associated focal segmental glomerulosclerosis [5]. Similarly, albumin increases ET-1 production by tubular epithelium which in turn elicits an inflammatory response [7]. Finally, ET-1 promotes activation of myeloid cells in the kidney [9].

Numerous preclinical studies indicate that ET_A receptor blockade reduces renal injury in in vivo models of diabetic nephropathy [11]. ET receptor blockade decreases proteinuria and glomerulosclerosis in established streptozotocin-induced and atherosclerotic diabetic nephropathy [12, 13]. In these experiments, ET receptor blockade also reduced BP, making it difficult to tease out pressure-independent effects. It is unclear if ET_A receptor antagonism is of comparable efficacy in diabetic nephropathy as that seen with RAS blockers; however, a recent study suggests that ET_A receptor antagonists and RAS blockers may exert a synergistic effect on preventing renal deterioration in diabetic nephropathy [14]. Another recent study found that ET receptor blockade prevented the onset of albuminuria and macrophage influx into the kidneys in streptozotocin-induced diabetes [15]. Taken together, the above findings suggest that blockade of ET_A receptors is a rational strategy in slowing renal disease progression in diabetes. Whether combined ET_A and ET_B receptor blockade would be effective is uncertain; however, given the potential problems associated with blockade of ET_B receptors (including fluid retention and reduced nitric oxide production), ensuing clinical trials have attempted to focus on selective blockade of ET_A receptors in diabetic nephropathy.

Clinical Studies in Diabetic Nephropathy

The first clinical indication for a possible renoprotective effect of ET blockade in diabetic nephropathy was obtained in a small exploratory study in patients with type 1 diabetes and albuminuria, where atrasentan was compared to placebo. Atrasentan effectively reduced albuminuria, which appeared to be independent of changes in renal blood flow, glomerular filtration rate (GFR), and BP [16]. The first phase 3 trial examining the effect of ET receptor antagonists on renal function in patients with diabetic nephropathy was a randomized double-blind study on the effect of the ET_A receptor antagonist avosentan on urinary albumin excretion rate (UAER) [17]. This study, sponsored by Speedel Pharma AG, involved 286 European patients recruited between 2003 and 2004 with type 2 (88%) and type 1 diabetic nephropathy, with the following (approximate) mean values: age 60 years, BP 144/83 mm Hg, creatinine clearance (ClCr) 80 ml/min, serum creatinine concentration 1.1–1.2 mg/dl, and UAER ranging between 1,135 and 1,743 mg/day. Patients with a ClCr <30 ml/min were excluded. Only

6 patients were not receiving RAS blockers. Patients received either 5, 10, 25, or 50 mg/day of avosentan or placebo for 12 weeks. The primary efficacy outcome measure was an absolute change in 12-hour UAER after 12 weeks and secondary efficacy outcome measures included changes in serum creatinine, ClCr, and BP.

The results of this trial were first reported in 2005 and the investigators noted that mean UAER significantly decreased with all doses of avosentan. Avosentan at doses of 5, 10, 25, and 50 mg/d decreased mean UACR by 20.9, 16.3, 25.0 and 29.9%, respectively; the placebo group had an unexpected substantial increase in mean UACR (35.5%). The authors stated that this was a dose-dependent response; however, this is not convincing based on the small numerical differences between groups. Further, due to the high variability, the data was also reported as median relative UAER levels, in which case avosentan at doses of 5, 10, 25, and 50 mg/day decreased median UACR by 28.7, 42.2, 44.8, and 40.2%, respectively, as compared to a 12.1% increase with placebo. Using the median values, there was no significant reduction in UAER with the 5-mg dose and there was no dose-dependence. There were no detectable effects on serum creatinine, ClCr, or systolic and diastolic BP.

ET receptor antagonists can have a prominent side effect profile, particularly with regard to liver toxicity (some ET receptor antagonists) and fluid retention (all ET receptor antagonists tested thus far). Liver enzymes were stable. There was no apparent difference between groups in weight gain; however, there was a dose-dependent increase in the incidence of fluid retention episodes (11.9, 21.1, 15.0, and 32.1% in the 5, 10, 25, and 50 mg/d groups, respectively). In addition, the proportion of patients who withdrew consent or failed to finish the study due to adverse events was dose-dependent: 18.9% in the 50-mg group, 5.1% in the 5-mg group, and 3.5% in the placebo group. Blood hemoglobin concentration tended to decrease in all the avosentan groups, most likely due to hemodilution from fluid retention.

While it is always easier to analyze data with the advantage of foreknowledge, it is notable that when this study was first reported in 2005, the investigators felt avosentan was well tolerated at all doses and that there was a dose-dependent reduction in UAER. In fact, there was evidence that there was not a clear dose response with respect to efficaciousness in reducing albuminuria, while the incidence of clinically significant fluid retention paralleled avosentan dosage. Nonetheless, Speedel used this phase 2 study as the basis to launch a phase 3 trial that met with quite unfortunate results.

ASCEND was Speedel's follow-up phase 3 trial that examined the effect of avosentan on the composite primary endpoint of time to doubling of serum creatinine, ESRD, or death in patients with type 2 diabetes and overt diabetic nephropathy [18]. Secondary outcomes included changes in albumin-to-creatinine ratio (ACR), change in estimated GFR (eGFR), and cardiovascular outcomes. This double-blind randomized study enrolled 1,392 participants from 36 countries beginning in July 2005; patients were given either 25 or 50 mg/d of avosentan or placebo. Enrollees had an ACR ≥309 mg/g and serum creatinine between approximately 1.2 and 3.0; patients were excluded if eGFR was ≤15. Mean parameters at baseline were (approximately): age 61 years, 32% female, BP 137/78, serum creatinine concentration 2.1 mg/dl, and eGFR 33 ml/min. The median ACR at baseline was approximately 1,460 across groups. All patients were on RAS blockade unless they were intolerant.

The trial was prematurely terminated in December 2006 after a median follow-up of 4 months due to an excess of cardiovascular events, primarily congestive heart failure and fluid overload in the avosentan groups. Despite this, avosentan significantly reduced ACR; the median decrease in ACR was 44.3, 49.3, and 9.7% for the 25-mg, 50-mg, and placebo groups, respectively. There was no difference in primary outcome; however, fewer individuals taking avosentan reached ESRD, while more died of cardiovascular causes (approx. 74% greater than placebo). eGFR declined similarly in all groups.

Adverse events leading to medication discontinuation occurred significantly more frequently in the 25- and 50-mg avosentan groups than in the placebo groups (19.6, 18.2, and 11.5%, respectively); these were mainly related to fluid overload. Fluid overload symptoms occurred in 44.8, 45.8, and 30.7% of patients taking 25 mg, 50 mg, or placebo, respectively. Both doses of avosentan were associated with a 6% incidence of diagnosed congestive heart failure, compared with 2% in those on placebo. Anemia was more frequent in patients taking avosentan, averaging about a 1.1 g/dl decrease in blood hemoglobin concentration, with no change in those on placebo (presumably due to hemodilution). BP decreased slightly in those taking avosentan (approx. 5 mm Hg systolic and approx. 3.5 mm Hg diastolic reduction) and hypotension was reported more frequently in these groups. Liver function tests were unchanged.

In retrospect, it is unclear why the 25- and 50-mg avosentan doses were chosen given that lower doses of avosentan may have comparable efficacy in reducing albuminuria, yet cause less fluid retention. The reasons why the higher doses of avosentan caused such significant fluid retention are not completely understood. One possibility is that these doses of avosentan block the ET_B receptor. In vitro studies suggest that avosentan has a relatively modest selectively for ET_A receptors (50:1 ET_A:ET_B binding selectivity) [19]; hence, higher doses of avosentan may also block natriuretic and diuretic ET_B receptors in the kidney. Another possibility is that blockade of ET_A receptors may also cause fluid retention which is also dose-dependent. Finally, the ASCEND trial involved patients with more advanced kidney disease (eGFR approx. 33 ml/min vs. approx. 80 mg/min in the earlier study); therefore, these individuals may have had a great propensity for fluid retention.

Despite the discontinuation of the ASCEND trial, both of the previous studies with avosentan were encouraging in that they showed a substantial

ET receptor antagonist-induced reduction in albuminuria. Given the possibility of activation of the ET_B receptor by avosentan, and that ET_A receptor blockade is likely of the greatest therapeutic benefit in slowing kidney disease progression [14, 15, 20] a trial using a more highly ET_A receptor-selective blocker was conducted. This ET receptor antagonist was atrasentan, with a reported ET_A:ET_B receptor selectivity of 1,800 [21]. The randomized double-blind trial sponsored by Abbott involved subjects with diabetic nephropathy who were receiving stable doses of RAS inhibitors [22]. Patients from the United States and Puerto Rico were studied from June 2009 to June 2010. A total of 89 subjects with an eGFR of 20 ml/min and an ACR of 100–3,000 mg/g were given 0.25, 0.75, or 1.75 mg/day of atrasentan or placebo for 8 weeks. The primary efficacy measure was change from baseline to each weekly postbaseline observation in ACR. The secondary efficacy measures included the proportion of subjects achieving at least a 40 and 25% reduction from baseline in the last on-treatment ACR level, and mean change in eGFR. The range of baseline values among groups were: 59–83% male, 50–64% Hispanic, ACR 350–515 mg/g, eGFR 48–61 ml/min, and serum creatinine 1.3–1.8 mg/dl. The mean BP was 137/75. Thus, these patients had less severe albuminuria and eGFR was of intermediate impairment as compared to the avosentan trials.

Atrasentan significantly reduced ACR in the 0.75-mg and 1.75-mg groups as compared to placebo. The geometric mean reduction in ACR was 21, 42, and 35% in the 0.25, 0.75, and 1.75 mg groups versus 11% in placebo. The reduction in ACR was seen as early as the first week of treatment. Only the 0.75-mg dose had significantly more subjects with ≥40% reduction in ACR (50% subjects). Systolic BP decreased significantly over the 8 weeks only in the 0.75-mg atrasentan group (by approx. 9 mm Hg), while diastolic BP decreased significantly only in the 1.75 mg atrasentan group (by approx. 7 mm Hg). Thus, it is not possible from these studies to separate out the BP-lowering effect from the antiproteinuric action of atrasentan.

The most common adverse event was peripheral edema, which was mostly mild or moderate, and occurred in 14, 18, 46, and 9% in the 0.25-atrasentan, 0.5-atrasentan, 1.75-mg atrasentan, and placebo groups, respectively. There was no change in body weight, while blood hemoglobin concentration decreased by 0.4–0.9 g/dl and not in a dose-dependent manner. Diuretic use was similar in all groups. Notably, the presence or absence of edema did not correlate with ACR reduction. There were no changes in liver function tests.

Thus, atrasentan was well tolerated and reduced albuminuria in patients taking RAS blockers. Moving forward and learning from the avosentan trials, it will be important to identify a dose of atrasentan that is antiproteinuric while causing minimal fluid retention in patients with reduced renal function. This phase 2A trial provides some optimism that such a dose can be identified.

Conclusion

There is a strong rationale for the involvement of the ET system in the pathogenesis of diabetic nephropathy. Moreover, diabetic nephropathy is frequently part of a spectrum of inter-related clinical disorders that include atherosclerosis, dyslipidemia, and obesity-induced inflammation, where blockade of ET has been shown to be beneficial as well. The first clinical studies indicate that ET receptor blockade reduces albuminuria potently. Since albuminuria is a direct reflection of glomerular injury in this setting, such treatment effects hold the promise of renoprotection in the long run. A serious concern in the clinical development of these drugs are off-target effects that particularly seem to be related to blockade of the ET_B receptor. As this receptor is essential for normal water and sodium balance, its blockade may result in edema formation. This shows the need for more ET_A selective treatment strategies, although it remains possible that ET_A receptor blockade also causes fluid retention. Thus, it is likely that future treatment strategies would benefit by including diuretics. Given the lack of new drugs to prevent diabetic nephropathy and the large proportion of patients that progress to end-stage renal failure, and despite adequate BP control during pharmacological blockade of the RAS, ET blockade is a path worth developing to prevent this deleterious condition.

References

1 Brenner BM, Cooper ME, de Zeeuw D, et al: Effects of losartan on renal and cardiovascular outcomes in patients with type 2 diabetes and nephropathy. N Engl J Med 2001;345:861–869.
2 Fioretto P, Mauer M, Brocco E, et al: Patterns of renal injury in NIDDM patients with microalbuminuria. Diabetologia 1996;39:1569–1576.
3 Pfister F, Feng Y, vom HF, et al: Pericyte migration: a novel mechanism of pericyte loss in experimental diabetic retinopathy. Diabetes 2008;57:2495–2502.
4 Rabelink TJ, de Boer HC, van Zonneveld AJ: Endothelial activation and circulating markers of endothelial activation in kidney disease. Nat Rev Nephrol 2010;6:404–414.
5 Ortmann J, Amann K, Brandes RP, et al: Role of podocytes for reversal of glomerulosclerosis and proteinuria in the aging kidney after endothelin inhibition. Hypertension 2004;44:974–981.
6 Morigi M, Buelli S, Zanchi C, et al: Shigatoxin-induced endothelin-1 expression in cultured podocytes autocrinally mediates actin remodeling. Am J Pathol 2006;169:1965–1975.
7 Zoja C, Benigni A, Remuzzi G: Cellular responses to protein overload: key event in renal disease progression. Curr Opin Nephrol Hypertens 2004;13:31–37.
8 Verhaar MC, Strachan FE, Newby DE, et al: Endothelin-A receptor antagonist-mediated vasodilatation is attenuated by inhibition of nitric oxide synthesis and by endothelin-B receptor blockade. Circulation 1998;97:752–756.
9 Amiri F, Paradis P, Reudelhuber TL, Schiffrin EL: Vascular inflammation in absence of blood pressure elevation in transgenic murine model overexpressing endothelin-1 in endothelial cells. J Hypertens 2008;26:1102–1109.

10 Morigi M, Buelli S, Angioletti S, et al: In response to protein load podocytes reorganize cytoskeleton and modulate endothelin-1 gene: implication for permselective dysfunction of chronic nephropathies. Am J Pathol 2005;166:1309–1320.
11 Barton M: Reversal of proteinuric renal disease and the emerging role of endothelin. Nat Clin Pract Nephrol 2008;4:490–501.
12 Hocher B, Schwarz A, Reinbacher D, et al: Effects of endothelin receptor antagonists on the progression of diabetic nephropathy. Nephron 2001;87:161–169.
13 Gross ML, Heiss N, Weckbach M, et al: ACE-inhibition is superior to endothelin A receptor blockade in preventing abnormal capillary supply and fibrosis of the heart in experimental diabetes. Diabetologia 2004;47:316–324.
14 Gagliardini E, Corna D, Zoja C, et al: Unlike each drug alone, lisinopril if combined with avosentan promotes regression of renal lesions in experimental diabetes. Am J Physiol Renal Physiol 2009;297:F1448–F1456.
15 Sasser JM, Sullivan JC, Hobbs JL, et al: Endothelin A receptor blockade reduces diabetic renal injury via an anti-inflammatory mechanism. J Am Soc Nephrol 2007;18:143–154.
16 ABT-627, a selective ET_A-receptor antagonist, reduces proteinuria in patients with diabetes mellitus; in: Regulation of Vascular Tone in Humans by Endothelium-Derived Mediators; thesis, University of Utrecht, Utrecht, 2000, pp 89–102.
17 Wenzel RR, Littke T, Kuranoff S, et al: Avosentan reduces albumin excretion in diabetics with macroalbuminuria. J Am Soc Nephrol 2009;20:655–664.
18 Mann JF, Green D, Jamerson K, et al: Avosentan for overt diabetic nephropathy. J Am Soc Nephrol 2010;21:527–535.
19 Neuhofer W, Pittrow D: Endothelin receptor selectivity in chronic kidney disease: rationale and review of recent evidence. Eur J Clin Invest 2009;39(suppl 2):50–67.
20 Dhaun N, Ferro CJ, Davenport AP, et al: Haemodynamic and renal effects of endothelin receptor antagonism in patients with chronic kidney disease. Nephrol Dial Transplant 2007;22:3228–3234.
21 Opgenorth TJ, Adler AL, Calzadilla SV, et al: Pharmacological characterization of A-127722: an orally active and highly potent ET_A-selective receptor antagonist. J Pharmacol Exp Ther 1996;276:473–481.
22 Kohan D, Pritchett Y, Molitch M, Wen S, Garimella T, Audhya P, Andress D: Addition of atrasentan to renin-angiotensin system blockade reduces albuminuria in diabetic nephropathy. J Am Soc Nephrol 2011;22:763–772.

Ton J. Rabelink, MD, PhD
LUMC Department of Nephrology
Room C3–28, PO Box 9600
NL–2300 RC Leiden (The Netherlands)
Tel. +31 71 526 2169, E-Mail t.rabelink@lumc.nl

Barton M, Kohan DE (eds): Endothelin in Renal Physiology and Disease.
Contrib Nephrol. Basel, Karger, 2011, vol 172, pp 243–254

Endothelin Antagonism in Patients with Nondiabetic Chronic Kidney Disease

Neeraj Dhaun[a,b] · Jane Goddard[a,b] · David J. Webb[a]

[a]Clinical Pharmacology Unit, British Heart Foundation (BHF) Centre of Research Excellence (CoRE), University of Edinburgh, Queen's Medical Research Institute, [b]Department of Renal Medicine, Royal Infirmary of Edinburgh, Edinburgh, UK

Abstract

The incidence of chronic kidney disease (CKD) is increasing worldwide. Cardiovascular disease is strongly associated with CKD and constitutes one of its major causes of morbidity and mortality. Although current treatments for CKD focus on blood pressure and proteinuria reduction, many CKD patients have ongoing hypertension and residual proteinuria. Newer treatments are needed that not only act on these parameters, but also slow the progression of CKD and improve the cardiovascular risk profile of CKD patients. The endothelins (ETs) are a family of related peptides of which ET-1 is the most powerful endogenous vasoconstrictor and the predominant isoform in the cardiovascular and renal systems. The ET system has been widely implicated in both cardiovascular disease and CKD. ET-1 contributes to the pathogenesis and maintenance of hypertension and arterial stiffness, as well endothelial dysfunction and atherosclerosis. By reversal of these effects, ET antagonists may reduce cardiovascular risk. In CKD patients, antagonism of the ET system may be of benefit in improving renal hemodynamics and reducing proteinuria. ET is likely also involved in the progression of renal disease, and data are emerging that suggest a synergistic role for ET receptor antagonists with angiotensin-converting enzyme inhibitors in slowing CKD progression.

The Endothelin System

Since its discovery in 1988 [1], endothelin (ET) has been widely implicated in the pathophysiology of renal disease. ETs are a family of three 21 amino acid peptides, each with distinct genes and tissue distributions, with powerful vasoconstrictor and pressor properties [1–3]. Of the three peptides, ET-1 is the major endothelial isoform and, in the human kidney, the only one so far shown to be expressed at the protein level [4]. Its main site of vascular production is the endothelial cell, but it is also produced by other cell types, including vascular smooth muscle cells and epicardial cells [5]. Within the renal system, it is produced by glomerular epithelial and mesangial cells, and renal tubular and medullary collecting duct cells [6]. Furthermore, the renal medulla is not only an important site of ET-1 generation, but also contains some of the highest concentrations of immunoreactive ET-1 of any organ [7].

Regulation of ET synthesis occurs at the level of gene transcription, with the gene product being the 212 amino acid preproET-1. Enhanced gene transcription occurs with a wide range of stimuli [8, 9]. Those pertinent to chronic kidney disease (CKD) include other vasoactive hormones, such as angiotensin and vasopressin, the cytokine interleukin-1, oxidized LDL, reduced extracellular pH, and cyclosporin A. In contrast, prostacyclin, nitric oxide, and the natriuretic peptides all inhibit gene transcription. PreproET-1 is cleaved to big ET-1 (38 amino acids), which is largely biologically inactive [10]. ET-converting enzyme then splits big ET-1 to the biologically active ET-1 and C-terminal fragment. Once synthesized, the secretion of mature ET-1 from endothelial cells is largely abluminal [11], towards the adjacent vascular smooth muscle, suggesting an autocrine or paracrine mechanism of action.

ET-1 acts by binding to two cognate receptors: ET_A and ET_B [12, 13]. Within blood vessels, both receptors are found on smooth muscle cells and their activation results in vasoconstriction. ET_B receptors are, however, predominantly found on the vascular endothelium where their activation results in vasodilation via prostacyclin and nitric oxide [14]. Because most ET-1 is released abluminally, plasma concentrations of ET-1 do not accurately reflect ET-1 production. However, some is released into the circulation and the ET_B receptor also acts as a clearance receptor for this circulating ET-1. Therefore, reductions in ET_B numbers, or ET_B receptor blockade, may reduce ET-1 clearance, increasing plasma concentrations without altering production. ET receptors are widely distributed within the human kidney, with the ET_A subtype localized to vascular smooth muscle, notably in the glomeruli, vasa recta, and arcuate arteries, whereas ET_B receptors are more numerous (ET_B to ET_A ratio 2:1) and more widespread, with a high concentration in the collecting system [4, 15].

Plasma ET-1 is elevated in CKD [16, 17]. This is likely due to a combination of reduced clearance and increased production. Evidence exists for renal and vascular ET-1 acting as two independent systems [18]. After systemic infusion of radiolabelled ET-1, labeled compound makes up less than 1% of total urinary ET-1 [19]. Therefore, neither glomerular filtration nor tubular secretion of plasma ET-1 accounts for urinary ET-1, which is therefore assumed to be primarily of renal origin. Urinary excretion of ET-1 is thus thought to reflect renal ET-1 production. Renal ET-1 is thought to have a role in the paracrine/autocrine regulation of renal and intrarenal blood flow, glomerular hemodynamics, sodium and water homeostasis [20], and acid-base balance [21].

Glomerulopathies, Nondiabetic Proteinuric Nephropathies and Chronic Kidney Disease

Glomerular injury is frequently associated with progressive CKD. This process typically involves glomerular sclerosis and interstitial fibrosis and occurs regardless of the nature of the initial renal insult. The mechanisms responsible for this continued renal deterioration are not fully understood, but likely involve a number of common pathways and may be distinct from those responsible for the original injury. Glomerular hypertension, glomerular cell hypertrophy, and extracellular matrix accumulation are all involved. Significant proteinuria, a marker of CKD, has emerged as a powerful predictor of renal disease progression [22, 23] regardless of underlying diagnosis, and proteinuria reduction is an important strategy to retard or prevent loss of renal function [24]. Furthermore, reduction of proteinuria confers cardiovascular protection [25]. Once filtered through the glomerulus, excess protein is tubulotoxic and excess protein reabsorption in the tubules can lead to activation of tubular-dependent pathways of interstitial inflammation and fibrosis, with progressive renal scarring [26]. The ET system has been implicated in these processes [6]. Much of the preclinical data are discussed elsewhere (see the chapter by Gagliardini et al., this vol., pp. 171–184, and [27]); however, based on these findings, a role for ET system antagonism appears logical in preventing, or at the very least slowing, the early pathological effects of protein overload on glomerular structure and function. At present there are only a few studies, which are detailed in a recent review by Barton [28], that support this hypothesis.

In a mouse model of IgA nephropathy, the most common glomerulonephritis worldwide, both ET-1 and the ET receptor message increased alongside progression of the nephritis, and treatment with a selective ET_A receptor antagonist ameliorated the histopathological lesions and proteinuria observed in this model [29]. In a rat model of proliferative nephritis, upregulation of ET-1 protein and receptors has been shown at the level of the glomerulus [30]. In this case nonselective ET receptor blockade reduced both the mesangial expansion and proteinuria characteristics of this model. Similarly, in a murine model of lupus nephritis, again defined by abnormal glomerular proliferation, selective blockade of the ET_A receptor improved glomerular structure and function, translating to an overall benefit in renal function [31]. Finally, in rats with accelerated passive Heymann nephritis, a model of membranous glomerulonephritis, combined blockade of the ET and angiotensin systems showed greater renoprotective benefit than either treatment alone [32].

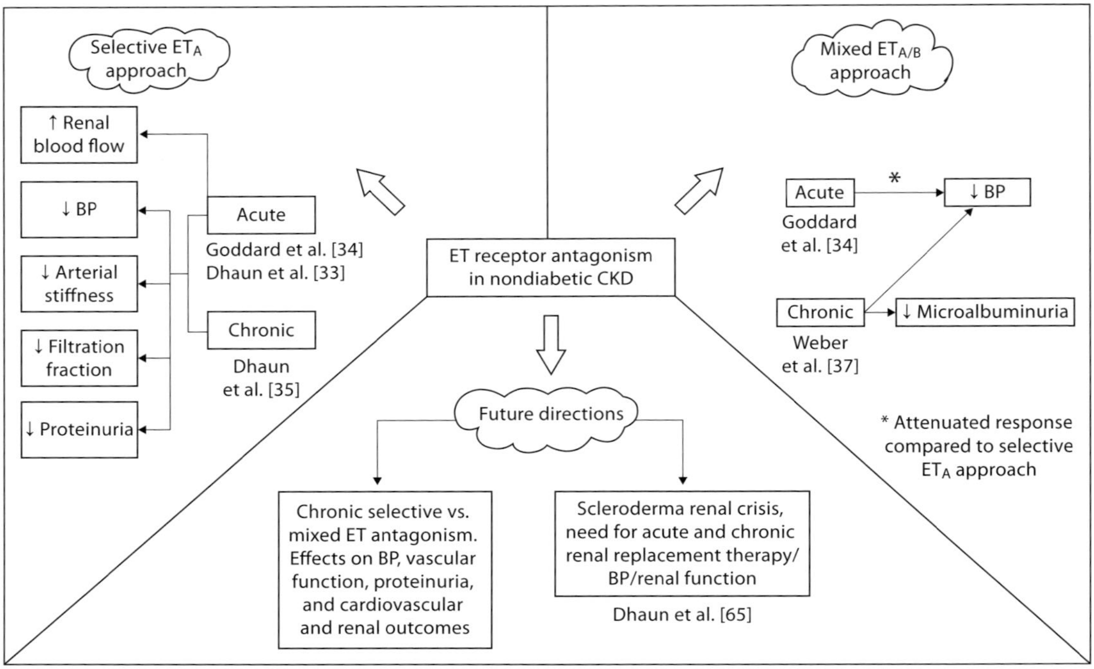

Fig. 1. Studies with selective ET_A and mixed $ET_{A/B}$ antagonism in nondiabetic proteinuric CKD.

Endothelin Antagonism in Nondiabetic Chronic Kidney Disease – Patient Studies

The management of CKD in the clinic focuses on blood pressure (BP) and proteinuria reduction, both key parameters that limit CKD progression and reduce the associated cardiovascular risk (fig. 1). In addition, there are a number of other areas of CKD where there may be a role for the ET system and where ET receptor antagonism may offer beneficial effects. These include endothelial dysfunction, atherosclerosis, arterial stiffness, and mineral bone disease. Clinical trials with ET receptor antagonists using these parameters as endpoints in nondiabetic CKD are limited.

Hypertension

In nondiabetic hypertensive CKD patients, the systemic vasodilation seen with acute ET_A receptor blockade (associated with a reduction in BP of approx. 10 mm Hg) [33] is attenuated by concomitant ET_B receptor antagonism [34], suggesting that, at least in this disease state, vasoconstrictor ET_B receptor activity is less important than ET_B vasodilatory function. Acute ET_A receptor antagonism also improves renal hemodynamics with a fall in renal vascular resistance and an increase in renal blood flow [33]. These changes are not accompanied by a measurable effect on glomerular filtration rate. There is a concomitant fall in the filtration fraction suggesting that ET-1 induces an ET_A receptor-mediated vasoconstriction, preferentially affecting the efferent arterioles, although not excluding effects on mesangial cells and the filtration coefficient [33].

Chronic dosing with a selective ET_A receptor antagonist also reduces BP, albeit to a lesser extent [35]. In this study a similar fall in systolic, diastolic, and mean BP of approximately 4 mm Hg was observed. Of note, in this trial patients already had good BP control (approx. 125/75 mm Hg) conforming to current CKD guidelines. Interestingly, in both these acute and chronic studies [33, 35], the majority of patients studied were already taking angiotensin-converting enzyme (ACE) inhibitors. Data from healthy subjects suggest a synergy between ET_A receptor antagonism and ACE inhibition that is not only dependent on an unblocked ET_B receptor, but also associated with significant natriuresis [36]. Clinically, this is important because patients with CKD are generally prescribed ACE inhibitors, not only for BP control but also for their renoprotective effects. With regards to mixed $ET_{A/B}$ antagonism, subgroup analysis of the DORADO study, which looked at the mixed antagonist darusentan in resistant hypertension, suggests that a combined $ET_{A/B}$ blocking strategy may also be of benefit in lowering BP in nondiabetic CKD [37].

Proteinuria

Proteinuria is one of the manifestations of glomerular hypertension and has its cause in the structural destruction of the glomerular filtration barrier. Albuminuria is incrementally associated with increased cardiovascular risk in both individuals with pre-existing risk (such as hypertensive patients) [25], and in individuals with no known risk factors [38]. This is true even in the presence of normal renal function [39]. Importantly, in patients with hypertension, reduction of albuminuria confers cardiovascular protection [25].

Both acute [33, 34] and chronic [35] selective ET_A blockade have been shown to reduce proteinuria in patients with nondiabetic proteinuric CKD. In the acute studies, these effects were abolished by concomitant ET_B receptor antagonism [34]. As for BP reduction, mixed $ET_{A/B}$ antagonism has also been shown to reduce proteinuria in nondiabetic CKD [37]. However, all of these subjects had, at most, microalbuminuria compared to the chronic dosing study with the selective ET_A antagonist [35], which included not only a wide range of nondiabetic proteinuric renal diagnoses (e.g. membranous glomerulopathy, IgA nephropathy, focal segmental glomerulosclerosis), but also varying proteinuria (0.3–8 g/day).

It still remains unclear as to what extent these effects on proteinuria are explained by BP reduction alone. In both the acute and chronic dosing studies using a selective ET_A receptor blocking approach, there was, as expected, a correlation between the changes in BP and proteinuria, with a greater fall in urinary protein seen in those patients with the greater fall in BP. However, both studies also had an active control which matched the reduction in BP seen with ET_A antagonism. Despite a similar fall in BP with both the active control and the ET_A antagonist, proteinuria fell to a greater extent with the latter supporting a BP-independent effect. Furthermore, a renal hemodynamic study following chronic selective ET_A antagonism suggests that falls in both glomerular filtration rate and filtration fraction are in part responsible for the reduction in proteinuria [35]. This is similar to the effects seen with ACE inhibitors and therefore may translate to longer term renoprotection.

Broader Cardiovascular Risk: Endothelial Dysfunction, Atherosclerosis, and Arterial Stiffness

The endothelium is a crucial regulator of vascular tone [40] and its function is impaired in CKD with a shift towards reduced vasodilation, associated with a proinflammatory and prothrombotic state. Endothelial dysfunction is recognized as a key early determinant in the progression to atherosclerosis, and is now well established as being independently associated with increased cardiovascular risk [41] and glomerular filtration rate loss [42]. Animal models show that ET-1 contributes to endothelial dysfunction [43], and ET receptor antagonism, predominantly with selective ET_A antagonists, improves nitric oxide-mediated endothelial function [44–46]. In addition to its effects on BP, ET-1 is proinflammatory [27] and has been implicated in the development of atherosclerosis. Several animal models have shown benefit of both selective and mixed ET receptor antagonism in the development of atherosclerotic lesions [44, 47–49]. There are currently no patient studies that focus on endothelial function in CKD. There are data that support a role for the ET_A receptor in coronary vascular tone and endothelial dysfunction in coronary artery disease [50, 51], Furthermore, CKD is a feature of aging, and ET-1 contributes to endothelial dysfunction in older subjects in comparison to younger ones [52]. Also, long-term administration of a selective ET_A receptor antagonist improves coronary endothelial dysfunction in older subjects with early atherosclerosis [53].

Arterial stiffness is linked to endothelial dysfunction [54] and the two commonly coexist in patients with increased cardiovascular risk. A number of interventions that reduce arterial stiffness also improve endothelial function [54]. To date, there have only been a few studies addressing the relationship between these two markers of cardiovascular disease after treatment. However, both animal and human studies suggest that the endothelium is an important regulator of arterial stiffness. ET-1 increases arterial stiffness in animals [43] and humans [55]. There are only two studies – one acute [33] and one chronic [35] – that use a selective ET_A receptor antagonist approach in nondiabetic CKD. These show a reduction in arterial stiffness with treatment, but the extent to which this is independent of BP is unclear. Interestingly, in the acute study [33], the reduction in arterial stiffness seen with ET antagonism was greater in those taking dual ACE inhibitor/angiotensin receptor blocker treatment than in those receiving an ACE inhibitor alone [56]. Indeed, there was a similar effect for proteinuria in this study. These findings support a role for ET antagonists to be considered in the multimodal regimen of the remission clinic.

Future Directions

While much of the current clinical research using ET receptor antagonists in CKD is focused on diabetic renal disease, there are some data to support their use in other areas. Scleroderma (systemic sclerosis) is an autoimmune disease of the connective tissues characterized by chronic inflammation and fibrosis with associated vascular damage [57]. The kidney is commonly involved [58, 59], and almost all renal pathology due to scleroderma is related to vascular damage and reduced renal blood flow [60]. Scleroderma renal crisis (SRC) is a relatively common and life-threatening complication. It is characterized by accelerated arterial hypertension associated with a rapid increase in serum creatinine concentration. Current treatment of SRC revolves around the use of ACE inhibitors [60]. Although there have been no randomized controlled trials of ACE inhibitor use in SRC, mortality from SRC has fallen approximately from 40 to 6% following their introduction [61]. Despite this, however, morbidity and mortality from SRC remain unacceptably high with a 5-year survival of approximately 60% and approximately half of all patients requiring either temporary or long-term dialysis [57].

ET-1 is implicated in the development of scleroderma. Scleroderma patients exhibit raised plasma and tissue concentrations of ET-1 [62, 63]. Currently, data regarding the use of ET receptor antagonism in scleroderma is limited, with the majority of the studies focused on pulmonary arterial hypertension [64]. The few clinical data available – one case report [65] – support a role for ET receptor antagonism in SRC. Following introduction of a selective ET_A receptor antagonist, there was a significant fall in both BP and proteinuria. Although there was some decline in renal function, this stabilized and indeed improved in the longer term. There was no requirement for renal replacement therapy. Given these encouraging data, further studies in this area are needed.

Safety of Endothelin Receptor Antagonists

Side effects with ET receptor antagonists in clinical trials are common. The most frequently reported clinical adverse events are headache, dizziness, nausea, nasal congestion, and dyspnea. These appear to be a class effect and likely relate to vasodilation. Liver enzyme derangement toxicity is also common and is dose-dependent. This has been reported to occur particularly with sulfonamide-derived compounds such as avosentan and sitaxentan. Indeed, sitaxentan was recently withdrawn due to this side effect. Importantly, successful switching from sulfonamide ET receptor antagonists to nonsulfonamide ones as a consequence of liver enzyme derangement is an option for those patients requiring therapy as recently shown by Rubin's group [66].

The mechanism of peripheral edema with ET receptor antagonism remains unclear. In the ASCEND study, which looked at the mixed ET receptor antagonist avosentan in diabetic CKD, fluid retention was the major cause for discontinuation of trial medication [67]. Interestingly, in a dose-ranging study in a similar study population, lower doses of avosentan were as effective at reducing microalbuminuria, but with a lower incidence of fluid overload [68]. It may be argued that at lower doses of avosentan there is less ET_B receptor blockade, which may be implicated in fluid retention. ET-1 acts in the renal tubule via the ET_B receptor to promote natriuresis and diuresis. Thus, peripheral edema associated with vasodilation could be aggravated by mixed $ET_{A/B}$ antagonists due to the reported ET_B-mediated downregulation of the renal tubular epithelial sodium channel [69]. In clinical trials, however, fluid retention appears to occur with both selective ET_A and mixed $ET_{A/B}$ receptor antagonists. It is clearly of importance in studies with CKD patients. However, using the ET receptor antagonist darusentan in patients with resistant hypertension, the recent DORADO study has shown this to be manageable [37].

Finally, all ET antagonists are contraindicated in pregnancy as they are teratogenic, but this may be less of an issue in CKD. As these drugs have only been available for a relatively short period of time, their balance of risks and benefits remain incompletely understood. Longer-term observational studies of individual ET receptor antagonists will be of benefit in this regard.

Conclusion

ET has been widely implicated in renal physiology and pathophysiology. Studies suggest that ET receptor antagonism could be of therapeutic importance in CKD. In clinical CKD, ET receptor antagonism can successfully reduce BP and arterial stiffness and also reduce renal vascular tone and proteinuria. Animal studies suggest that there is further potential benefit in terms of prevention of proliferation and fibrosis as well as atherosclerosis, an important comorbidity in CKD. To date, only one longer term study in nondiabetic CKD has been reported, focusing on proteinuria, BP, and arterial stiffness as its outcomes. This suggests that ET receptor antagonism may be renoprotective. Hopefully, other such trials will report on the ability of this class of drugs to delay the progression of CKD and its cardiovascular complications.

References

1 Yanagisawa M, Kurihara H, Kimura S, et al: A novel potent vasoconstrictor peptide produced by vascular endothelial cells. Nature 1988;332:411–415.

2 Inoue A, Yanagisawa M, Kimura S, et al: The human endothelin family: three structurally and pharmacologically distinct isopeptides predicted by three separate genes. Proc Nat Acad Sci USA 1989;86:2863–2867.

3 Arinami T, Ishikawa M, Inoue A, et al: Chromosomal assignments of the human endothelin family genes: the endothelin-1 gene (EDN1) to 6p23-p24, the endothelin-2 gene (EDN2) to 1p34, and the endothelin-3 gene (EDN3) to 20q13.2-q13.3. Am J Hum Genet 1991;48:990–996.

4 Karet FE, Davenport AP: Localization of endothelin peptides in human kidney. Kidney Int 1996;49:382–387.

5 Eid H, de Bold ML, Chen JH, et al: Epicardial mesothelial cells synthesize and release endothelin. J Cardiovasc Pharmacol 1994;24:715–720.

6 Kohan DE: Endothelins in the normal and diseased kidney. Am J Kid Dis 1997;29:2–26.

7 Morita S, Kitamura K, Yamamoto Y, et al: Immunoreactive endothelin in human kidney. Ann Clin Biochem 1991;28:267–271.

8 Attina T, Camidge R, Newby DE, et al: Endothelin antagonism in pulmonary hypertension, heart failure, and beyond. Heart 2005;91:825–831.

9 Wesson DE, Simoni J, Green DF: Reduced extracellular pH increases endothelin-1 secretion by human renal microvascular endothelial cells. J Clin Invest 1998;101:578–583.

10 Haynes WG, Webb DJ: Contribution of endogenous generation of endothelin-1 to basal vascular tone. Lancet 1994;344:852–854.

11 Yoshimoto S, Ishizaki Y, Sasaki T, et al: Effect of carbon dioxide and oxygen on endothelin production by cultured porcine cerebral endothelial cells. Stroke 1991;22:378–383.

12 Arai H, Hori S, Aramori I, et al: Cloning and expression of a cDNA encoding an endothelin receptor. Nature 1990;348:730–732.

13 Sakurai T, Yanagisawa M, Takuwa Y, et al: Cloning of a cDNA encoding a non-isopeptide selective subtype of the endothelin receptor. Nature 1990;348:732–735.

14 DeNucci G, Thomas R, D'Orleans-Juste P, et al: Pressor effects of circulating endothelin are limited by its removal in the pulmonary circulation and by the release of prostacyclin and endothelium-derived relaxing factor. Proc Natl Acad Sci 1988;85:9797–9800.

15 Kuc R, Davenport AP: Comparison of endothelin-A and endothelin-B receptor distribution visualized by radioligand binding versus immunocytochemical localization using subtype selective antisera. J Cardiovasc Pharmacol 2004;44:224–226.

16 Koyama H, Tabata T, Nishzawa Y, et al: Plasma endothelin levels in patients with uraemia. Lancet 1989;333:991–992.

17 Dhaun N, Lilitkarntakul P, Macintyre IM, et al: Urinary endothelin-1 in chronic kidney disease and as a marker of disease activity in lupus nephritis. Am J Physiol 2009;296:1477–1483.

18 Serneri GGN, Modesti PA, Cecioni I, et al: Plasma endothelin and renal endothelin are two distinct systems involved in volume homeostasis. Am J Physiol 1995;268:1829–1837.

19 Benigni A, Perico N, Gaspari F, et al: Increased renal endothelin production in rats with reduced renal mass. Am J Physiol 1991;260:331–339.

20 Kohan DE, Padilla E: Osmolar regulation of endothelin-1 production by rat inner medullary collecting duct. J Clin Invest 1993;91:1235–1240.

21 Wesson DE: Endogenous endothelins mediate increased acidification in remnant kidneys. J Am Soc Nephrol 2001;12:1826–1835.

22 Mallick NP, Short CD, Hunt LP: How far since Ellis? The Manchester Study of glomerular disease. Nephron 1987;46:113–124.

23 Cameron JS, Turner DR, Ogg CS, et al: The long-term prognosis of patients with focal segmental glomerulosclerosis. Clin Nephrol 1978;10:213–218.

24 Ruggenenti P, Perna A, Remuzzi G: Retarding progression of chronic renal disease: the neglected issue of residual proteinuria. Kidney Int 2003;63:2254–2261.
25 Ibsen H, Olsen MH, Wachtell K, et al: Reduction in albuminuria translates to reduction in cardiovascular events in hypertensive patients: losartan intervention for endpoint reduction in hypertension study. Hypertension 2005;45:198–202.
26 Remuzzi G, Bertani T: Pathophysiology of progressive nephropathies. N Engl J Med 1998;339:1448–1456.
27 Dhaun N, Goddard J, Webb DJ: The endothelin system and its antagonism in chronic kidney disease. J Am Soc Nephrol 2006;17:943–955.
28 Barton M: Reversal of proteinuric renal disease and the emerging role of endothelin. Nat Clin Pract Nephrol 2008;4:490–501.
29 Nakamura T, Ebihara I, Fukui M, et al: Effect of a specific endothelin receptor A antagonist on glomerulonephritis of ddY mice with IgA nephropathy. Nephron 1996;72:454–460.
30 Yoshimura A, Iwasaki S, Inui K, et al: Endothelin-1 and endothelin B type receptor are induced in mesangial proliferative nephritis in the rat. Kidney Int 1995;48:1290–1297.
31 Nakamura T, Ebihara I, Tomino Y, et al: Effect of a specific endothelin A receptor antagonist on murine lupus nephritis. Kidney Int 1995;47:481–489.
32 Benigni A, Corna D, Maffi R, et al: Renoprotective effect of contemporary blocking of angiotensin II and endothelin-1 in rats with membranous nephropathy. Kidney Int 1998;54:353–359.
33 Dhaun N, Macintyre IM, Melville V, et al: Blood pressure-independent reduction in proteinuria and arterial stiffness after acute endothelin-a receptor antagonism in chronic kidney disease. Hypertension 2009;54:113–119.
34 Goddard J, Johnston NR, Hand MF, et al: Endothelin-A receptor antagonism reduces blood pressure and increases renal blood flow in hypertensive patients with chronic renal failure: a comparison of selective and combined endothelin receptor blockade. Circulation 2004;109:1186–1193.
35 Dhaun N, MacIntyre IM, Kerr D, et al: Selective endothelin-A receptor antagonism reduces proteinuria, blood pressure, and arterial stiffness in chronic proteinuric kidney disease. Hypertension 2011;57:772–779.
36 Goddard J, Eckhart C, Johnston NR, et al: Endothelin A receptor antagonism and angiotensin-converting enzyme inhibition are synergistic via an endothelin B receptor-mediated and nitric oxide-dependent mechanism. J Am Soc Nephrol 2004;15:2601–2610.
37 Weber MA, Black H, Bakris G, et al: A selective endothelin-receptor antagonist to reduce blood pressure in patients with treatment-resistant hypertension: a randomised, double-blind, placebo-controlled trial. Lancet 2009;374:1423–1431.
38 Wang TJ, Evans JC, Meigs JB, et al: Low-grade albuminuria and the risks of hypertension and blood pressure progression. Circulation 2005;111:1370–1376.
39 Freedman BI, Langefeld CD, Lohman KK, et al: Relationship between albuminuria and cardiovascular disease in type 2 diabetes. J Am Soc Nephrol 2005;16:2156–2161.
40 Endemann DH, Schiffrin EL: Endothelial dysfunction. J Am Soc Nephrol 2004;15:1983–1992.
41 Lerman A, Zeiher AM: Endothelial function: cardiac events. Circulation 2005;111:363–368.
42 Perticone F, Maio R, Perticone M, et al: Endothelial dysfunction and subsequent decline in glomerular filtration rate in hypertensive patients. Circulation 2010;122:379–384.
43 Amiri F, Virdis A, Neves MF, et al: Endothelium-restricted overexpression of human endothelin-1 causes vascular remodeling and endothelial dysfunction. Circulation 2004;110:2233–2240.
44 Barton M, Haudenschild CC, D'Uscio LV, et al: Endothelin ET(A) receptor blockade restores NO-mediated endothelial function and inhibits atherosclerosis in apolipoprotein E-deficient mice. Proc Nat Acad Sci USA 1988;95:14367–14372.
45 Best PJM, McKenna CJ, Hasdai D, et al: Chronic endothelin receptor antagonism preserves coronary endothelial function in experimental hypercholesterolaemia. Circulation 1999;99:1747–1752.

46 Bauersachs J, Fraccarollo D, Galuppo P, et al: Endothelin-receptor blockade improves endothelial vasomotor dysfunction in heart failure. Cardiovasc Res 2000;47:142–149.

47 Babaei S, Picard P, Ravandi A, et al: Blockade of endothelin receptors markedly reduces atherosclerosis in LDL receptor deficient mice: role of endothelin in macrophage foam cell formation. Cardiovasc Res 2000;48:158–167.

48 Kowala MC, Rose PM, Stein PD, et al: Selective blockade of the endothelin subtype A receptor decreases early atherosclerosis in hamsters fed cholesterol. Am J Pathol 1995;146:819–826.

49 d'Uscio LV, Barton M, Shaw S, et al: Chronic ET(A) receptor blockade prevents endothelial dysfunction of small arteries in apolipoprotein E-deficient mice. Cardiovasc Res 2002;53:487–495.

50 Halcox JP, Nour KR, Zalos G, et al: Coronary vasodilation and improvement in endothelial dysfunction with endothelin ET(A) receptor blockade. Circ Res 2001;89:969–976.

51 Halcox JP, Nour KR, Zalos G, et al: Endogenous endothelin in human coronary vascular function. Differential contribution of endothelin receptor types A and B. Hypertension 2007;49:1–8.

52 Westby CM, Weil BR, Greiner JJ, et al: Endothelin-1 vasoconstriction and the age-related decline in endothelium-dependent vasodilation in men. Clin Sci 2011;120:485–491.

53 Reriani M, Raichlin E, Prasad A, et al: Long-term administration of endothelin receptor antagonist improves coronary endothelial function in patients with early atherosclerosis. Circulation 2010;122:958–966.

54 Oliver JJ, Webb DJ: Noninvasive assessment of arterial stiffness and risk of atherosclerotic events. Arterioscler Thromb Vasc Biol 2003;23:554–566.

55 Vuurmans TJ, Boer P, Koomans HA: Effects of endothelin-1 and endothelin-1 receptor blockade on cardiac output, aortic pressure, and pulse wave velocity in humans. Hypertension 2003;41:1253–1258.

56 Dhaun N, Macintyre IM, Melville V, et al: Effects of endothelin receptor antagonism relate to the degree of renin-angiotensin system blockade in chronic proteinuric kidney disease. Hypertension 2009;54:e19–e20.

57 Penn H, Howie AJ, Kingdon EJ, et al: Scleroderma renal crisis: patient characteristics and long-term outcomes. QJM 2007;100:485–494.

58 Medsger TA Jr, Masi AT, Rodnan GP, et al: Survival with systemic sclerosis (scleroderma). A life-table analysis of clinical and demographic factors in 309 patients. Ann Intern Med 1971;75:369–376.

59 Medsger TA Jr, Masi AT: Survival with scleroderma. II. A life-table analysis of clinical and demographic factors in 358 male U.S. veteran patients. J Chronic Dis 1973;26:647–660.

60 Denton CP: Renal manifestations of systemic sclerosis – clinical features and outcome assessment. Rheumatology (Oxford) 2008;47:54–56.

61 Steen VD, Medsger TA: Changes in causes of death in systemic sclerosis, 1972–2002. Ann Rheum Dis 2007;66:940–944.

62 Vancheeswaran R, Magoulas T, Efrat G, et al: Circulating endothelin-1 levels in systemic sclerosis subsets – a marker of fibrosis or vascular dysfunction? J Rheumatol 1994;21:1838–1844.

63 Abraham DJ, Vancheeswaran R, Dashwood MR, et al: Increased levels of endothelin-1 and differential endothelin type A and B receptor expression in scleroderma-associated fibrotic lung disease. Am J Pathol 1997;151:831–841.

64 Dhaun N, Pollock DM, Goddard J, et al: Selective and mixed endothelin receptor antagonism in cardiovascular disease. Trends Pharmacol Sci 2007;28:573–579.

65 Dhaun N, Macintyre IM, Bellamy CO, et al: Endothelin receptor antagonism and renin inhibition as treatment options for scleroderma kidney. Am J Kidney Dis 2009;54:726–731.

66 McGoon MD, Frost AE, Rubin LJ, et al: Ambrisentan therapy in patients with pulmonary arterial hypertension who discontinued bosentan or sitaxsentan due to liver function test abnormalities. Chest 2009;135:122–129.
67 Mann JF, Green D, Jamerson K, et al: Avosentan for overt diabetic nephropathy. J Am Soc Nephrol 2010;21:527–535.
68 Wenzel RR, Littke T, Kuranoff S, et al: Avosentan reduces albumin excretion in diabetics with macroalbuminuria. J Am Soc Nephrol 2009;20:655–664.
69 Kohan DE, Rossi NF, Inscho EW, et al: Regulation of blood pressure and salt homeostasis by endothelin. Physiol Rev 2011;91:1–77.

Dr. Neeraj Dhaun
Queen's Medical Research Institute, 3rd Floor East, Room E3.23
47 Little France Crescent
Edinburgh EH16 4TJ (UK)
Tel. +44 131 242 9210, E-Mail bean.dhaun@ed.ac.uk

Barton M, Kohan DE (eds): Endothelin in Renal Physiology and Disease.
Contrib Nephrol. Basel, Karger, 2011, vol 172, pp 255–260

Endothelin Antagonists in Clinical Trials: Lessons Learned

Matthias Barton[a] · Donald E. Kohan[b]

[a]Molecular Internal Medicine, University of Zürich, Zürich, Switzerland;
[b]Division of Nephrology, University of Utah, Health Sciences Center, Salt Lake City, Utah, USA

Abstract

Over the past 20 years, a large number of endothelin receptor antagonists (ERAs) have been developed, many of which underwent clinical testing. Yet, today only two ERAs are approved for the treatment of only two clinical indications, namely pulmonary arterial hypertension and scleroderma-related digital ulcers. Clinical development of ERAs has been hampered by problems with dosing, with the makeup of study cohorts, and adverse events. More recently, a number of studies evaluated the potential of ERA treatment for proteinuric renal disease. Although several ERAs were found to reduce proteinuria in patients with nephropathy, clinical testing was stopped for all but one drug. The reasons for the failure of some of these studies and general considerations about how to move forward with clinical studies involving ERAs in renal disease are discussed in this chapter.

Endothelin Receptor Antagonists: Few Remaining Trials in Kidney Disease

Endothelin receptor antagonists (ERAs) have been investigated in clinical trials as treatment for patients with kidney injury based on strong preclinical evidence of ERA-mediated nephroprotection in models of chronic renal disease [1–3]. Indeed, basic research has provided a good understanding of the role of endothelin (ET) in kidney and cardiovascular function, and has also shown that ET is involved in metabolism, inflammation, nociception, and other systems (reviewed in [4–6]). Initial studies in renal patients looked promising [7], and a number of investigators conducted clinical studies on ERAs in renal disease or hypertension using different ERAs (including atrasentan, avosentan, darusentan, and sitaxentan), all showing significant reductions in proteinuria [Dhaun et al., this vol., pp. 243–254; 7–11]. Unfortunately, only one ERA (atrasentan) continues to be studied in patients with kidney disease [12]. This is even more disappointing since several hundred ERAs have been developed [Davenport and Maguire, this vol., pp. 1–17]; to this day, only two of them are approved for the treatment of patients [Davenport and Maguire, this vol., pp. 1–17; 12], albeit for nonrenal indications (pulmonary arterial hypertension and scleroderma-related digital ulcers) [12].

Early Failures of Endothelin Receptor Antagonist Clinical Trials

The availability of orally active drugs within 4 years of the discovery of ET [4] represents a relatively unique case in biomedicine. Only 3 years after oral ERAs became available, the first clinical study on ERA treatment (in patients with congestive heart failure) was published [13, 14]. Not surprisingly, dosing parameters needed refining [15, 16]. Significant drug-related side effects arose that were not observed or were more severe than found in healthy volunteers, including volume overload and hepatic toxicity of sulfonamide-based ERAs. Finally, adhering to FDA regulations, investigators in heart failure studies had to perform treatment on top of standard therapies, which may have reduced the likelihood of detecting effects of ERA treatment [17, 18]. Unfortunately, none of the trials performed in patients with congestive heart failure or cancer reported positive results with regard to the primary outcomes [12, 17]. Whether the failure of these trials was due to incomplete knowledge of ERAs and the biology of the ET system under disease conditions or to other factors remains speculative, but they cast a negative light on this class of agents.

The Need for Peer Review and Publication of All Clinical Data

Clinical drug development requires scientific transparency and careful peer review [19–21]. Kelland and Webb [22] previously proposed that drug companies release and publish all data generated from the ERA heart failure trials, a view recently emphasized by Martine Clozel [23] discussing the need for publication in the context of the interests of the pharmaceutical industry. The importance of this is underscored by the failure of avosentan (an ERA) in the ASCEND trial. This phase 3 trial investigated the effects of avosentan on proteinuria and glomerular filtration rate in patients with advanced chronic kidney disease (CKD) stages 3 and 4 diabetic nephropathy [9]. Notably, the trial was undertaken in the absence of any published peer-reviewed preclinical or clinical data with the exception of pharmacokinetic and pharmacodynamic studies in healthy subjects. Despite the substantial reduction in proteinuria, avosentan treatment was associated with marked fluid retention (close to 50% of the patients in the avosentan groups experienced some form of fluid retention, including congestive heart failure, as compared to about 33% in the placebo-treated individuals). As a result of these adverse events, the trial was prematurely terminated. While preclinical and clinical data may well have been available to the investigators, this study raises the important question as to whether the study would have proceeded, particularly with the doses employed, if rigorous peer review had occurred.

Safety, Side Effects, and Drug Toxicity of Endothelin Receptor Antagonists

ERAs have a significant array of potential adverse effects in addition to fluid retention. We have previously discussed the potential side effects related to the chemical nature of certain ERAs [24]. Elevations of LFTs are a known problem of all sulfonamides and sulfonamide-derived drugs [25]. Hepatic toxicity is particularly important in patients prone to liver dysfunction due to fatty liver disease (in obese individuals), or in patients with right heart failure and hepatic congestion (pulmonary arterial hypertension or congestive heart failure). Several cases of liver failure in patients receiving sulfonamide ERAs have been reported [26, 27]. Switching of patients with drug-induced LFT elevations from sulfonamide to nonsulfonamide ERAs has proven to be successful [28]. Sitaxentan, a sulfonamide-based ERA, was under investigation as an antiproteinuric drug and showed promising initial clinical data [Dhaun et al., this vol., pp. 243–254]. Unfortunately, in 2010, several cases of fatal liver failure in sitaxentan-treated patients were reported. Among the 90,000 patients on either bosentan or ambrisentan, no deaths due to hepatic failure have been reported; however, there were 4 deaths among the 2,000 patients on sitaxentan [25]. Consequently, at the end of 2010, sitaxsentan was withdrawn from the market.

Another side effect associated with ERA therapy is potential testicular toxicity. While this has been described in drug company product literature, to our knowledge, no peer-reviewed study on ERA testicular toxicity has ever been published.

Clinical Endothelin Receptor Antagonist Trials: Methodological Issues

In 2006, we discussed the renoprotective potential of ERAs in patients with kidney disease and/or hypertension [3]. Darusentan is a propanoic acid, nonsulfonamide ERA that had been shown to exert vascular and renal protection in preclinical studies. In a phase 3 clinical trial in resistant hypertension [11], darusentan was found to reduce pre-existent proteinuria in patients with renal disease. In addition, darusentan was effective in reducing blood pressure. In a second trial, darusentan reduced blood pressure to a greater degree than the active control, the α_2-antagonist guanfacine [Lazich and Bakris, this vol., pp. 223–234; 29, 30]. However, after completion of both of these trials (DORADO and DORADO-AC), Gilead Sciences announced at the end of 2009 that because the DORADO coprimary endpoints were not met and additional phase 3 trials would be needed to establish darusentan as an antihypertensive, further development of darusentan for this clinical indication would be abandoned [31]. Subsequent reviews have discussed why this trial failed to show statistical significance despite an 'apparent' benefit [Lazich and Bakris, this vol., pp. 223–234; 29], noting different results obtained between ambulatory and clinic blood pressures and other issues. Hence, another ERA that had potential therapeutic benefit in patients with CKD was no longer available.

Where Do We Go from Here with Endothelin Receptor Antagonist Trials in Kidney Disease?

Knowledge of the physiological and pathophysiological mechanisms, pharmacological profiles, and side effects obviously must be applied to the design of future clinical studies using ERAs. Patient selection is also a key issue. Data from preclinical studies in CKD suggest that there appears to a 'point of no return' with regard to disease progression [3–5]. If this also applies to patients, we still need to determine which patients can still benefit from ERA treatment. Factors aggravating overall health may also be important (almost all patients in the renal ERA studies were obese by WHO definition, patients were included in ASCEND despite a history of myocardial infarction or unstable angina at least 3 months prior, and patients up to 80 years of age were also included in ASCEND). The well-known side effects of all ERAs are fluid retention and nitrate-like headaches, both of which are dose-dependent. Careful up-titration of ERAs with concomitant diuretic therapy to counteract fluid retention has been successfully used in a recent study [8] of patients with diabetic nephropathy, and represents a rational strategy to mitigate the most common adverse events.

The recently published phase 2A trial with atrasentan in diabetic nephropathy [8] suggests that if ERAs are used carefully in patients with CKD, with close attention to potential adverse effects, careful patient selection, and peer-reviewed data-driven decision-making, then there could be substantial hope for this class of drugs in slowing disease progression in patients with CKD where targeted therapy does not currently exist. Moving forward, we recognize that the financial incentive to develop ERAs for additional clinical applications is diminishing due to reduced patent life, even considering the 5-year exclusivity. Nonetheless, the therapeutic potential of the ERA class of drugs for treating human disease in general, and renal disease in particular, remains substantial and is important to pursue.

Acknowledgements

This work was supported by the Swiss National Science Foundation grants No. 3200-108258 and K33KO-122504 (to Dr. Barton) and U.S. National Institutes of Health grants DK-96392 and P01 HL96974 (to Dr. Kohan).

References

1 Remuzzi G, Perico N, Benigni A: New therapeutics that antagonize endothelin: promises and frustrations. Nat Rev Drug Discov 2002;1:986–1001.

2 Kohan DE: Endothelin, hypertension and chronic kidney disease: new insights. Curr Opin Nephrol Hypertens 2010;19:134–139.

3 Barton M, Mullins JJ, Bailey MA, Kretzler M: Role of endothelin receptors for renal protection and survival in hypertension: waiting for clinical trials. Hypertension 2006;48:834–837.

4 Barton M, Yanagisawa M: Endothelin: 20 years from discovery to therapy. Can J Physiol Pharmacol 2008;86:485–498.

5 Barton M, Nett PC, Amann K, Teixeira MM: Anti-inflammatory effects of endothelin receptor antagonists and their importance for treating human disease; in Chaudhary I, Ur-Rahman A (eds): Frontiers in Cardiovascular Drug Discovery. Oak Parks, Bentham Science, 2010, vol 1, pp 236–258.

6 Kohan DE, Rossi NF, Inscho EW, Pollock DM: Regulation of blood pressure and salt homeostasis by endothelin. Physiol Rev 2011;91:1–77.

7 Honing MLH, Bouter PK, Ballard DE, Padley P, Morrison PJ, Rabelink TJ: ABT-627, a selective ET_A-receptor antagonist, reduces proteinuria in patients with diabetes mellitus; in: Regulation of Vascular Tone in Humans by Endothelium-Derived Mediators (doctorate thesis). University of Utrecht, 2000, pp 89–102.

8 Kohan DE, Pritchett Y, Molitch M, Wen S, Garimella T, Audhya P, Andress DL: Addition of atrasentan to renin-angiotensin system blockade reduces albuminuria in diabetic nephropathy. J Am Soc Nephrol 2011;22:763–772.

9 Mann JF, Green D, Jamerson K, Ruilope LM, Kuranoff SJ, Littke T, Viberti G: Avosentan for overt diabetic nephropathy. J Am Soc Nephrol 2010;21:527–535.

10 Wenzel RR, Littke T, Kuranoff S, Jurgens C, Bruck H, Ritz E, Philipp T, Mitchell A: Avosentan reduces albumin excretion in diabetics with macroalbuminuria. J Am Soc Nephrol 2009;20:655–664.

11 Weber MA, Black H, Bakris G, Krum H, Linas S, Weiss R, Linseman JV, Wiens BL, Warren MS, Lindholm LH: A selective endothelin-receptor antagonist to reduce blood pressure in patients with treatment-resistant hypertension: a randomised, double-blind, placebo-controlled trial. Lancet 2009;374:1423–1431.

12 Barton M: Endothelin receptor antagonists in cardiovascular medicine: challenges and opportunities; in Black CM (ed): Cardiovascular Medicine: Pathogenesis and Translational Therapies. Springer, Heidelberg, 2011 in press.

13 Kiowski W, Sutsch G, Hunziker P, Muller P, Kim J, Oechslin E, Schmitt R, Jones R, Bertel O: Evidence for endothelin-1-mediated vasoconstriction in severe chronic heart failure. Lancet 1995;346:732–736.

14 Love MP, McMurray JJ: Endothelin in chronic heart failure: current position and future prospects. Cardiovasc Res 1996;31:665–674.

15 Kelland NF, Webb DJ: Clinical trials of endothelin antagonists in heart failure: a question of dose? Exp Biol Med (Maywood) 2006;231:696–699.

16 Krum H: Sympathetic activation and the role of beta-blockers in chronic heart failure. Aust NZ J Med 1999;29:418–427.

17 Battistini B, Berthiaume N, Kelland NF, Webb DJ, Kohan DE: Profile of past and current clinical trials involving endothelin receptor antagonists: the novel '-sentan' class of drug. Exp Biol Med (Maywood) 2006;231:653–695.

18 Seed A, Love MP, McMurray JJ: Clinical experience with endothelin receptor antagonists in chronic heart failure. Heart Fail Rev 2001;6:317–323.

19 Mello MM, Joffe S: Compact versus contract – industry sponsors' obligations to their research subjects. N Engl J Med 2007;356:2737–2743.

20 Bodenheimer T: Uneasy alliance – clinical investigators and the pharmaceutical industry. N Engl J Med 2000;342:1539–1544.

21 Blumenthal D: Academic-industrial relationships in the life sciences. N Engl J Med 2003;349:2452–2459.

22 Kelland NF, Webb DJ: Clinical trials of endothelin antagonists in heart failure: publication is good for the public health. Heart 2007;93:2–4.
23 Clozel M: Between confidentiality and scientific exchange: the place of publication in drug discovery and pharmaceutical research. Sci Transl Med 2011;3:67cm2.
24 Barton M: Therapeutic potential of endothelin receptor antagonists for chronic proteinuric renal disease in humans. Biochim Biophys Acta 2010;1802:1203–1213.
25 Galie N, Hoeper MM, Simon J, Gibbs R, Simonneau G: Task Force for the Diagnosis and Treatment of Pulmonary Hypertension of the European Society of Cardiology (ESC) and the European Respiratory Society (ERS): Liver toxicity of sitaxentan in pulmonary arterial hypertension. Eur Heart J 2011;32:386–387.
26 Hoeper MM: Liver toxicity: the Achilles' heel of endothelin receptor antagonist therapy? Eur Respir J 2009;34:529–530.
27 Hoeper MM, Olsson KM, Schneider A, Golpon H: Severe hepatitis associated with sitaxentan and response to glucocorticoid therapy. Eur Respir J 2009;33:1518–1519.
28 McGoon MD, Frost AE, Oudiz RJ, Badesch DB, Galie N, Olschewski H, McLaughlin VV, Gerber MJ, Dufton C, Despain DJ, Rubin LJ: Ambrisentan therapy in patients with pulmonary arterial hypertension who discontinued bosentan or sitaxsentan due to liver function test abnormalities. Chest 2009;135:122–129.
29 Bakris GL, Lindholm LH, Black HR, Krum H, Linas S, Linseman JV, Arterburn S, Sager P, Weber M: Divergent results using clinic and ambulatory blood pressures: report of a darusentan-resistant hypertension trial. Hypertension 2010;56:824–830.
30 Webb DJ: DORADO: opportunity postponed: lessons from studies of endothelin receptor antagonists in treatment-resistant hypertension. Hypertension 2010;56:806–807.
31 Second pivotal phase III study of Gilead's darusentan for resistant hypertension misses primary endpoints. December 2009. http://www.pharmiweb.com/pressreleases/pressrel.asp?ROW_ID=12716&channel=7 (accessed April 20, 2011).

Matthias Barton, MD
Molecular Internal Medicine, University of Zürich, LTK Y44 G22
Winterthurerstrasse 190
CH–8057 Zürich (Switzerland)
Tel. +41 77 439 5554, E-Mail barton@access.uzh.ch

Author Index

Subject Index